# Geometry Made Easy

*Ultimate Study Guide and Test Prep with Key Points, Examples, and Practices. The Best Tutor for Beginners and Pros + Two Practice Tests*

**Dr. Abolfazl Nazari**



PUBLISHED BY EFFORTLESS MATH EDUCATION

EFFORTLESSMATH.COM

# Welcome to Geometry Made Easy

# 2024

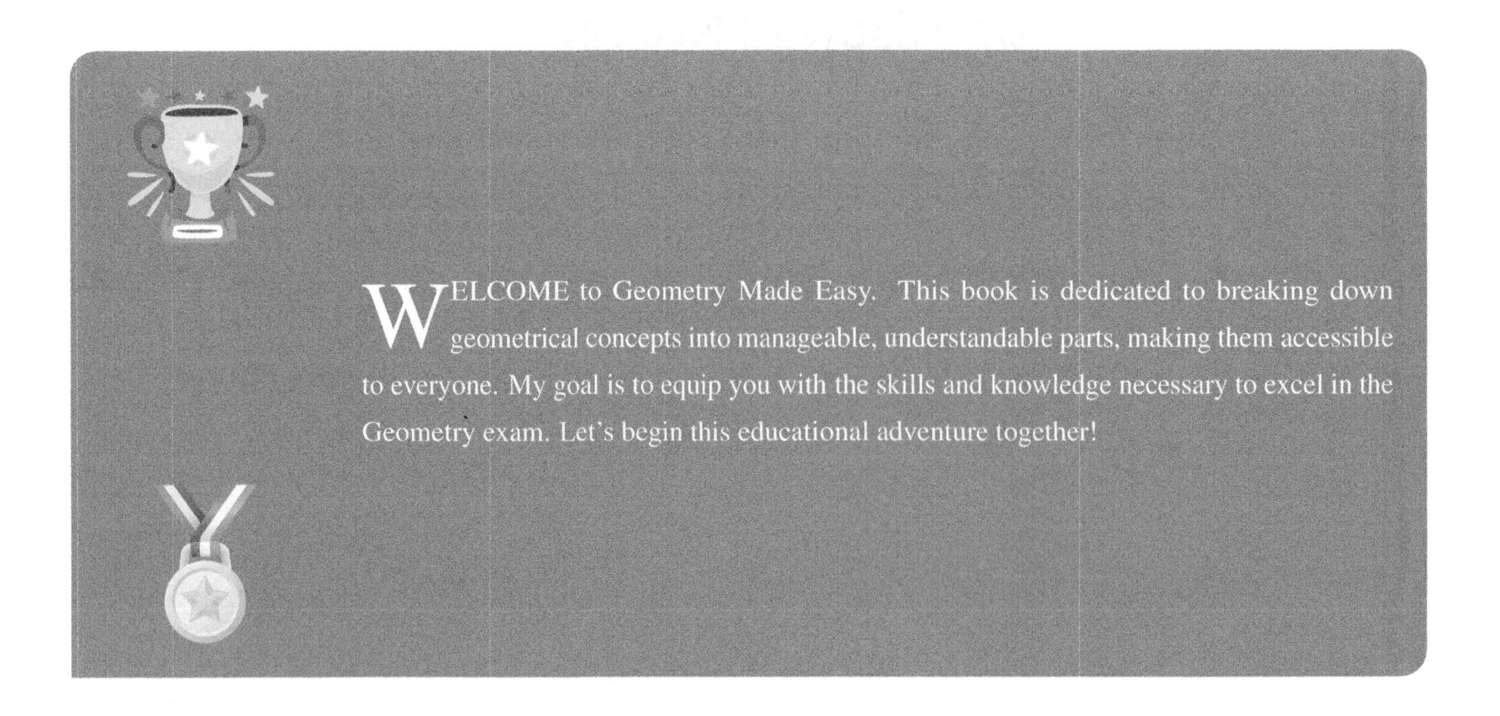


**Geometry Made Easy** provides comprehensive coverage of the key geometrical topics needed for the Geometry exam. The book is structured into detailed chapters covering all topics of Geometry. Each chapter starts with basic concepts and gradually moves to more complex ones, ensuring you gain a complete understanding of each topic. The content is tailored to not only prepare you for the Geometry test but also to apply these skills in real-life situations.

In keeping with the *Math Made Easy* series' philosophy, this book adopts an interactive and practice-oriented approach to learning. Each mathematical concept is introduced in a clear and straightforward manner, accompanied by examples to help illustrate its application. A variety of practice problems are provided to mirror the style and challenges of the Geometry test, enabling you to test your knowledge and strengthen your understanding. I am excited to show you what the book contains.

## What is included in this book

- ☑ Online resources for additional practice and support.
- ☑ A guide on how to use this book effectively.
- ☑ All Geometry concepts and topics you will be tested on.
- ☑ End of chapter exercises to help you develop the basic math skills.
- ☑ 2 full-length practice tests with detailed answers.

## Effortless Math's Geometry Online Center

Effortless Math Online Geometry Center offers a complete study program, including the following:

- ☑ *Step-by-step instructions on how to prepare for the Geometry test*
- ☑ *Numerous Geometry worksheets to help you measure your math skills*
- ☑ *Complete list of Geometry formulas*
- ☑ *Video lessons for all Geometry topics*
- ☑ *Full-length Geometry practice tests*

Visit `EffortlessMath.com/Geometry` to find your online Geometry resources.

Scan this QR code

**(No Registration Required)**

# Tips for Making the Most of This Book

This book is all about making mathematics easy and approachable for you. Our aim is to cover everything you need to know, keeping it as straightforward as possible. Here is a guide on how to use this book effectively: First, each math topic has a core idea or concept. It's important to understand and remember this. That's why we have highlighted key points in every topic. These are like mini-summaries of the most important stuff.

Examples are super helpful in showing how these concepts work in real problems. In every topic, we've included a couple of examples. If you feel very smart, you can try to solve them on your own first. But they are meant to be part of the teaching; They show how key concepts are applied to the problems. The main thing is to learn from these examples.

And, of course, practice is key. At the end of each chapter, you will find problems to solve. This is where you can really sharpen your skills.

To wrap it up:

- ***Key Points***: Don't miss the key points. They boil down the big ideas.
- ***Examples***: Try out the examples. They show you how to apply what you're learning.
- ***Practices***: Dive into the practice problems. They're your chance to really get it.

In addition to the material covered in this book, it is crucial to have a solid plan for your test preparation. Effective test preparation goes beyond understanding concepts; it involves strategic study planning and practice under exam conditions.

- **Begin Early.** Start studying well before the exam to avoid rushing, allowing for a thorough review.
- **Daily Study Sessions.** Study regularly for 30 to 45 minutes each day to enhance retention and reduce stress.
- **Active Note-Taking.** Write down key points to internalize concepts and improve focus. Review notes regularly.
- **Review Challenges.** Spend extra time on difficult topics for better understanding and performance.
- **Practice.** Engage in extensive practice using end-of-chapter problems and additional workbooks.

Explore other guides, workbooks, and tests in the series to complement your study, offering extra practice and enhancing understanding, problem-solving skills, and academic preparation.

# Contents

# 1. Geometric Tools and Concepts

## 1.1 Points, Lines, and Planes

We represent a *point* using a dot and label it with a capital letter. For example, we may refer to a point as Point $A$.

Point $A$

**Key Point**

A point is a specific location in space. It is dimensionless, meaning it has no width, no length, and no depth.

We can name a *line* by any two points on the line, such as $\overleftrightarrow{AB}$ or $\overleftrightarrow{BA}$, or by a single lowercase letter, such as Line $l$.

A B

Line $l = \overleftrightarrow{AB}$

**Key Point**

A line is one-dimensional, has no thickness, and extends infinitely in both directions. It is determined by at least two points on it.

We usually name a *plane* by three non-collinear points on it, symbolized as plane $ABC$, or by a capital letter Plane $P$.

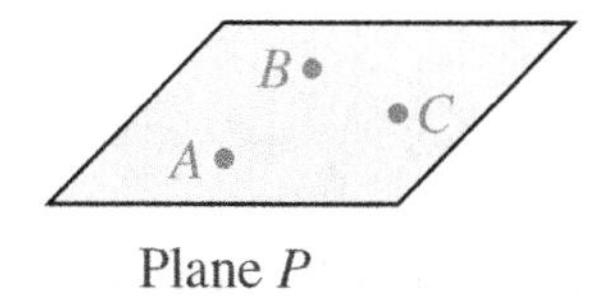


Plane $P$

**Key Point**

A plane is a flat two-dimensional surface that extends infinitely in all directions.

Based on the figure, answer the following questions:

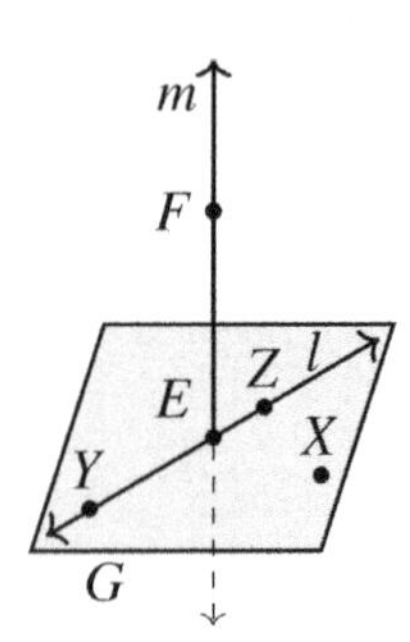


-What are two other ways to name $\overleftrightarrow{EF}$?

-What are three other ways to name Plane $G$?

-What are the names of three collinear points?

-What are the names of four coplanar points?

**Solution:** For each one:

-This line can also be named as $\overleftrightarrow{FE}$ or Line $m$.

-Plane $G$ may also be referred to as Plane $EXY$, Plane $XYZ$, or Plane $EXZ$, using any three non-collinear points in the plane.

-Points $Y$, $E$, and $Z$ are collinear.

-Points $X$, $Y$, $Z$, and $E$ are coplanar.

## 1.2 Line Segments and Measurements

A line segment in geometry is uniquely defined by its two endpoints. If you take any two points, say $A$ and $B$, the line segment connecting them is denoted as segment $AB$. Unlike a line, line segment $AB$ does not extend forever; it starts at $A$ and ends at $B$, containing all the points located on the line that lie between these endpoints.

**Key Point**

A line segment length is fixed and measurable, contrasting with an infinitely extending line.

To measure a line segment, you can use geometric tools like a ruler or a measuring tape. Simply align the ruler so that one endpoint of the segment coincides with the zero mark, then read the length at the other endpoint.

Sometimes, we come across scenarios where a line segment is part of a larger segment. If a point $C$ lies between points $A$ and $B$ on a line segment, the total length of segment $AB$ is the sum of the lengths of segments $AC$ and $CB$. This is known as the Segment Addition Postulate.

**Key Point**

Segment Addition Postulate: If $C$ lies between $A$ and $B$, then $AB = AC + CB$.

**Example** Given a line segment $AB$ with a point $C$ between $A$ and $B$, if $AC$ measures 4 units and $CB$ measures 3 units, find the length of $AB$.

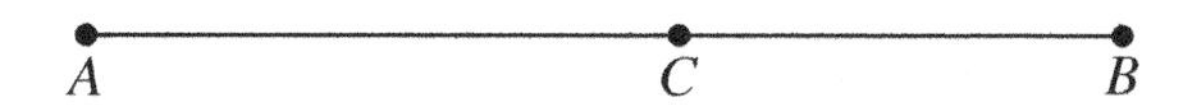


**Solution:** Using the Segment Addition Postulate, $AB = AC + CB$. So, $AB = 4 + 3 = 7$ units.

Measure the line segment $FE$.

**Solution:** To measure with a ruler, first align point $F$ with the zero mark and ensure the ruler is parallel to segment $FE$. The reading at point $E$ is approximately 3 cm.

## 1.3 Midpoint and Distance

The *midpoint* refers to the exact center point between two given points. It is particularly useful for dividing a line segment into two equal parts. On a number line, finding the midpoint involves simple arithmetic.

**Key Point**

Given two numbers $a$ and $b$ on the number line, the midpoint $M$ can be calculated using the formula $M = \frac{a+b}{2}$.

For instance, if $a = 2$ and $b = 8$, the midpoint $M$ would be calculated as follows: $M = \frac{2+8}{2} = 5$.

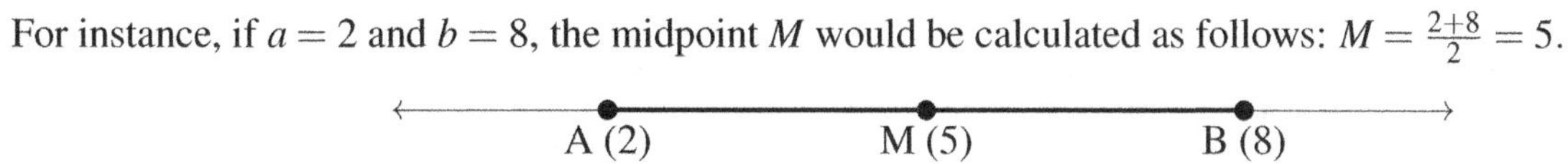


The *distance* between two points on a number line is the absolute value of the difference between their coordinates. This ensures that distance is a non-negative value, as it represents the length of the shortest path between two points.

**Key Point**

The distance $d$ between two numbers $a$ and $b$ is given by the formula $d = |b - a|$.

For example, the distance between points 3 and 8 on a number line would be $|8 - 3| = 5$.

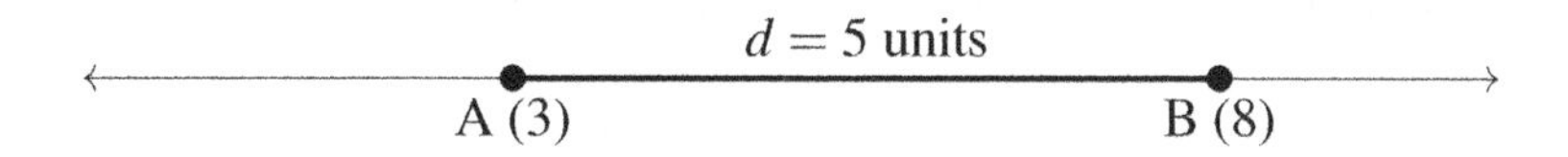

If the length of the line segment $AB$ is 18 $cm$, find the midpoint of this line segment.

**Solution:** The midpoint $M$ of the line segment $AB$ is found by dividing the total length of the segment by 2:

$$M = \frac{18\ cm}{2} = 9\ cm.$$

So, if you start at one end of the segment and measure 9 $cm$, you will reach the midpoint. The midpoint of this line segment is 9 $cm$ from either endpoint. Here is a visual representation of this situation:

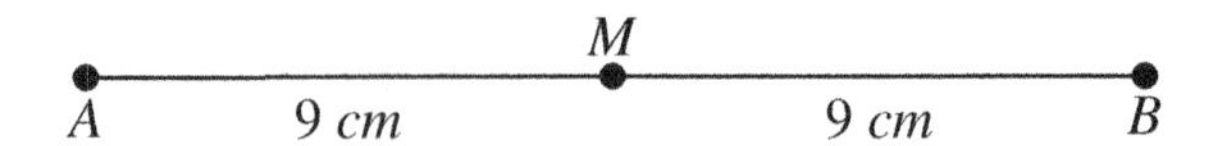


## 1.4 Parallel lines and Transversals

Parallel lines are lines in a plane that, no matter how far they extend, will never intersect or meet. They remain constantly equidistant from one another.

A transversal is a line that intersects two or more lines. When it crosses parallel lines, it creates a total of eight angles.

When a transversal crosses parallel lines, four pairs of congruent angles are formed. These angles are critical for solving many geometric problems.

The relationships between these angles are particularly important. Let us illustrate this with a graph:

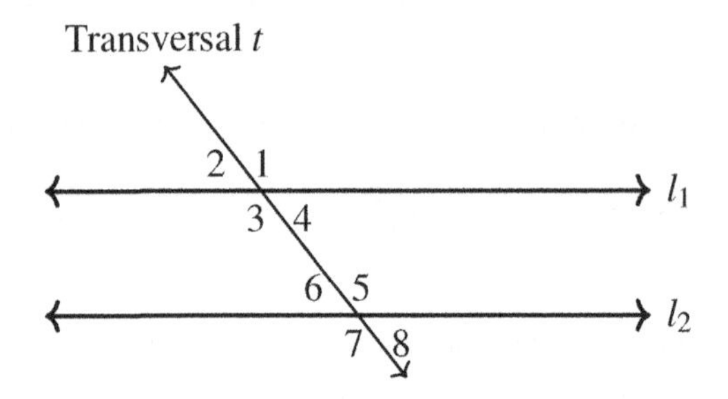


The angles formed have special properties: Angles 1, 3, 5, and 7 are congruent. Angles 2, 4, 6, and 8 are also congruent. Additionally, the following angle pairs are supplementary (their sum is 180°): Angles 1 and 2, 3 and 4, 5 and 6, 7 and 8.

**Key Point**

When a transversal intersects two parallel lines, each pair of corresponding angles is congruent. Moreover, each pair of interior angles on the same side of the transversal is supplementary.

Determine the value of $x$ in the diagram where two parallel lines are intersected by a transversal.

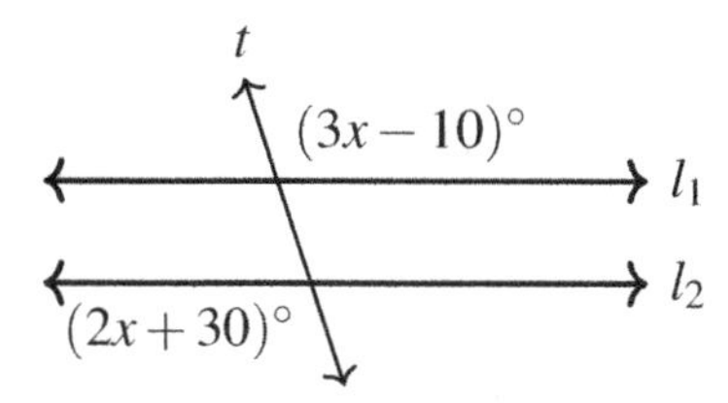


**Solution:** Since the lines are parallel, the angles $3x-10$ and $2x+30$ are congruent. Setting the two expressions equal to each other gives us:

$$3x-10=2x+30 \Rightarrow 3x-2x=30+10 \Rightarrow x=40.$$

## 1.5 Perpendicular Lines

A line is defined as *perpendicular* to another line if they intersect to form a right angle, quantified as $90^\circ$.

Perpendicular lines meet at exactly $90^\circ$, a right angle, and are denoted using the symbol $\perp$.

To construct a line perpendicular to line $QR$ that passes through point $S$ (not necessarily on the line $QR$), follow these steps:

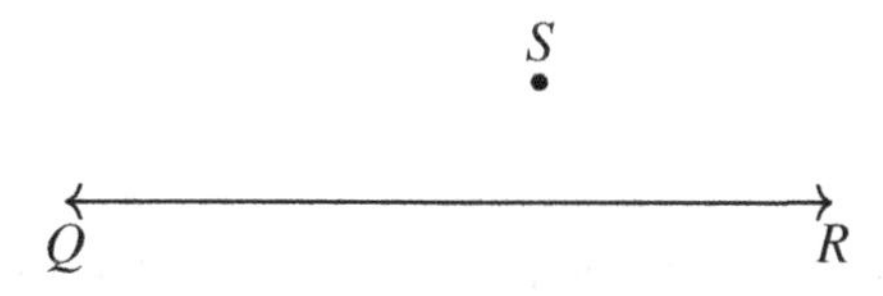


**Choose the Point.** Place point $S$ outside line segment $QR$ as the base point for the perpendicular line.
**Drawing an Arc from Point** $S$. With point $S$ as the center and a suitable radius, draw an arc that intersects the line $QR$ at two points, which we will call $A$ and $B$.
**Drawing Additional Arcs.** Without adjusting the compass width, place the compass point on $A$ and draw an arc either above or below line $QR$. Repeat the same process from point $B$ to create another arc that intersects the first arc.
**Marking the Intersection Point.** Label the point where the two arcs intersect closest to $S$ as $T$.
**Drawing the Perpendicular Line.** Draw a straight line from point $S$ through point $T$. This line, labeled $ST$, is perpendicular to line $QR$ and passes through point $S$.

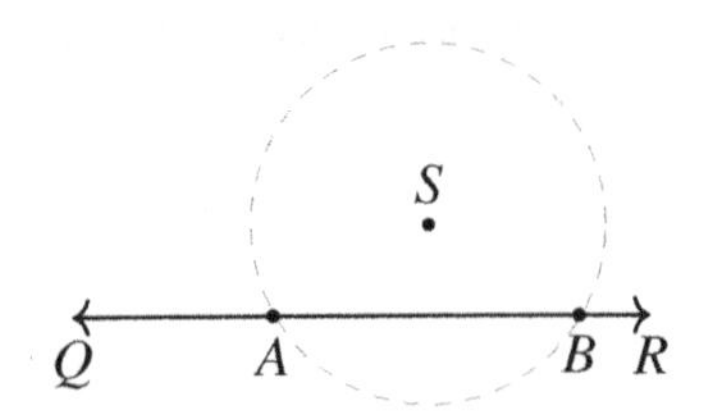


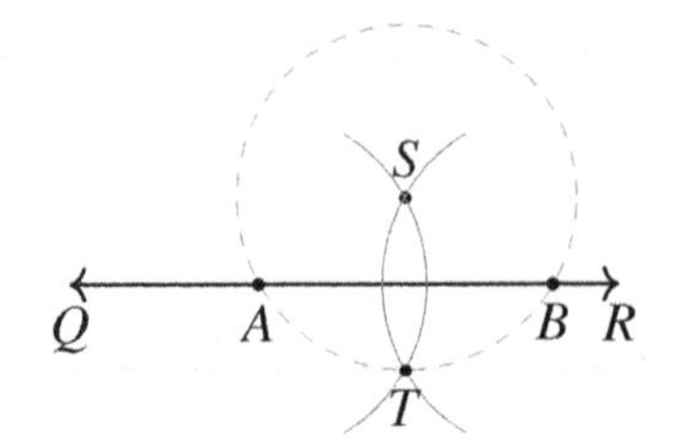


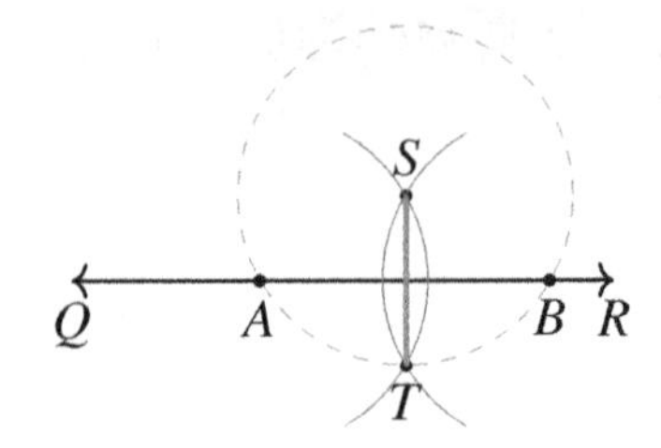

Draw a perpendicular line from point $O$ on line $XY$.

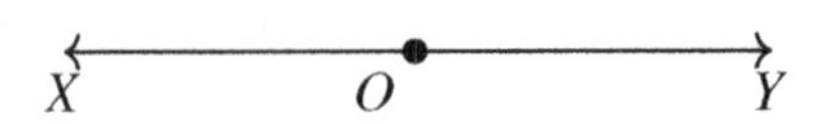


**Solution:** At $O$, draw a semicircle intersecting $XY$ at points $C$ and $D$.

With the same compass width, draw arcs from $C$ and $D$ to intersect above and below $XY$, marking the intersection points as $F$ and $G$. Finally, draw a straight line through $F$, $O$, and $G$ to form the perpendicular to $XY$ at $O$.

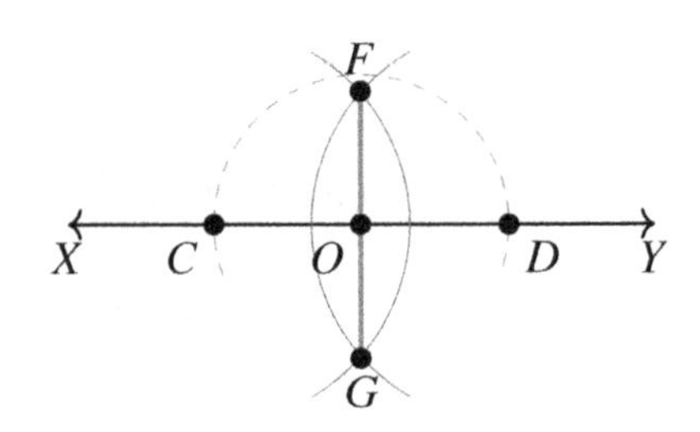


## 1.6 Lines, Rays, and Angles

A *line* is an endless one-dimensional figure, which means it stretches out infinitely in both directions with no endpoints. For instance, a line passing through points $A$ and $B$ is denoted as $\overleftrightarrow{AB}$.

$\overleftrightarrow{AB}$

$A$ $B$

**Key Point**

Lines are infinite in both directions and are denoted by a line over the two points, like $\overleftrightarrow{AB}$.

A *ray* is part of a line that begins at a specific point and extends infinitely in one direction. For example, a ray with an endpoint at $A$ that passes through another point $B$ is represented as $\overrightarrow{AB}$.

$\overrightarrow{AB}$

$A$ $B$

**Key Point**

Rays start at a point and extend infinitely in one direction, depicted as $\overrightarrow{AB}$ for a ray that starts at $A$ and passes through $B$.

*Angles* are formed by two rays sharing a common endpoint, or vertex. The measure of an angle represents the rotation needed to align these rays. For instance, the angle formed by rays $\overrightarrow{BA}$ and $\overrightarrow{BC}$ is denoted as $\angle ABC$, with $B$ as the vertex.

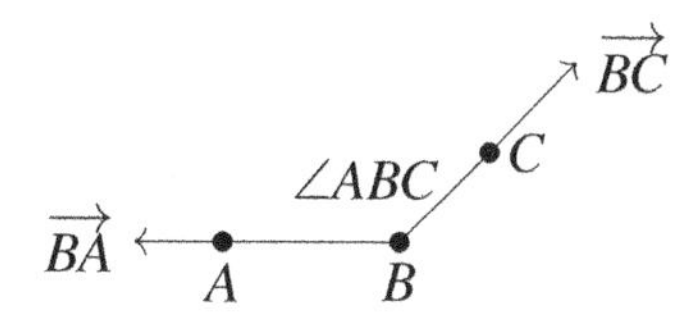


**Key Point**

The total sum of angles around a point is always 360°, useful for solving problems involving several angles.

**Example** Consider a point $M$ with rays extending to points $L$ and $N$.

If $\angle OMN = (x+10)°$, $\angle LMO = 30°$, and $\angle LMN = (3x-10)°$, find the numerical value of $\angle LMN$.

**Solution:** Knowing an angle sum around point $M$, we have $\angle LMN = \angle LMO + \angle OMN$. Solving for $x$: $3x - 10 = 30 + (x+10) \Rightarrow 2x = 50 \Rightarrow x = 25$.
Substituting $x$: $\angle LMN = 3(25) - 10 = 75 - 10 = 65°$. So, $\angle LMN = 65°$.

## 1.7 Types of Angles

In geometry, when two rays connect at a common endpoint, known as the vertex, they form an angle. The amount of turn between each ray is measured in degrees, and based on this measurement, we classify angles into several types.

**Key Point**

Angles are measured in degrees, and their type is determined by the size of their measurement.

An *acute angle* is formed when the measure of an angle is less than 90°. These angles appear quite often in various geometric figures, especially in triangles.
A *right angle* is exactly 90°. It is one of the most basic angle types and plays a crucial role in definitions, theorems, and properties across geometry.
An *obtuse angle* measures more than 90° but less than 180°. These angles are common in geometric shapes like obtuse triangles.

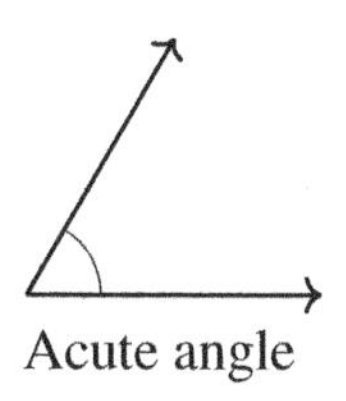
Acute angle

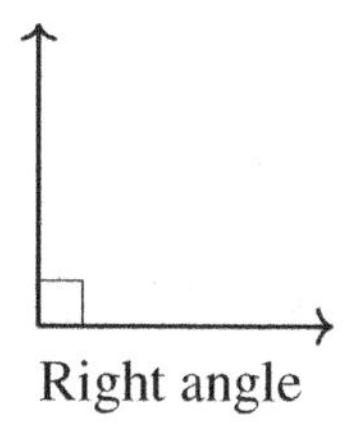
Right angle

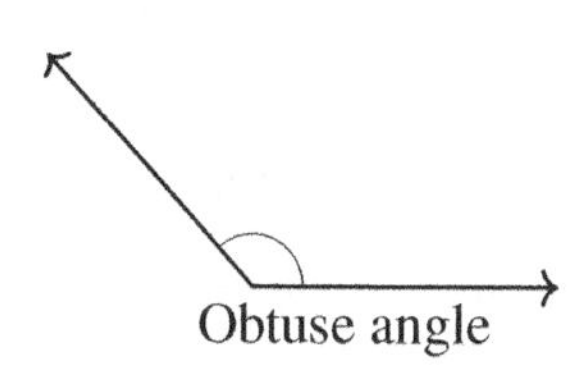
Obtuse angle

A *straight angle* measures exactly 180°, forming a straight line. This type of angle serves as a bridge between obtuse angles and reflex angles.

A *reflex angle* is an angle greater than 180° but less than 360°. It is generally the larger angle that one can trace when looking past a line or along the outside of a simple shape.

A *full rotation* or full angle measures 360°, signifying a complete turn around a point.

Given an angle that measures 60°, what type of angle is it?

**Solution:** Since 60° is less than 90°, it is an acute angle.

Given an angle that measures 95°, what type of angle is it?

**Solution:** Since 95° is more than 90° but less than 180°, the angle is an obtuse angle.

## 1.8 Complementary and Supplementary Angles

*Complementary angles* are two angles whose measures add up to 90°. These pairs of angles are significant in various proofs and problem-solving situations.

**Key Point**

Two angles are considered *complementary* if their measures add up to 90°, which is the measure of a right angle.

*Supplementary angles* are two angles whose sum is 180°, which can often be found along a straight line or dealing with polygons, especially rectangles and squares.

**Key Point**

Two angles are *supplementary* if their measures sum to 180°, which is the measure of a straight angle.

If one angle measures 70° and the other angle measures 20°, are the two angles complementary?

**Solution:** The sum of the angles is $70° + 20° = 90°$. Since their sum is 90°, the two angles are complementary.

 Find the measure of angle $x$ if it forms a complementary angle with a $25^\circ$ angle.

**Solution:** Since $x$ and the $25^\circ$ angle are complementary, we know that their sum should be $90^\circ$. Therefore, $x = 90 - 25 = 65$.

 If angles $A$ and $B$ are supplementary and angle $A$ measures $125^\circ$, what is the measure of angle $B$?

**Solution:** Since angles $A$ and $B$ are supplementary, they must add up to $180^\circ$. Thus, we have: $B = 180 - 125 = 55$. Therefore, the measure of angle $B$ is $55^\circ$.

## 1.9 Bisecting an Angle

Bisecting an angle is the process of dividing an angle into two equal parts.

To bisect an angle, use a compass and straightedge to create two congruent angles from the original angle.

Here is a step-by-step guide to bisecting an angle using just a compass and a straightedge:

**Locate the Vertex:** Begin with an angle, for example, $\angle ABC$ where $B$ is the vertex.

**Draw an Arc:** Place the compass point on vertex $B$ and select an arbitrary radius to draw an arc intersecting both rays $BA$ and $BC$ at points $D$ and $E$, respectively.

**Draw Another Arc:** Without altering the compass width, set the compass on $D$ and draw a semi-arc within the angle. Repeat this step with the compass on $E$ to create another semi-arc until they intersect at a new point $F$.

**Draw the Bisector:** Now, use the straightedge to draw a new ray from vertex $B$ through point $F$. This new line, $BF$, is the bisector of $\angle ABC$, splitting it into two angles of equal measure.

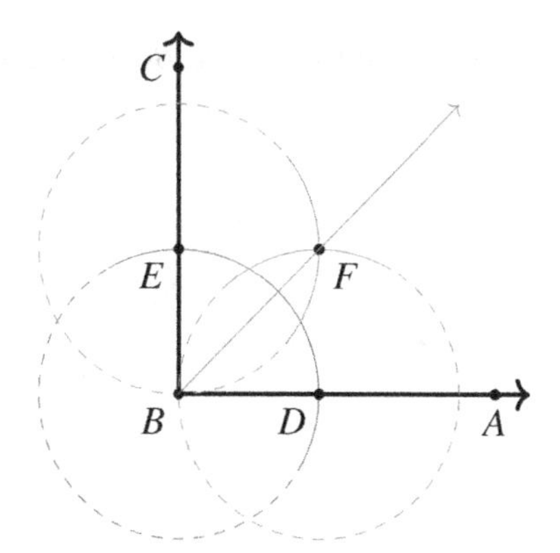

 Given $\angle XYZ$ with vertex $Y$, if the angle measures $70^\circ$, what will be the measure of the angles formed after bisecting it?

**Solution:** After bisecting $\angle XYZ$, each of the two congruent angles formed will measure half of 70°. That is, $\frac{70^\circ}{2} = 35^\circ$. Thus, both angles will measure 35° each.

 Given angle $\angle ABC$ with its bisector $BD$.

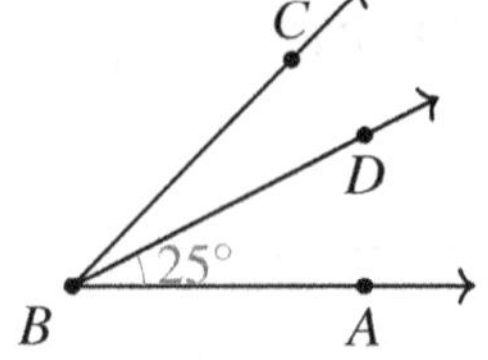


If the measure of angle $ABD$ is 25°, what is the measure of angle $DBC$?

**Solution:** Since $BD$ is the bisector, the two angles $ABD$ and $DBC$ are congruent. Thus, angle $DBC$ also measures 25°.

## 1.10 Constructing a Triangle Given Its Sides

A triangle can be constructed using the lengths of its three sides, known as the SSS (Side-Side-Side) criterion.

Triangle construction using the Side-Side-Side (SSS) criterion involves determining a unique triangle from three known side lengths.

The steps for constructing a triangle using the SSS Criterion are as follows:

**Draw the Base:** Choose a side as the base, say of length $a$. Draw a line segment $AB$ of length $a$.

**Construct the Second Side:** With the compass set to length $b$, draw an arc from $B$, indicating potential locations for the third vertex.

**Construct the Third Side:** Set the compass to length $c$, draw an arc from $A$. The intersection with the previous arc marks the third vertex.

**Complete the Triangle:** Mark the intersection as $C$. Draw $AC$ and $BC$ to finalize triangle $ABC$.

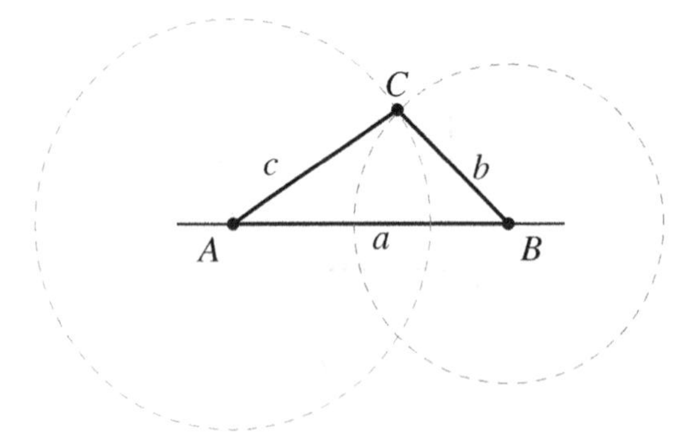

 Construct a triangle with sides measuring 4 $cm$, 2 $cm$, and 3 $cm$.

**Solution:**

Draw a line segment $AB$ of 4 $cm$ (our chosen base). With the compass set to 3 $cm$ (the second side), draw an arc centered at $A$. Similarly, with the compass set to 2 $cm$ (the third side), draw an arc centered at $B$. The intersection of these arcs determines point $C$. Join $A$ to $C$ and $B$ to $C$ to complete the triangle.

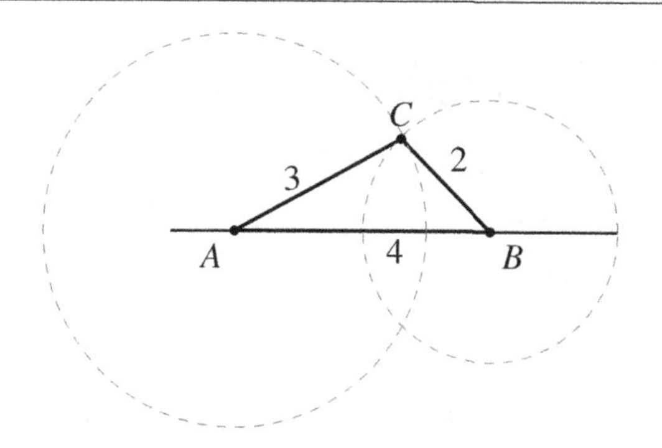


## 1.11 The Circumscribed Circle

A polygon is said to have a *circumcircle*, or *circumscribed circle*, if there exists a circle that passes through all of its vertices. In the case of a triangle, which is the simplest polygon, the construction of its circumcircle is a fascinating application of basic geometric tools such as a compass and straightedge.

**Key Point**

The circumcenter, equidistant from all the vertices of a triangle, is also the center of the circumscribed circle.

To construct the circumcircle of a triangle, follow these steps:

1. Find the *perpendicular bisectors* for each side of the triangle. Identify the midpoint of each side and with a compass and straightedge, draw the line that is perpendicular to the side and passes through the midpoint.
2. Locate the *circumcenter* by finding the point of intersection of all three perpendicular bisectors. Depending on the type of triangle, the circumcenter will be located in different regions; within the triangle for acute triangles, on the hypotenuse for right triangles, and outside the triangle for obtuse triangles.
3. Finally, draw the *circumcircle* by placing the compass point on the circumcenter, adjusting its width to reach one of the vertices, and drawing a circle that will naturally pass through the other two vertices.

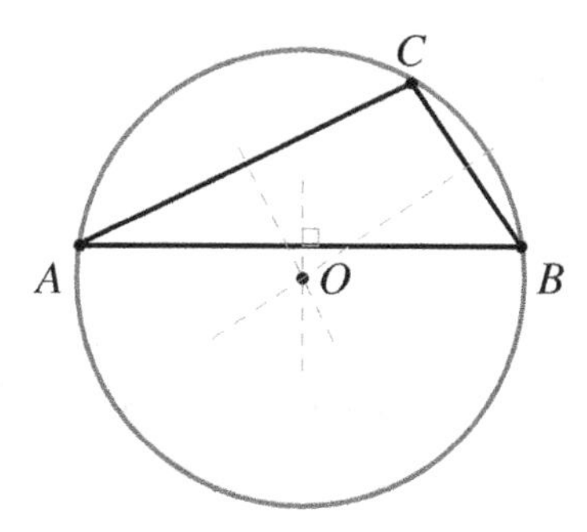


**Key Point**

The circumcenter of a triangle, equidistant from all vertices and defined as the circumradius, serves as the center of the circumscribed circle and is the concurrent point of the sides' perpendicular bisectors.

In triangle ABC with a circumcircle, if the circumradius is 3 cm, what is the distance from the circumcenter to vertex $A$

**Solution:** As the circumradius is the uniform distance from the circumcenter to all vertices, the distance to vertex $A$ is 3 cm.

## 1.12 The Inscribed Circle of a Triangle

The inscribed circle of a triangle, also known as the incircle, is the largest circle that fits entirely within the triangle and touches all three sides.

The incenter of a triangle is equidistant from all three sides of the triangle, making it the ideal center for the incircle.

The construction of the incircle of a triangle involves the following steps:

**Construct the Angle Bisectors:** First, use a compass and straightedge to bisect each angle of the triangle.

**Determine the Incenter:** The point at which all three angle bisectors of the triangle intersect is known as the incenter.

**Find the Inradius:** Use a compass to draw a perpendicular line from the incenter to any side of the triangle. The length of this line is the inradius.

**Draw the Incircle:** Place the compass at the incenter, adjust it to the length of the inradius, and draw the circle. This is your incircle which will touch all three sides of the triangle.

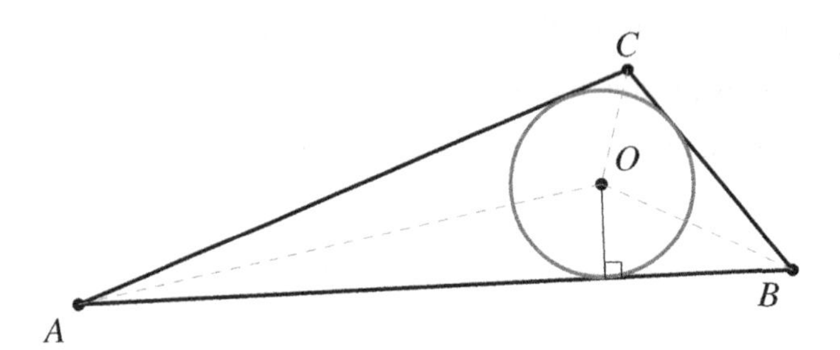


Key Point

A triangle's incircle has several useful properties:

- The incenter, the center of the incircle, is equally distant from all three sides of the triangle.
- The incenter is located at the point where the angle bisectors of the triangle intersect.
- The line segments from the incenter to the points where the incircle touches the triangle's sides are all of equal length.
- You can calculate the area of the triangle by multiplying its inradius (the radius of the incircle) by its semiperimeter (half the triangle's perimeter).

 Given a triangle $DEF$, after constructing the incircle, if the inradius is 5 $cm$ and the triangle's perimeter is 30 $cm$, what is the area of triangle $DEF$?

**Solution:** By the properties of the incircle, we know that the area of the triangle can be determined by using the inradius and semiperimeter: Area of triangle = inradius × semiperimeter,

$$\Rightarrow \text{Area} = 5\ cm \times \frac{30\ cm}{2} = 5\ cm \times 15\ cm = 75\ cm^2.$$

Therefore, the area of the triangle $DEF$ is 75 $cm^2$.

## 1.13 Inscribing Regular Polygons

Inscribing a polygon means drawing it inside a circle so that all of its vertices lie on the circumference of the circle. A polygon is regular when all its sides and angles are equal.

A regular polygon can be inscribed in a circle if all of its vertices lie on the circumference of the circle, creating an equal arc between adjacent vertices.

The general steps to inscribe a regular polygon in a circle are:

**Draw the Circumscribing Circle:** Begin the process by drawing a circle of the desired radius. This will be the outer limit within which the polygon will be inscribed.

**Divide the Circle:** Find the central angle for the polygon using the formula

$$\text{Central Angle} = \frac{360^\circ}{n},$$

where $n$ is the number of sides of the regular polygon. This angle will be used to partition the circle into equal arcs.

**Draw the Polygon:** Select any point on the circle as the first vertex. Then, using a compass or a protractor, step off the calculated central angle around the circle to locate subsequent vertices. After finding all the vertices, connect them with straight lines using a straightedge—this forms the inscribed polygon.

 Draw a regular hexagon inscribed inside a circle.

**Solution:** First, draw a circle with a compass. Since a hexagon has 6 sides, the central angle we should divide the circle into is $\frac{360^\circ}{6} = 60^\circ$. Starting at any point on the circle, this will be the first vertex labeled $A$. Measuring $60^\circ$ around the circle from vertex $A$, find and label the second vertex as $B$. Continue this process until all six vertices, $A, B, C, D, E$, and $F$, are located. Then connect these vertices with straight lines to form the inscribed hexagon (see the following graph).

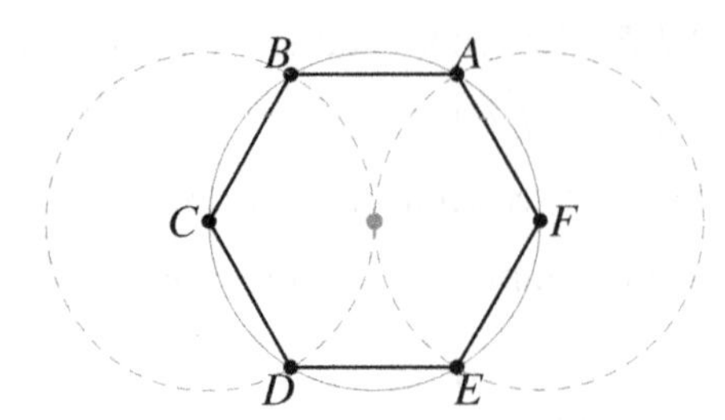


Note that the vertices of the hexagon are equally spaced around the circle, and the dashed circles show the consistent radius from points $C$ and $F$ to their surrounding vertices.

## 1.14 Practices

**1)** Classify each statement as true or false.

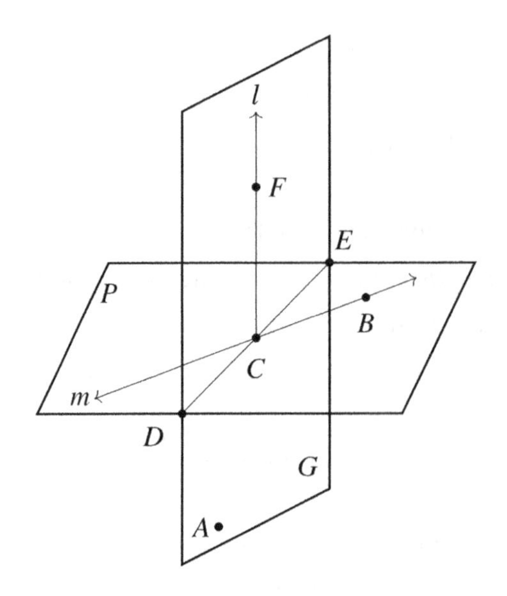


1-1) $CF$ is in plane $P$.

1-2) $E$ is in plane $P$.

1-3) $B$, $C$, and $A$ are coplanar.

1-4) Plane $G$ contains $C$, $D$, $E$, and $F$.

1-5) Line $m$ intersects $CF$ at $C$.

**2)** Find the length of $AB$.

2-1)

14 *cm*

$A$ $B$ 5 *cm* $C$

2-2)

20 *cm*

$C$ 13 *cm* $B$ $A$

2-3)

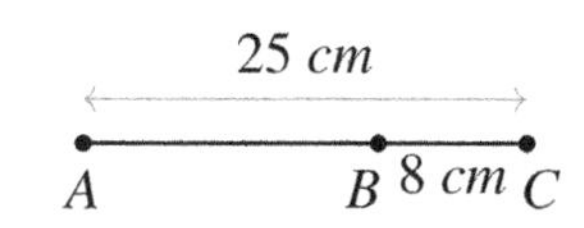


2-4)

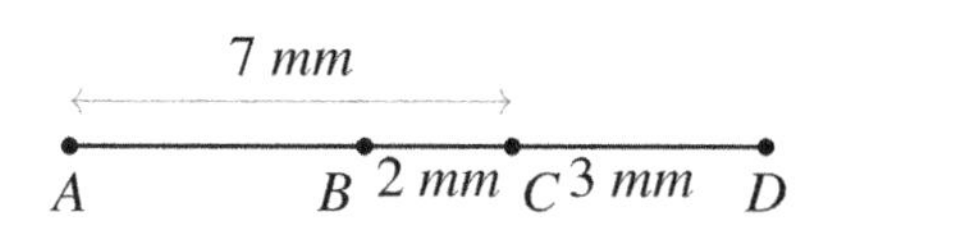


2-5)

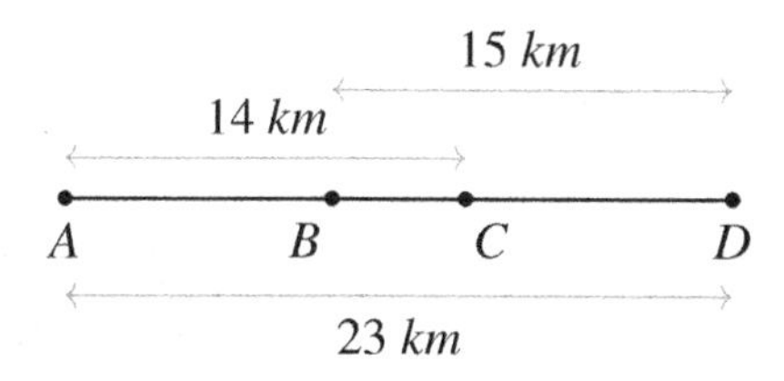


2-6)

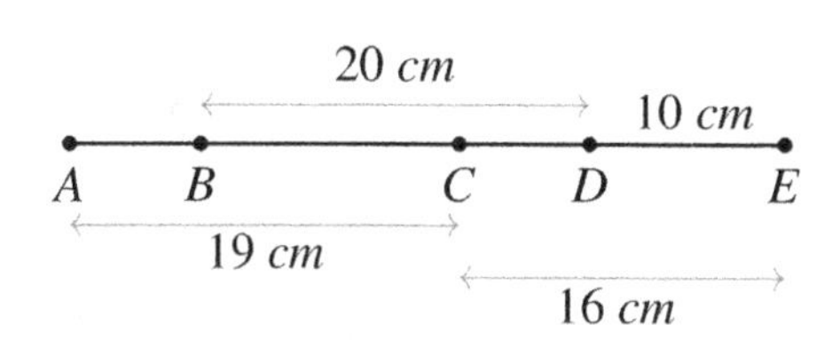


**3)** Points $A$, $B$ and $C$ are collinear. Point $B$ is between $A$ and $C$. find the length indicated.

3-1) Find $AC$ if $AB = 44$ and $BC = 21$.

3-2) Find $AC$ if $AB = 4$ and $BC = 9$.

**4)** Points $A$, $B$ and $C$ are collinear. Point $B$ is between $A$ and $C$. Solve for $x$.

4-1) $AC = x + 29$, $AB = -3 + 4x$, and $BC = 23$.

4-2) $AC = 24$, $BC = x + 4$, and $AB = x + 8$.

**5)** Locate the midpoint of each line segment.

5-1)

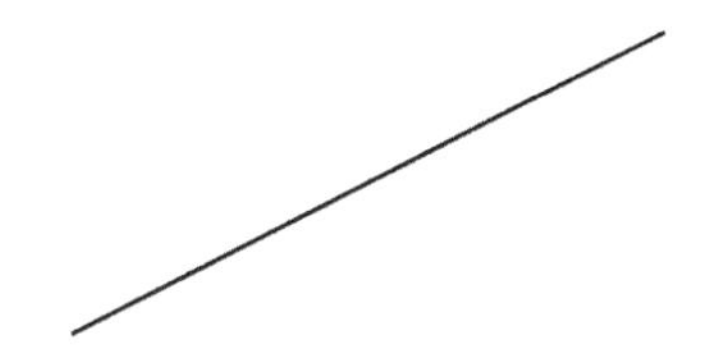

5-2)

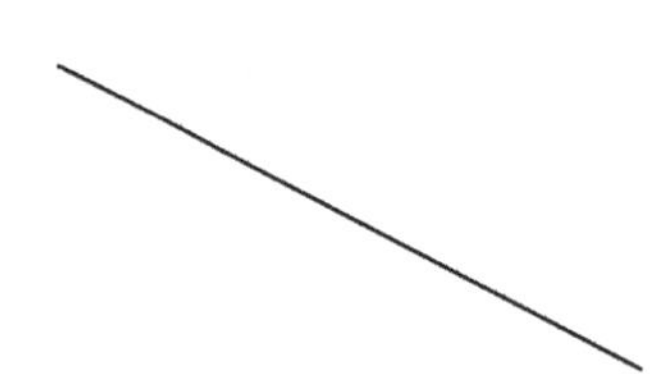

**6)** Solve.

6-1) 18) Construct a line segment half as long as the given line segment.

6-2) Divide the line segment into the 3 equal parts.

**7)** Find the distance between each two points.

7-1) $-2$, 12

7-2) 5, 9

7-3) $-14$, $-3$

7-4) 6, $-7$

7-5)

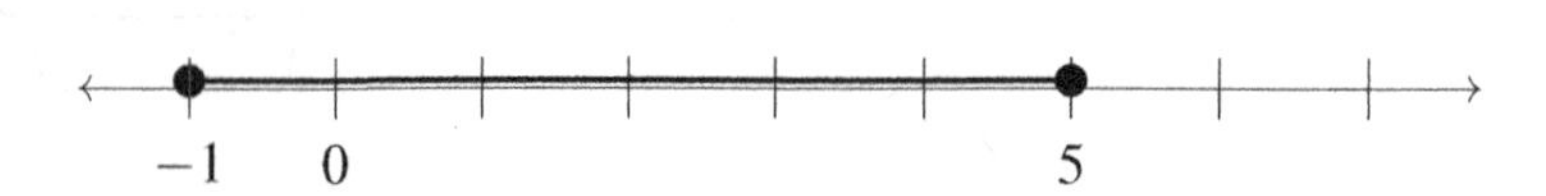


7-6)

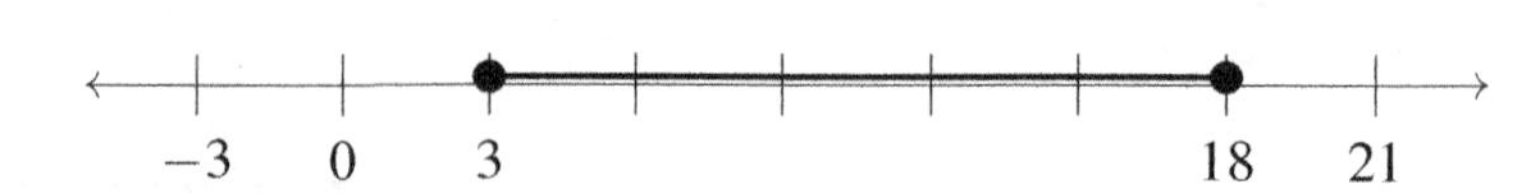


**8)** Identify each pair of angles as corresponding, alternate interior, alternate exterior, or consecutive interior and find the measure of each angle indicated.

8-1)

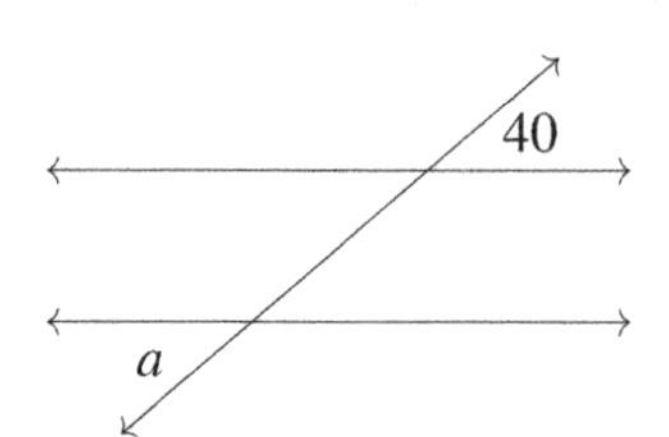


8-2)

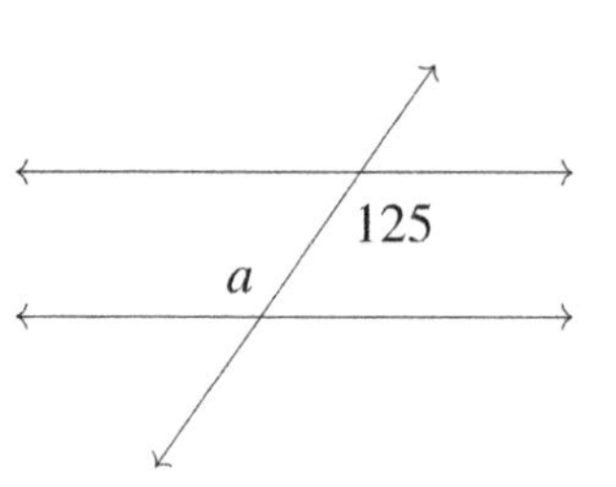


8-3)

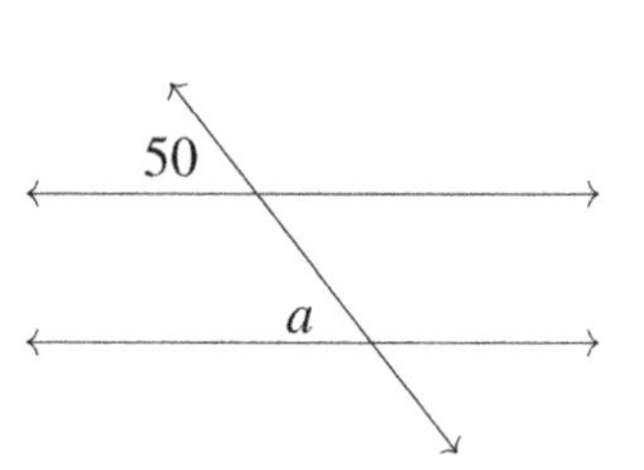


8-4)

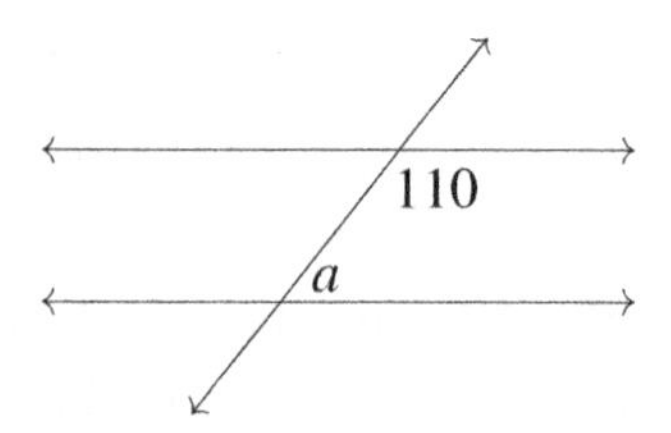


8-5)

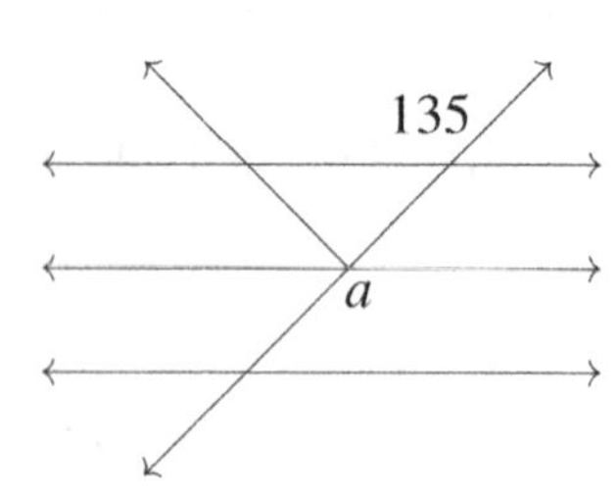


8-6)

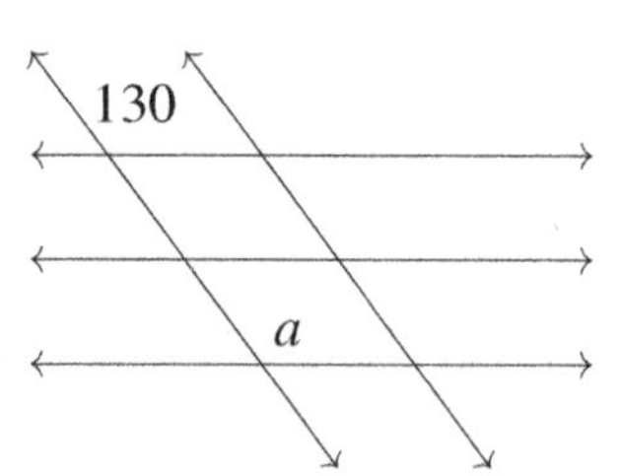


**9)** Draw a perpendicular line from point $E$ on line $AB$.

9-1)

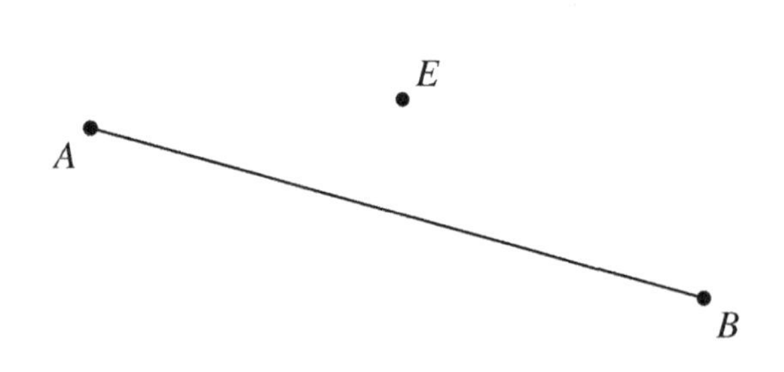


9-2)

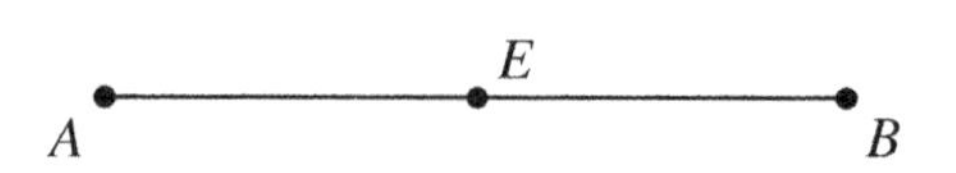

9-3)

9-4)

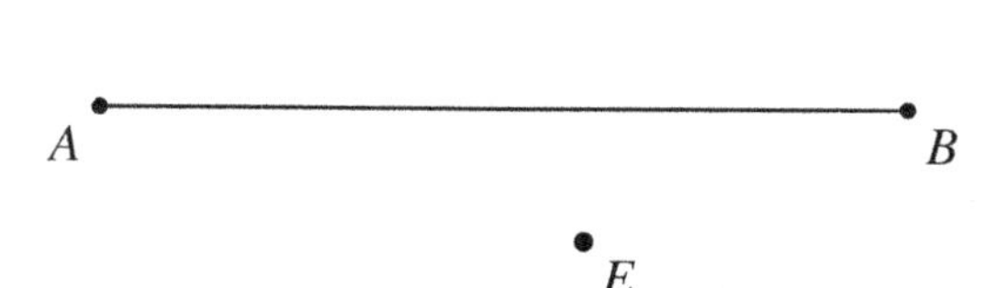


**10)** Find the measure of each angle to the nearest degree.

10-1)

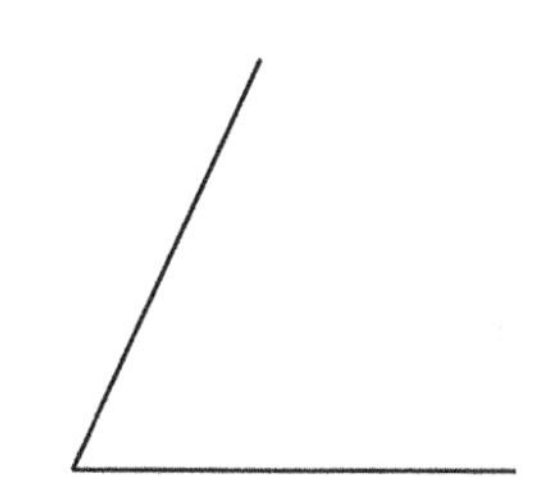

10-2)

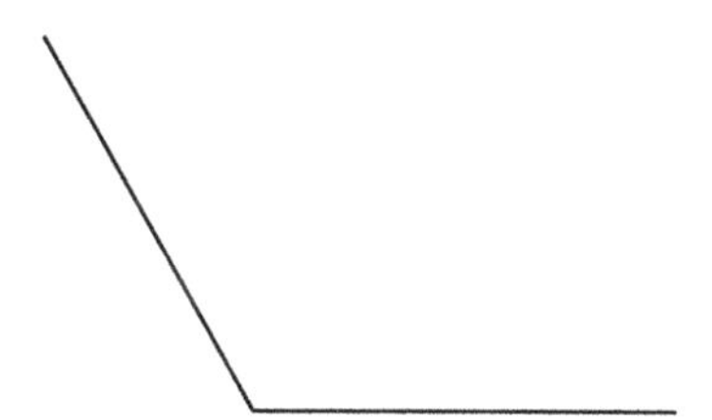

10-3)

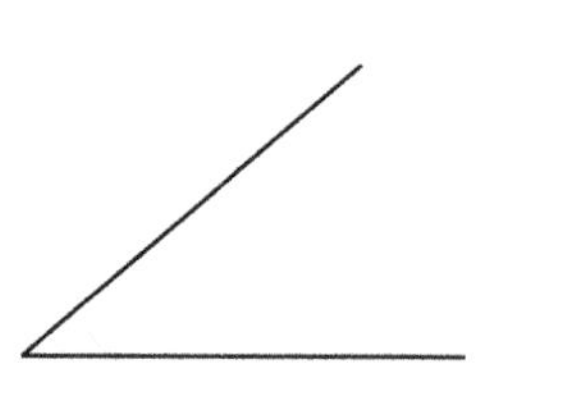

10-4)

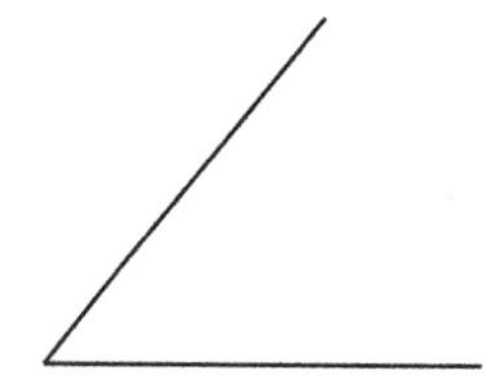

10-5)

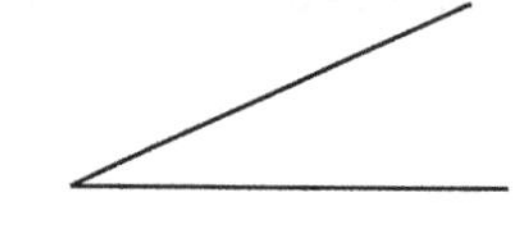

10-6)

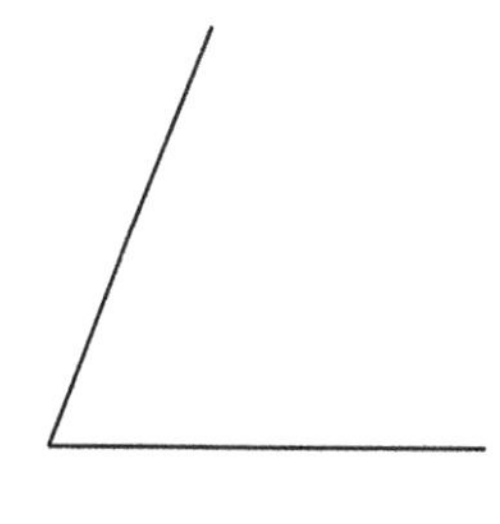

10-7)

10-8)

**11)** Classify each angle as acute, obtuse, right, or straight.

11-1)

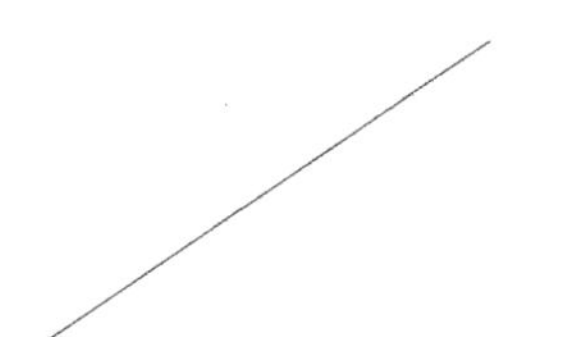

11-2)

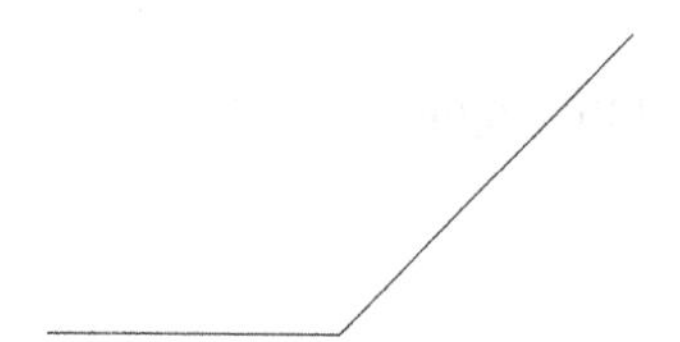

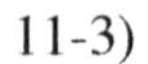
11-3)

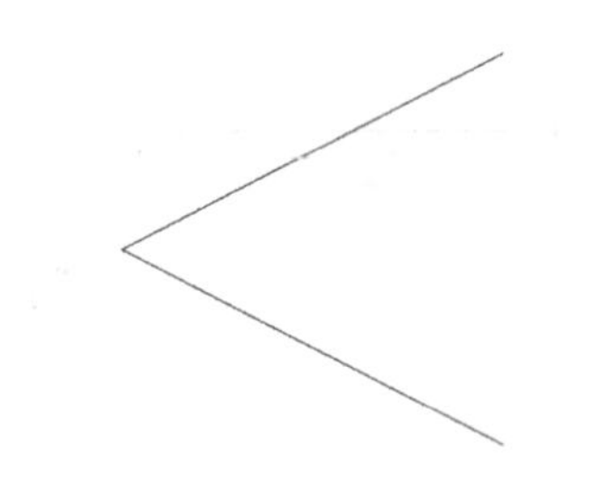

11-5)

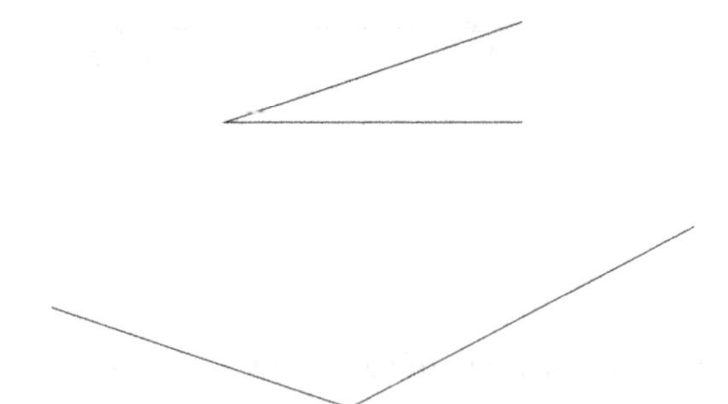

11-6)

11-4)

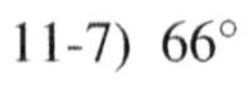
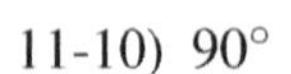
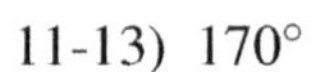

11-7) 66°

11-8) 122°

11-9) 180°

11-10) 90°

11-11) 10°

11-12) 45°

11-13) 170°

11-14) 91°

11-15) 89°

**12)** Name the vertex and sides of each angle.

12-1)

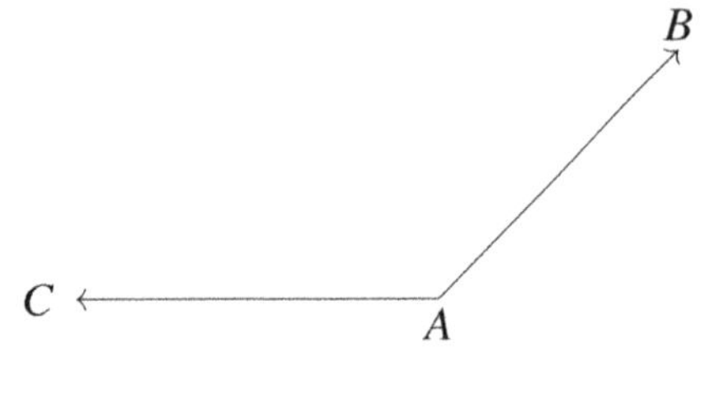


12-3)

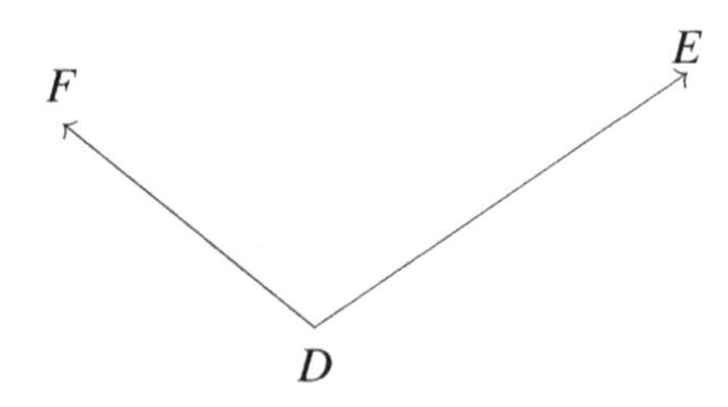


12-2)

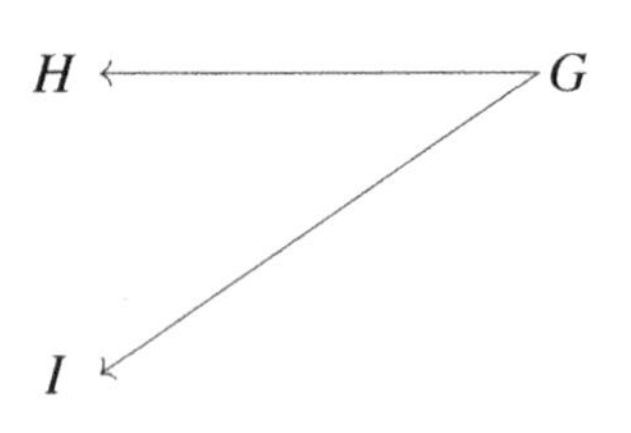


12-4)

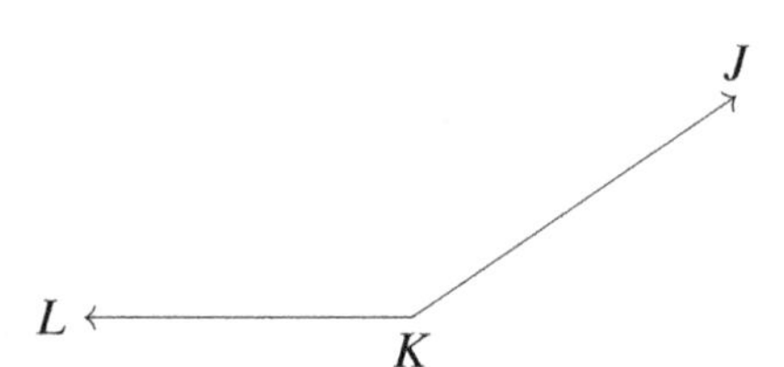


**13)** Draw and label an angle to fit each description.

13-1) An acute angle, $\angle A$

13-2) A right angle, $\angle B$

13-3) An obtuse angle, $\angle ABC$

13-4) A straight angle, $\angle DEF$

**14)** Name all the angles that have $V$ as a vertex.

14-1)

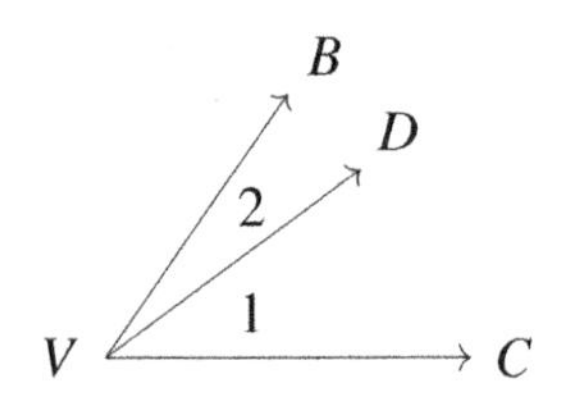

14-2)

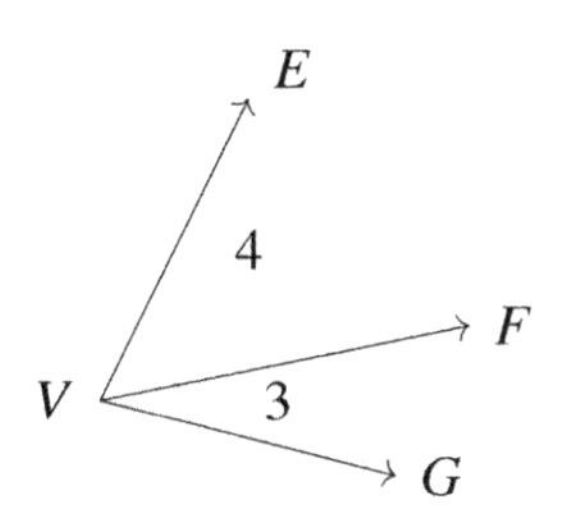


**15)** Draw and label an angle to fit each description.

15-1) $\angle GDF = 35^\circ$ and $\angle FDE = 70^\circ$. Find $\angle GDE^\circ$.

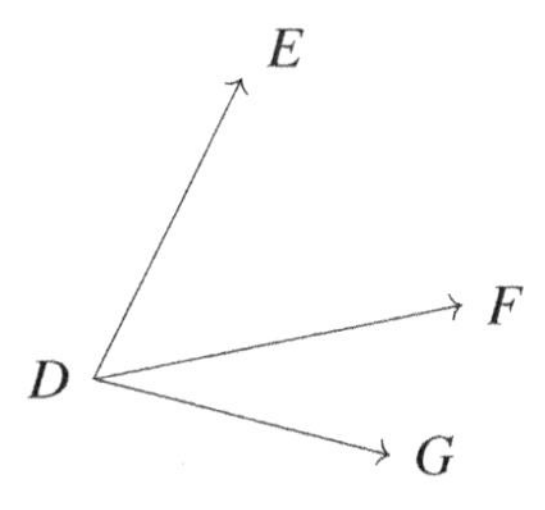


15-2) Find $\angle KHJ$, if $\angle KHI = 145^\circ$ and $\angle JHI = 45^\circ$.

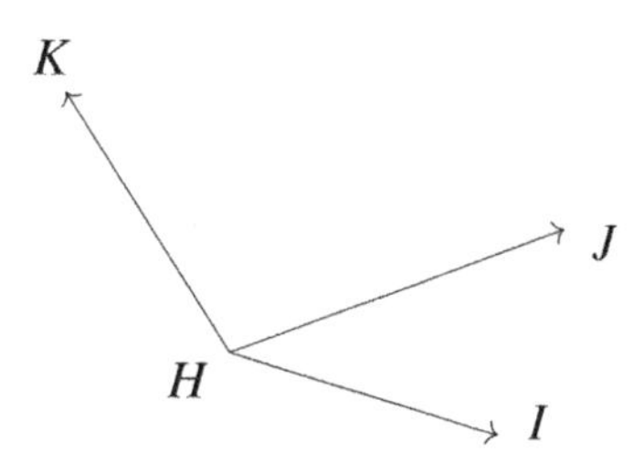


15-3) $\angle OLN = 50^\circ$ and $\angle OLM = 180^\circ$. Find $\angle NLM$.

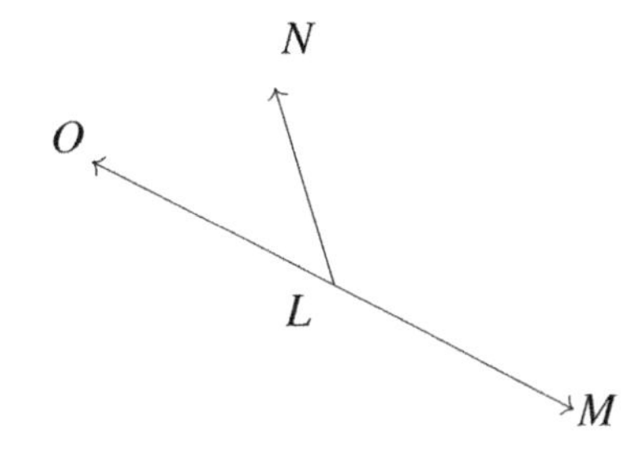

15-4) Find $\angle RPO$, if $\angle SPR = 90^\circ$ and $\angle SPQ = 120^\circ$.

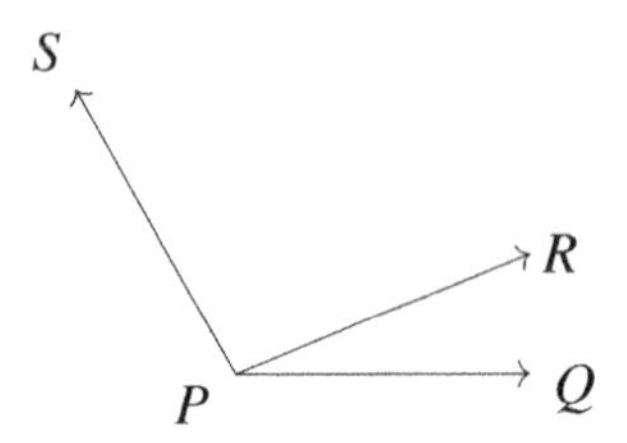


**16)** Construct the bisector of each angle.

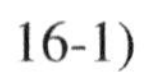

16-1)

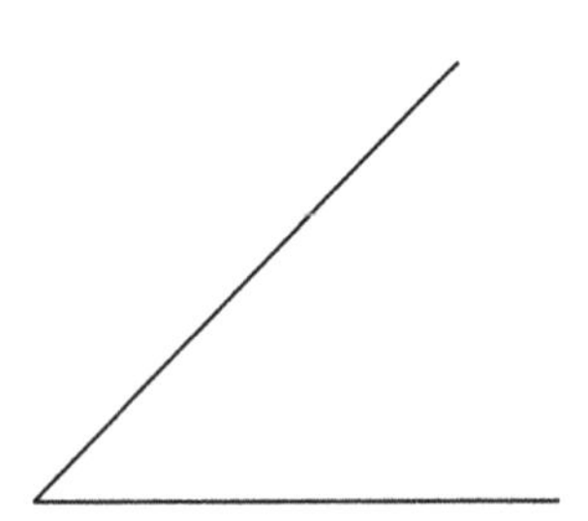

16-2)

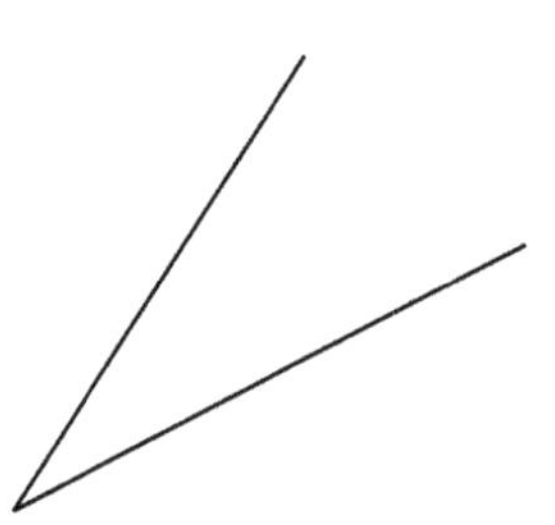

16-3)

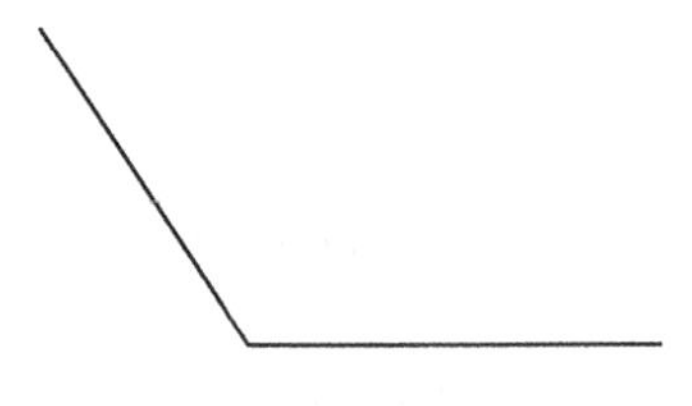

16-4)

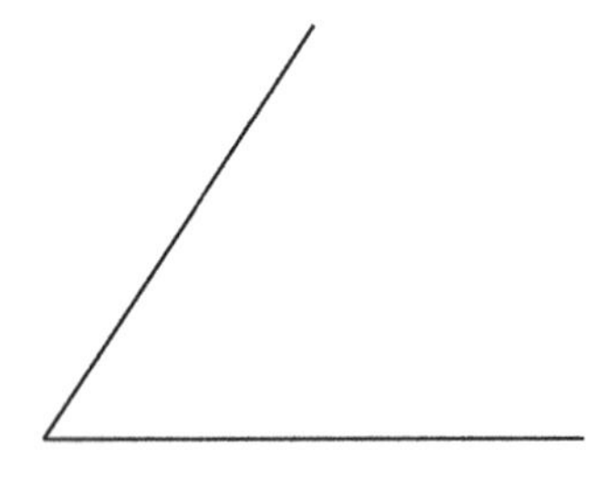

**17)** Construct the bisector of angle $M$.

17-1)

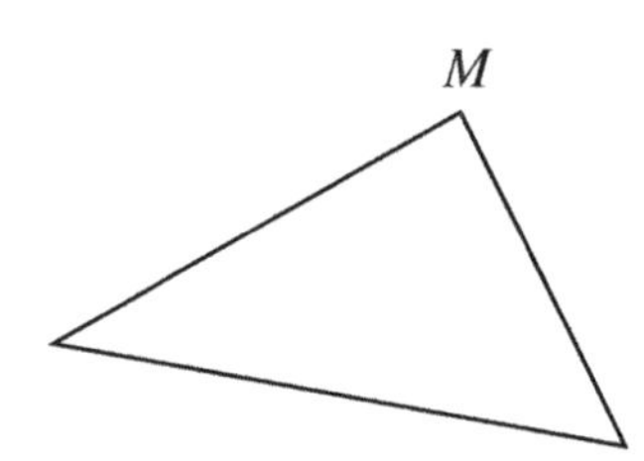


17-2)

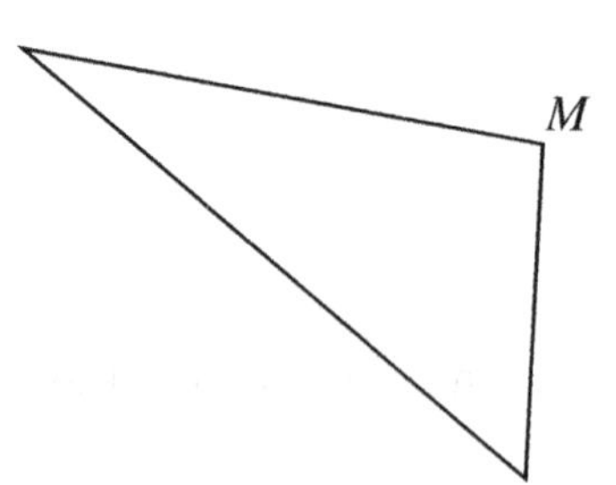

17-3)

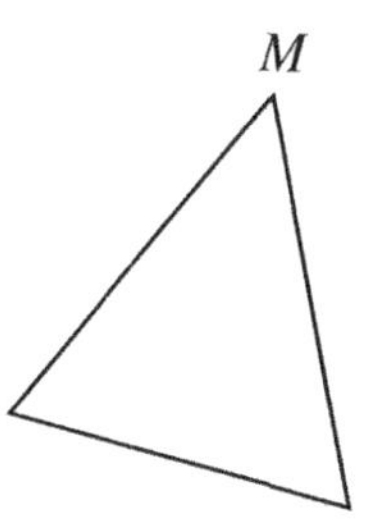


17-4)

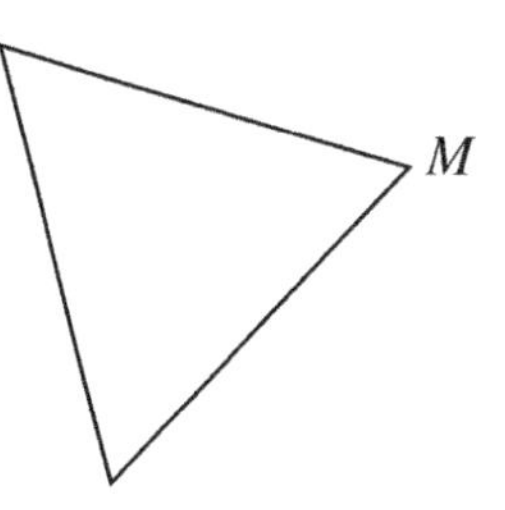


**18)** Construct a triangle whose sides are half as long as the sides of the given triangle.

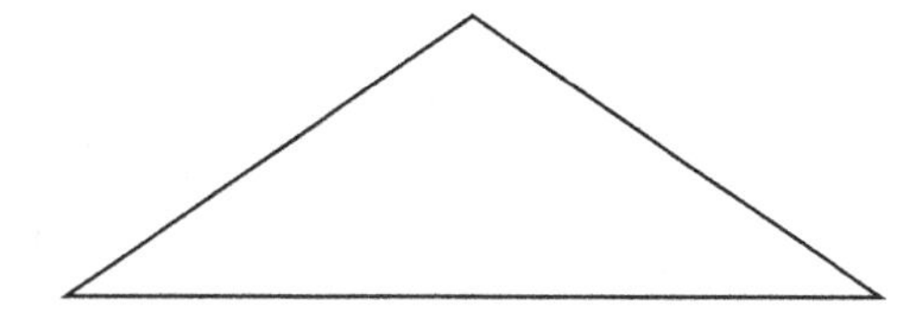

**19)** Construct an isosceles triangle according to the given lengths.

19-1) Base: 3, Side: 4.

19-2) Base: 6, Altitude: 4

**20)** Construct a right triangle with a hypotenuse of 5 cm and one leg measuring 3 cm.

**21)** Locate the circumcenter of each triangle.

21-1)

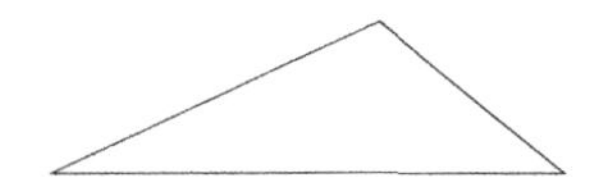

21-2)

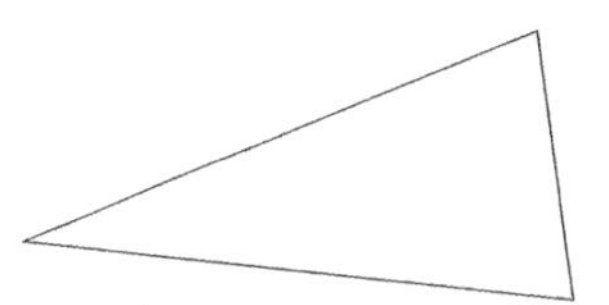

**22)** Circumscribe a circle about each triangle.

22-1)

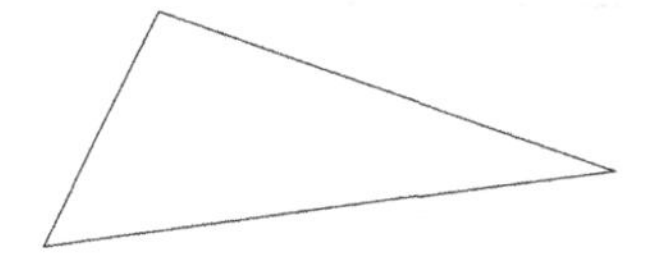

22-2)

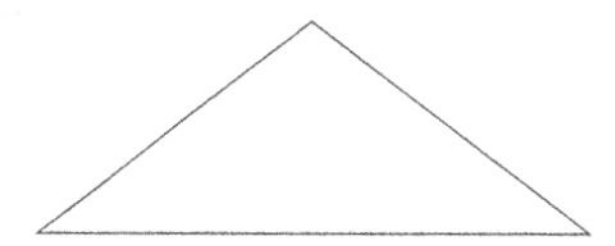

**23)** Find the radius of the inscribed circle in the triangle. Round your answer to nearest tenth.

23-1)

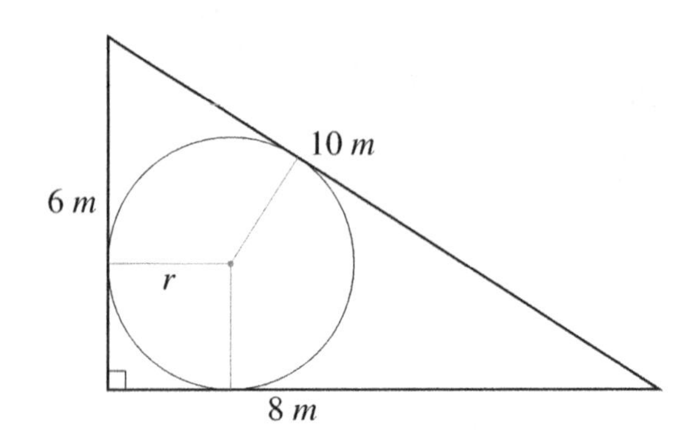


23-2)

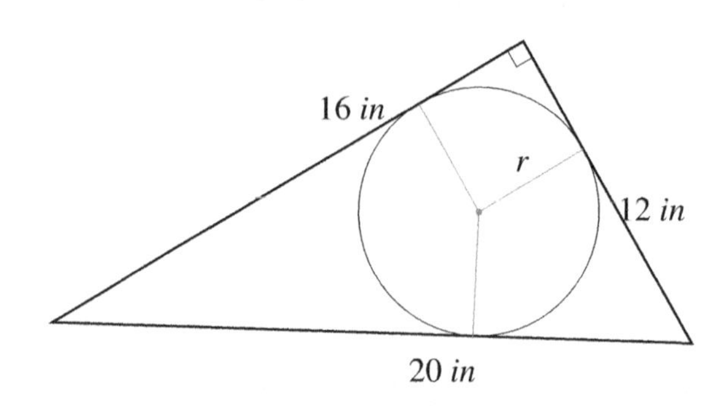


23-3) $h = 13.9\ in$

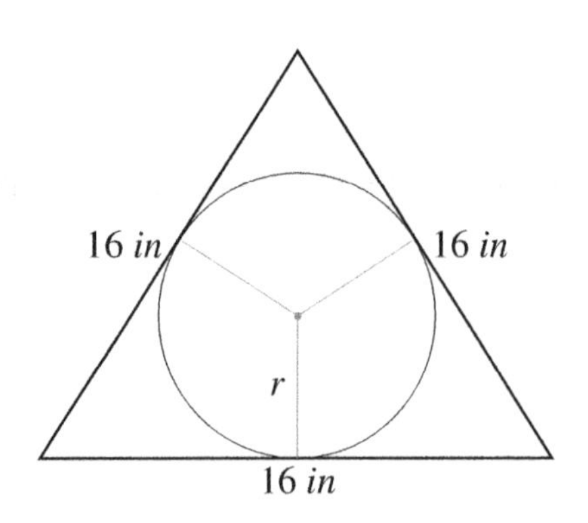


23-4) Area= 8.2 $cm^2$

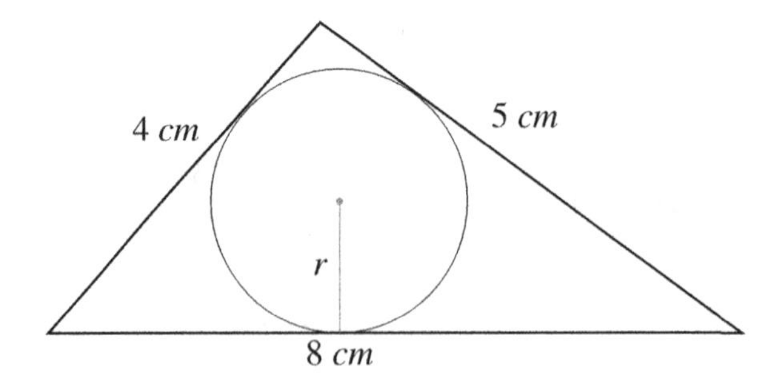


**24)** Draw the given Regular polygon, inside a circle with compass.

24-1) Heptagon

24-2) Pentagon

24-3) Octagon

## 1.15 Answers

**1)**

1-1) False

1-2) True

1-3) False

1-4) True

1-5) True

**2)**

2-1) 9 *cm*

2-2) 7 *cm*

2-3) 17 *m*

2-4) 2 *mm*

2-5) 8 *km*

2-6) 5 *cm*

**3)**

3-1) 65

3-2) 13

**4)**

4-1) 3

4-2) 6

**5)**

5-1)

5-2)

**6)**

6-1)

6-2)

**7)**

7-1) 14

7-2) 4

7-3) 11

7-4) 13

7-5) 6

7-6) 15

**8)**

8-1) Alternate exterior, $a = 40^\circ$

8-2) Alternate interior, $a = 125^\circ$

8-3) Corresponding, $a = 50^\circ$

8-4) Consecutive interior, $a = 70^\circ$

8-5) Alternate exterior, $a = 135^\circ$

8-6) Corresponding, $a = 130^\circ$

**9)**

9-1)

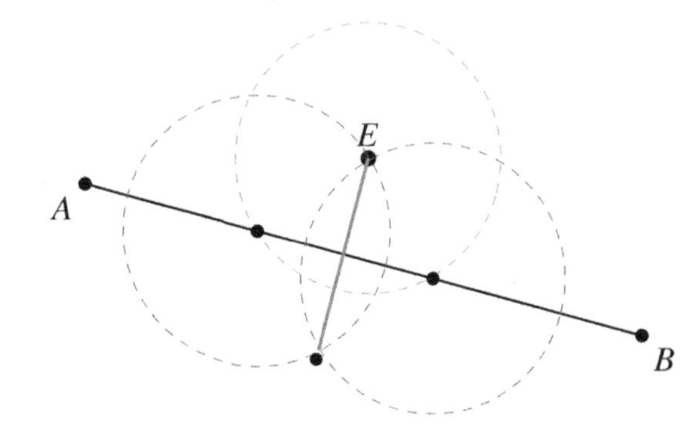


9-2)

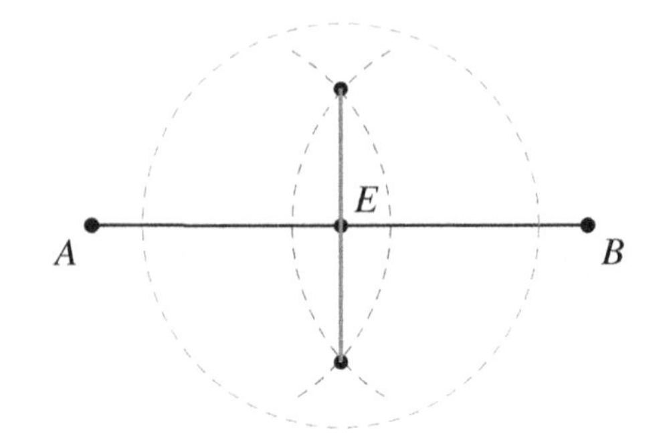

9-3)

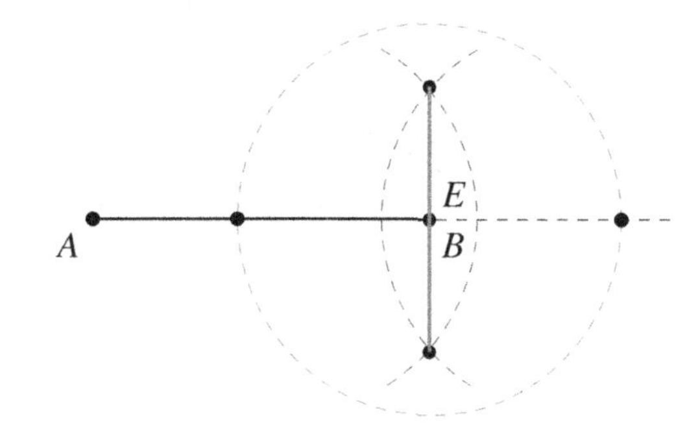


9-4)

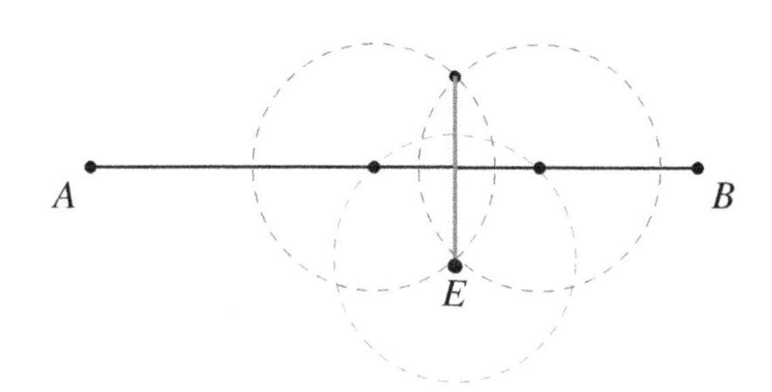


**10)**

10-1) 65°

10-2) 120°

10-3) 40°

10-4) 50°

10-5) 24°

10-6) 68°

10-7) 10°

10-8) 150°

**11)**

11-1) Acute

11-2) Obtuse

11-3) Acute

11-4) Straight

11-5) Acute

11-6) Obtuse

11-7) Acute

11-8) Obtuse

11-9) Straight

11-10) Right

11-11) Acute

11-12) Acute

11-13) Obtuse

11-14) Obtuse

11-15) Acute

**12)**

12-1) Vertex: point $A$
Sides: $\overrightarrow{AB}$, $\overrightarrow{AC}$

12-2) Vertex: point $G$
Sides: $\overrightarrow{GH}$, $\overrightarrow{GI}$

12-3) Vertex: point $D$
Sides: $\overrightarrow{DE}$, $\overrightarrow{DF}$

12-4) Vertex: point $K$
Sides: $\overrightarrow{KL}$, $\overrightarrow{KJ}$

**13)**

13-1)

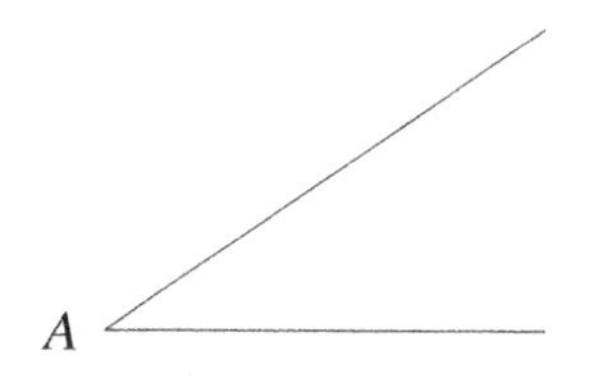


13-2)

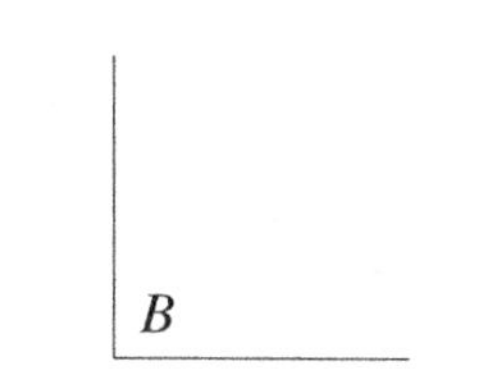


13-3)

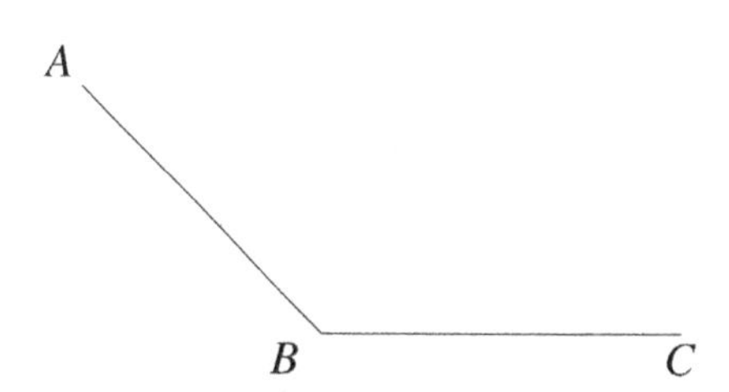


13-4)

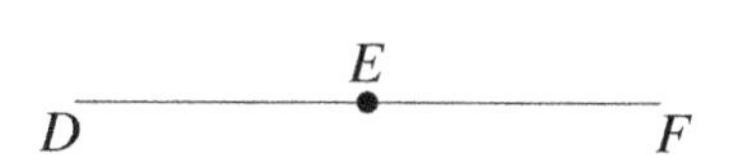


**14)**

14-1) $\angle BVD$, $\angle BVC$, and $\angle DVC$

14-2) $\angle EVF$, $\angle EVG$, and $\angle FVG$

**15)**

15-1) 105°

15-2) 100°

15-3) 130°

15-4) 30°

**16)**

16-1)

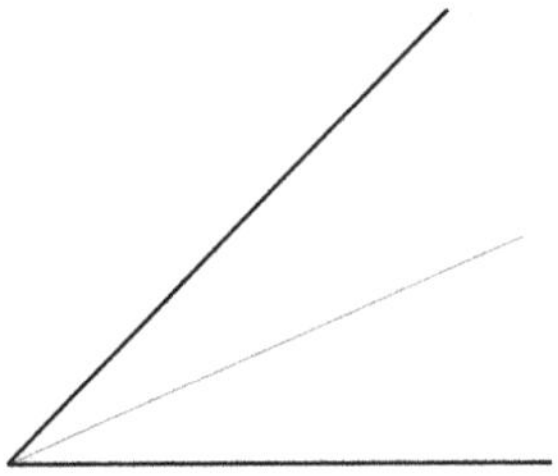

16-2)

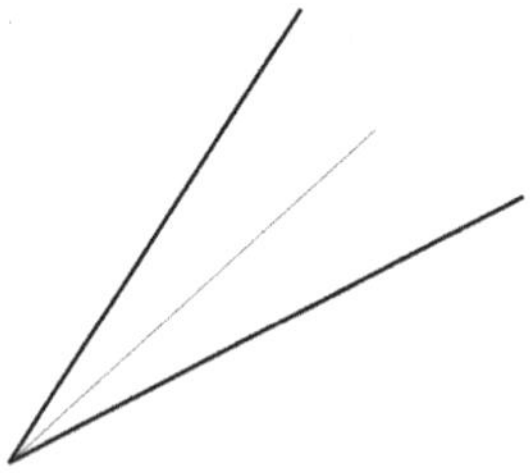

16-3)

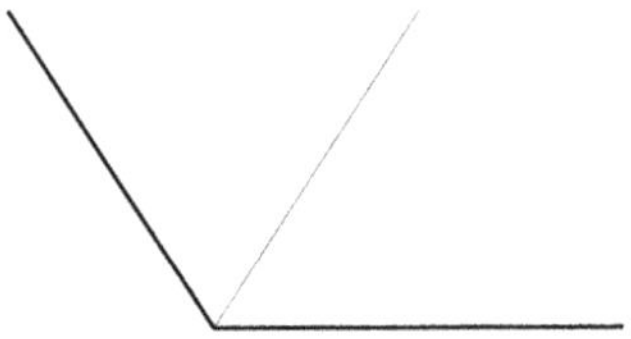

16-4)

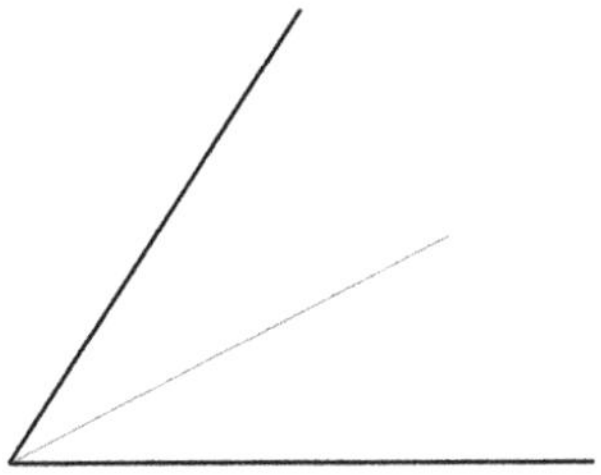

**17)**

17-1)

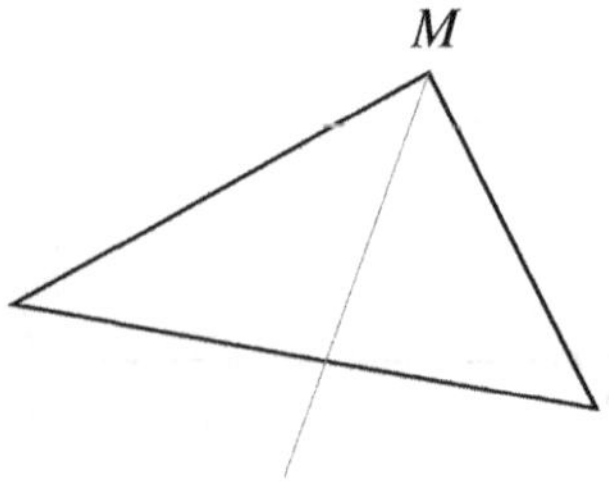

M

17-2)

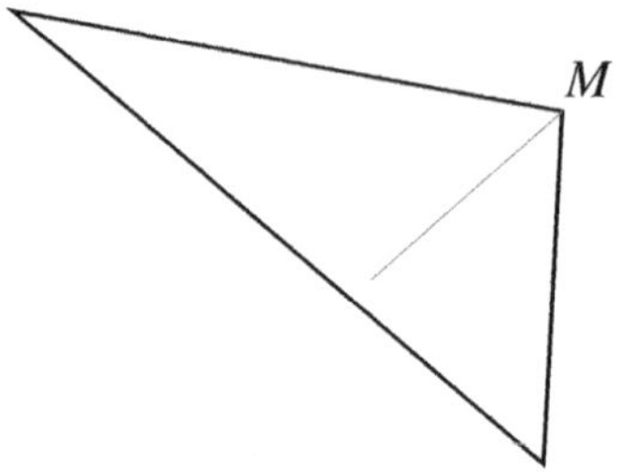

M

17-3)

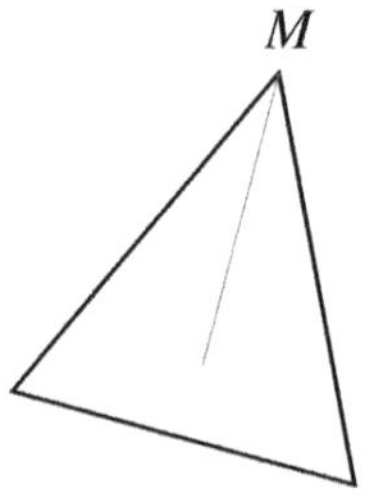

M

17-4)

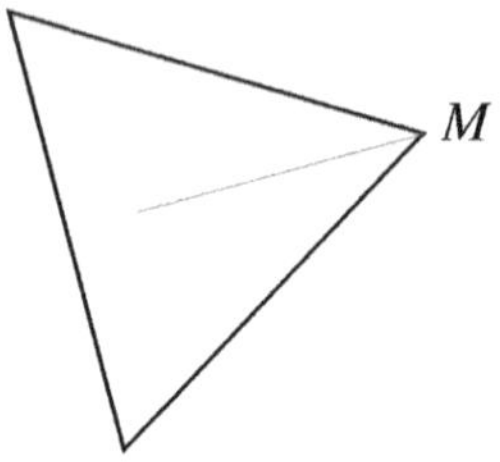

M

**18)**

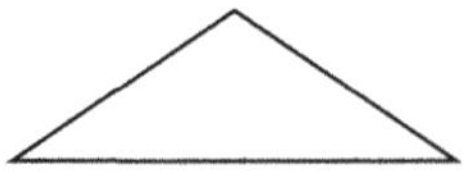

**19)**

19-1)

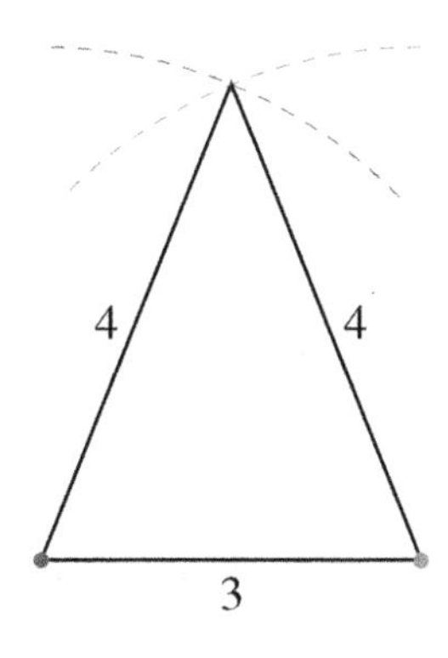


19-2)

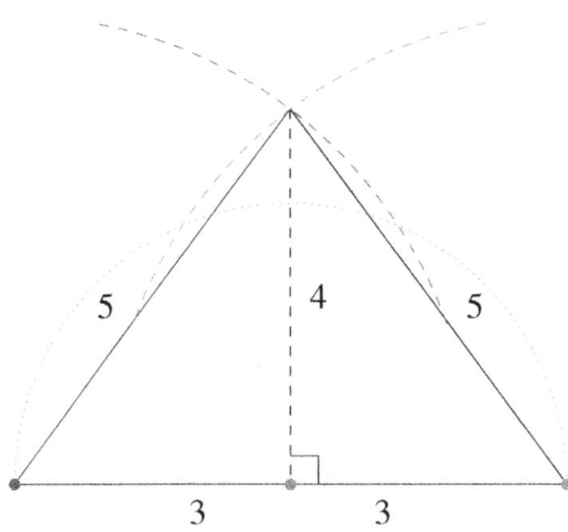


**20)**

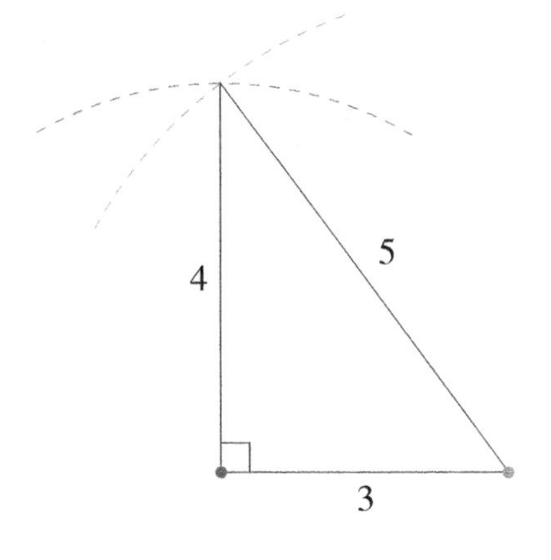


**21)**

21-1)

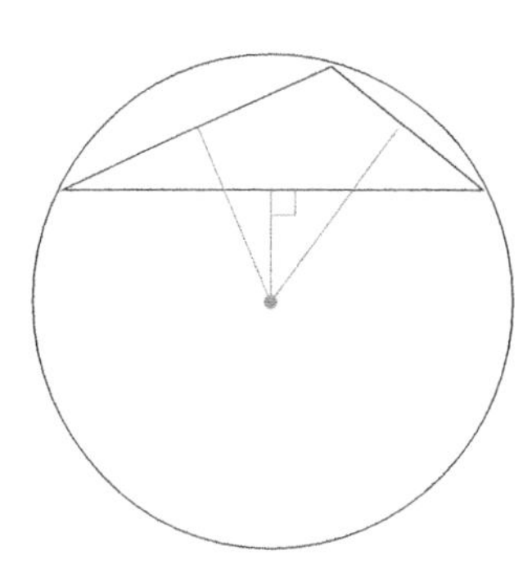

21-2)

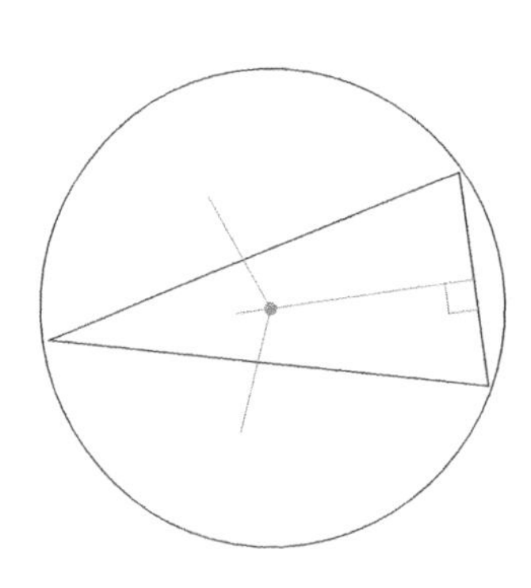

**22)**

22-1)

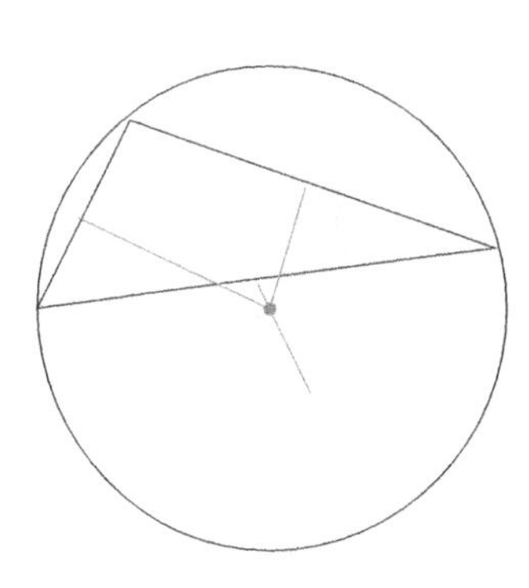

22-2)

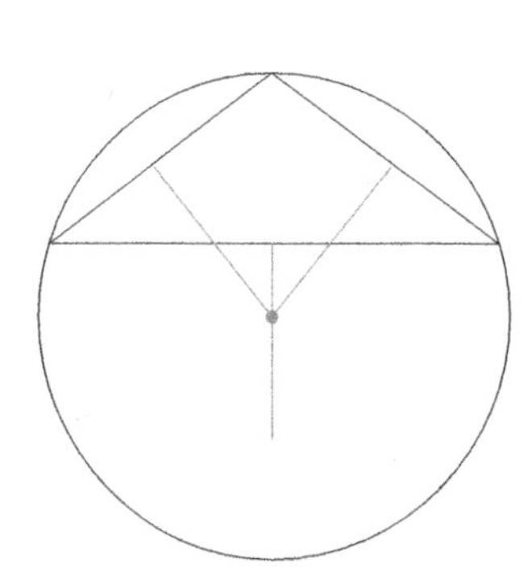

**23)**

23-1) $r = 2\ m$

23-2) $r = 4\ in$

23-3) $r \approx 4.6\ in$

23-4) $r \approx 0.96\ cm$

**24)**

24-1)

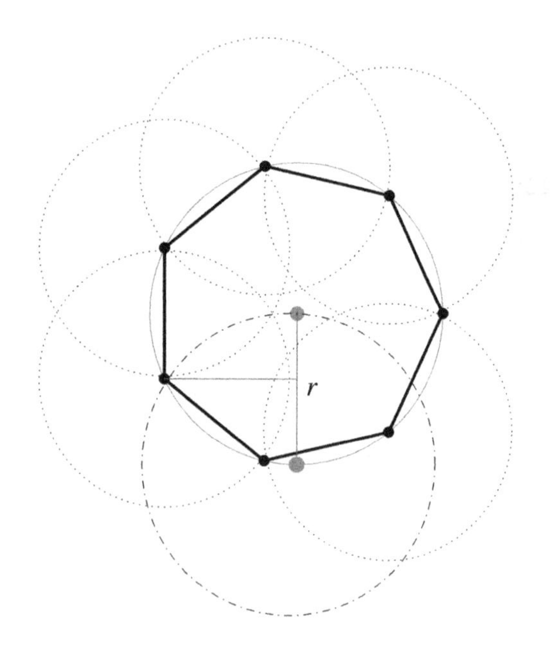


24-2)

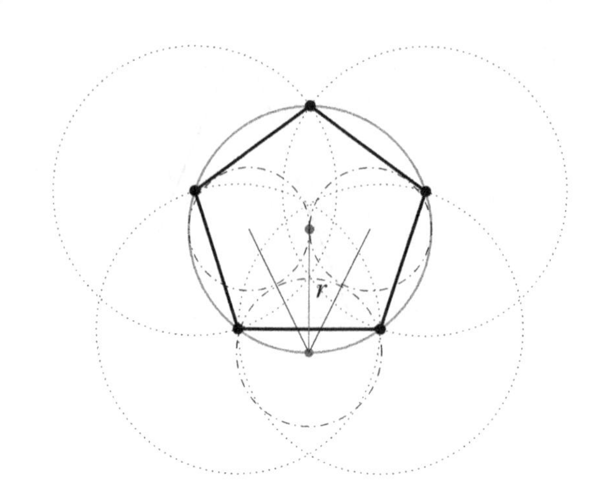


24-3)

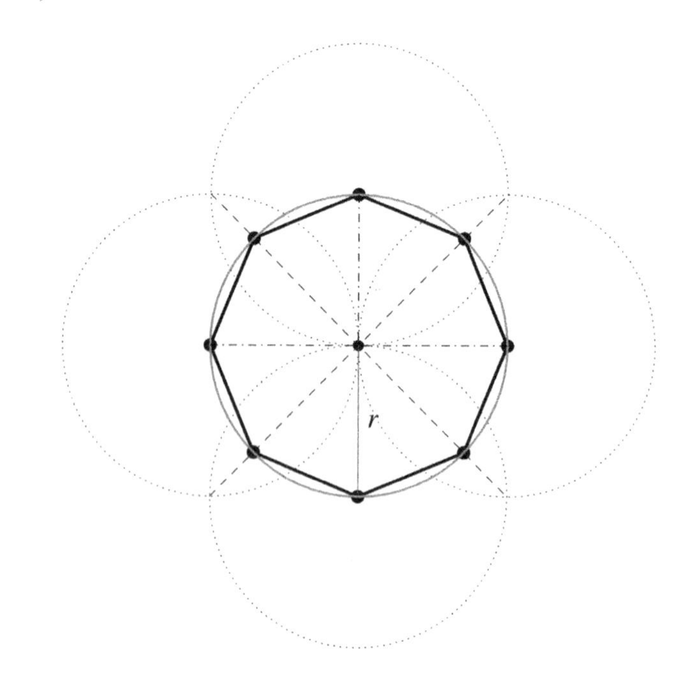

# 2. Reasoning and Proofs

## 2.1 Conjectures and Counterexamples

A *conjecture* is essentially an educated guess about a mathematical relationship or property that has not yet been verified formally.

A *counterexample* is a single instance where the conjecture does not hold true. The discovery of a counterexample is critical because it disproves the conjecture.

**Key Point**

A single *counterexample* is sufficient to disprove a conjecture. Finding a counterexample is a powerful tool in mathematical argumentation.

**Key Point**

Proving a conjecture requires establishing its truth in all possible cases. This goes beyond simply accumulating examples that support it. Rigorous mathematical proofs are necessary.

Now, let us explore some examples to understand conjectures and counterexamples better.

**Example** Conjecture: The sum of any two odd numbers is odd. Consider this conjecture and determine its validity.

**Solution:** Counterexample: Take two odd numbers, for example, 3 and 5. Their sum is $3+5=8$, which is an even number. This single counterexample shows that the conjecture is false.

**Example** Conjecture: All even numbers are divisible by 2.

**Solution:** This is a true statement, and no counterexample can be found because, by definition, an even number is any integer that can be divided by 2 without leaving a remainder.

Conjecture: All prime numbers are odd.

**Solution:** Counterexample: The number 2 is a prime number, as it has only two distinct positive divisors: itself and 1. However, it is an even number. This single counterexample disproves the conjecture.

## 2.2 Inductive Reasoning from Patterns

Inductive reasoning is a method of thinking where we derive general principles from specific observations. The conclusions we draw from this reasoning may be probable, but they are not guaranteed to be certain.

**Key Point**

Inductive reasoning can lead to probable conclusions that are based on observed evidence, but these conclusions are not absolutely certain.

**Observing Patterns:** The initial step in inductive reasoning involves identifying trends or patterns within certain cases or amongst related data points.

**Making Generalizations:** Once a pattern is apparent, we make a generalized statement or hypothesis that applies not only to the observed instances but also to other cases that fit within the pattern.

**Limitations:** Even though inductive reasoning can be very powerful, it is imperative to remember that the conclusions reached may not always be accurate. Exceptions can occur, and hence, conclusions must be treated as hypotheses rather than absolute truths.

**Application in Mathematics:** Inductive reasoning is frequently used in mathematical contexts, particularly when recognizing patterns in number sequences or geometric figures.

*Number Patterns:* When we see a sequence like $2,4,6,8,\ldots$, we use inductive reasoning to hypothesize that the pattern involves increasing by 2, hence predicting the next number would be 10.

*Geometric Patterns:* After measuring several triangles and finding that their angles sum to $180^\circ$, inductive reasoning leads us to conclude that all triangles have angle sums of $180^\circ$.

*Noticing Exceptions:* It is essential to test and validate the conclusions made by inductive reasoning. A pattern's existence does not confirm its universality.

**Key Point**

Inductive reasoning starts with observations and moves to general conclusions, in contrast to deductive reasoning, which begins with general statements to reach specific conclusions.

Every time you drop a pen, it falls to the ground. Use inductive reasoning to draw a conclusion.

**Solution:** Inductive Reasoning: When you drop an object, it falls to the ground.

Use the pattern of dots in the following sequence to answer the questions.

a) Draw the next figure in the pattern.

b) How does the number of dots in each shape relate to the figure number?

c) Use part b to determine a formula for the $n$th figure.

Shape1 Shape2 Shape3 Shape4

**Solution:** a) Following the current pattern, the next figure, Shape 5, will have 5 dots on top and 1 dot centered below, for a total of 6 dots.

b) Observing the pattern: Each figure number corresponds to the number of top dots, plus 1 for the center. Hence, each figure's number of dots is the figure number plus 1.

Shape 1: $1+1=2$ dots.

Shape 2: $2+1=3$ dots.

Shape 3: $3+1=4$ dots.

Shape 4: $4+1=5$ dots.

So the pattern is that each shape's number of dots is the figure number plus 1.

c) Using the observed pattern, the formula for the number of dots in the $n$-th figure is: $n+1$. For any figure number $n$, the number of dots it contains is $n$ plus 1.

## 2.3 Conditional Statements

Conditional statements, or "if-then" statements, are logical propositions asserting that if a particular condition, the hypothesis, is met, then a specific conclusion follows. In mathematical notation, a conditional statement is written as "if $p$, then $q$", where $p$ is the hypothesis and $q$ is the conclusion.

**Key Point**

A conditional statement is false only when the hypothesis is true, and the conclusion is false. Otherwise, the statement is true.

These statements can be structurally represented using the symbolic form $p \to q$. The following truth table encapsulates all possible truth values related to conditional statements:

| $p$ | $q$ | $p \to q$ |
|---|---|---|
| True | True | True |
| True | False | False |
| False | True | True |
| False | False | True |

Let us work through a couple of examples to understand how to identify the hypothesis and conclusion in a conditional statement.

Determine the hypothesis and conclusion in the following statement:

If a number is even, then it is divisible by 2.

**Solution:** *Hypothesis:*A number is even. *Conclusion:* It is divisible by 2.

Determine the hypothesis and conclusion in the following statement:

If a shape has four sides, then it is a quadrilateral.

**Solution:** *Hypothesis:* A shape has four sides. *Conclusion:* It is a quadrilateral.

If $3-1=4$, then snakes can fly.

**Solution:** *Hypothesis:* $3-1=4$. *Conclusion:* snakes can fly. The statement $3-1=4$ is false. Similarly, the statement "snakes can fly" is false under normal circumstances. However, according to the principles of classical logic, the conditional statement "If $3-1=4$, then snakes can fly" is considered true, even though both its parts are false.

## 2.4 Logic and Truth Tables

*Logic* is the study of the principles of valid reasoning and inference. It deals with statements that are either true or false, and their combinations through logical connectives. In logic, a statement must always be clear enough so that it cannot be both true and false at the same time.

**Key Point**

In the study of logic, different symbols are used to represent basic logical connectives:

- $\wedge$ denotes the *logical conjunction* (AND), where $p \wedge q$ is true only if both $p$ and $q$ are true.
- $\vee$ represents the *logical disjunction* (OR), where $p \vee q$ is true if at least one of $p$ or $q$ is true.
- $\neg$ stands for *negation* (NOT), where $\neg p$ is true if $p$ is false.
- $\rightarrow$ symbolizes the *conditional* (IMPLIES), where $p \rightarrow q$ is false only when $p$ is true and $q$ is false.
- $\leftrightarrow$ indicates the *biconditional* (IF AND ONLY IF), where $p \leftrightarrow q$ is true when both $p$ and $q$ have the same truth value.

The symbol $\therefore$ is used to denote "therefore" or "hence," indicating a conclusion derived from the preceding logical argument.

Basic logical is used to combine or modify statements. The *truth tables* for each of these connectives have their specific patterns which determine the truth value of compound statements based on the truth values of their components.

| $p$ | $q$ | $p \wedge q$ |
|---|---|---|
| True | True | True |
| True | False | False |
| False | True | False |
| False | False | False |

| $p$ | $q$ | $p \vee q$ |
|---|---|---|
| True | True | True |
| True | False | True |
| False | True | True |
| False | False | False |

| $p$ | $\neg p$ |
|---|---|
| True | False |
| False | True |

| $p$ | $q$ | $p \rightarrow q$ |
|---|---|---|
| True | True | True |
| True | False | False |
| False | True | True |
| False | False | True |

| $p$ | $q$ | $p \leftrightarrow q$ |
|---|---|---|
| True | True | True |
| True | False | False |
| False | True | False |
| False | False | True |

Key Point

A *truth table* displays all possible truth values that logical expressions can have by listing them under each possibility of their compound statements.

Create a truth table for the compound statement $p \vee \neg q$.

**Solution:** The truth table for $p \vee \neg q$ shows that the statement is true if at least one of $p$ or $\neg q$ is true:

| $p$ | $q$ | $\neg q$ | $p \vee \neg q$ |
|---|---|---|---|
| T | T | F | T |
| T | F | T | T |
| F | T | F | F |
| F | F | T | T |

where T stands for True and F stands for False.

## 2.5 Converse, Inverse, and Contrapositive

Let us explore key operations on conditional statements: the converse, inverse, and contrapositive. These concepts are essential for evaluating arguments and proofs in geometry.

**The Converse**. For the conditional statement $p \rightarrow q$, its converse is given by $q \rightarrow p$. It is important to note that the truth value of the converse may not be the same as the original statement.

The converse of a statement is formed by interchanging the hypothesis and the conclusion.

**The Inverse**. For $p \rightarrow q$, the inverse is $\neg p \rightarrow \neg q$. Like the converse, the inverse might not share the original statement's truth value.

**Key Point**

The inverse of a statement is formed by negating both the hypothesis and the conclusion of the original statement.

**The Contrapositive**. For $p \to q$, the contrapositive is $\neg q \to \neg p$. The contrapositive always has the same truth value as the original conditional statement.

**Key Point**

The contrapositive of a statement negates and then switches the hypothesis and conclusion of the original statement.

**Key Point**

A conditional statement and its contrapositive always share the same truth value, while the converse and inverse do not necessarily share this trait.

If both a conditional statement and its converse are true, they can be combined into a biconditional statement, written as $p \leftrightarrow q$, signifying $p$ if and only if $q$.

**Example** Consider the statement "If it is raining, then the ground is wet". Write down the converse, inverse, and contrapositive of the given conditional statement and analyze their truth values.

**Solution:** Let us analyze the converse, inverse, and contrapositive of the original statement to understand their logical implications:

- Original statement: If it is raining ($p$), then the ground is wet ($q$).
- Converse: If the ground is wet ($q$), then it is raining ($p$).
  The converse might not be true since there could be other reasons for the ground being wet, such as watering the lawn.
- Inverse: If it is not raining ($\neg p$), then the ground is not wet ($\neg q$). The inverse might not be true since the ground could be wet from earlier rainfall.
- Contrapositive: If the ground is not wet ($\neg q$), then it is not raining ($\neg p$).

The contrapositive shares the truth value with the original statement; hence, if the original statement is true, the ground not being wet implies that it is not raining.

## 2.6 Biconditionals and Definitions

A *biconditional statement* consists of two parts (the '*if*' part and the '*only if*' part) and can be considered a combination of a conditional statement and its converse. In simple terms, the phrase "*if and only if*," abbreviated as "*iff*," is used to connect these parts.

**Key Point**

The symbol '$\leftrightarrow$' represents a biconditional statement. The statement $p \leftrightarrow q$ is true if and only if both statements are true or both are false.

To be more precise, if we have a statement "A shape is a square *if and only if* it has four equal sides and four right angles," this creates a two-way link between both conditions.

**Key Point**

Usages of a biconditional statement include:

- For a definition to be precise in mathematics, it often takes the form of a biconditional statement.
- A biconditional is a succinct way to define equivalence between properties.

In practice, biconditionals help us understand that certain properties are not just necessary, but also sufficient for a definition.

Remember that for a definition to be strong and effective, it has to specify a condition that is both necessary and sufficient. It is this precise characteristic that biconditionals capture so well.

Write the biconditional for the statement: "A number is even if and only if it is divisible by 2."

**Solution:** Let $p$ be the statement "A number is even," and let $q$ be the statement "A number is divisible by 2." The biconditional $p \leftrightarrow q$ is written as "A number is even *if and only if* it is divisible by 2."
Here, the "*if*" part means: If a number is even, then it is divisible by 2. The "*only if*" part means: If a number is divisible by 2, then it is an even number.

Define a right angle using a biconditional statement.

**Solution:** Let $r$ represent the statement "An angle measures 90°," and $a$ represent the statement "An angle is a right angle." The biconditional $a \leftrightarrow r$ is then written as "An angle is a right angle *if and only if* it measures 90°."

## 2.7 Deductive Reasoning

Deductive reasoning, or "top-down" logic, is a method that starts with a set of premises and derives a certain conclusion from them.

**Key Point**

Deductive reasoning starts with a general statement and then applies it to a specific case to reach a conclusion.

A geometric proof is a classic example of deductive reasoning where one starts with known axioms, definitions, and previously proven theorems to prove a new theorem. This method is powerful because it establishes a logical argument whose validity does not depend on the author's credibility or intuition but on the structure of the argument itself.

**Key Point**

If the premises are true and the argument is valid, then the conclusion reached by deductive reasoning is certainly true. This means that the truth of the premises guarantees the truth of the conclusion.

Remember that whereas inductive reasoning is probabilistic and involves making generalizations from observations, deductive reasoning is about certainty and starts with a general statement before reaching a specific conclusion.

Determine the validity of the conclusion:

- **Premise 1**: If a figure is a rectangle, then it has four sides.
- **Premise 2**: Figure $X$ is a rectangle.

**Solution:** Using deductive reasoning, we can conclude that since figure $X$ is a rectangle, it must have four sides. This conclusion is valid and necessarily true if the premises are true.

**Example** Use deductive reasoning to identify the missing angle:

- **Premise 1**: The sum of the internal angles in a triangle is $180^\circ$.
- **Premise 2**: In triangle $ABC$, angles $A$ and $B$ measure $60^\circ$ each.

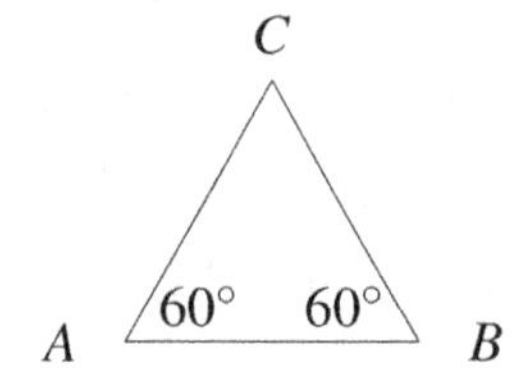


**Solution:** The conclusion is: Angle $C$ in triangle $ABC$ measures $180-(60+60)=60^\circ$.

## 2.8 Properties of Equality and Congruence

Equality is a relation that holds between two values when they are the same in value. There are several properties for equality.

The Reflexive Property of Equality states that for any number $a$, we have $a=a$. This seems obvious, but it is important for demonstrating that an object is equal to itself.

The Reflexive Property of Equality means that anything is equal to itself.

The Symmetric Property of Equality tells us that if $a=b$, then it must also be true that $b=a$. The position of equal values can be swapped without affecting their equivalence.

The Symmetric Property of Equality allows the reversal of equality.

For the Transitive Property of Equality, if $a = b$ and $b = c$, we can deduce that $a = c$. This property is extremely useful in proving that two things are equal by relating each to a third object.

**Key Point**

The Transitive Property creates a chain of equality.

The Substitution Property of Equality permits us to replace equal values within an expression or equation, maintaining the original value's equality.

**Key Point**

Equal values are interchangeable.

Equality has several operations-based properties. The Addition, Subtraction, Multiplication, and Division Properties of Equality allow us to perform the same operation on both sides of an equation without changing the truth value. These are essential when solving equations in algebra and geometry.

**Key Point**

Arithmetic operations performed on equals produce equals:

- Addition and Subtraction Properties of Equality: If $a = b$, then $a \pm c = b \pm c$ for any number $c$.
- Multiplication Property of Equality: If $a = b$, then $ac = bc$, for any number $c$.
- Division Property of Equality: If $a = b$ and $c \neq 0$, then $\frac{a}{c} = \frac{b}{c}$.

The Distributive Property, $a(b+c) = ab+ac$, is a bridge between addition and multiplication, often used in simplifying expressions and proving other geometric properties.

**Key Point**

The Distributive Property combines addition and multiplication.

In geometry, congruence denotes equal shape and size. The Reflexive, Symmetric, and Transitive Properties of Congruence are analogous to those for equality, but they apply to geometric figures.

Here are some examples demonstrating these properties.

If $x = 3$ and $3 = y$, which property of equality allows us to deduce $x = y$?

**Solution:** The property of equality that allows this deduction is the Transitive Property of Equality. Since $x = 3$ and $3 = y$, it follows that $x = y$.

Identify which property is being used: "If $a+3 = b$ and $a = 4$, then $7 = b$".

**Solution:** The property being used here is the Substitution Property of Equality. Since $a = 4$, substituting 4 for $a$ in $a+3 = b$ gives $4+3 = b$ or equivalently, $7 = b$.

 Given that $\angle A \cong \angle B$ and $\angle B \cong \angle C$, prove that $\angle A \cong \angle C$.

**Solution:** By applying the Transitive Property of Congruence, since $\angle A \cong \angle B$ and $\angle B \cong \angle C$, we can conclude that $\angle A \cong \angle C$.

## 2.9 Two Column Proofs

A two-column proof is an organized method used to demonstrate the logical progression from a set of given premises (hypothesis) to a conclusion within the realm of geometry. It is primarily crafted to enhance clarity and rigor in the proof process.

The format of a two-column proof is simple: it consists of two columns—on the left, you list the statements, and on the right, you provide reasons or justifications for each statement. This layout ensures that every claim made is substantiated by a definition, postulate, or theorem.

The left column of a two-column proof contains the statements that gradually lead to the conclusion; the right column contains the reasons, which include given information, definitions, postulates, and previously established theorems.

It is crucial to progress in a stepwise manner, ensuring the transition from one statement to the next is justified and logically sound. This practice not only helps us draw correct conclusions but also trains our logical thinking skills.

Let us apply this strategy to a practical scenario:

 Given: Lines $EF$ and $GH$ are parallel, and line $IJ$ is a transversal. Prove: $\angle EIJ \cong \angle HJI$.

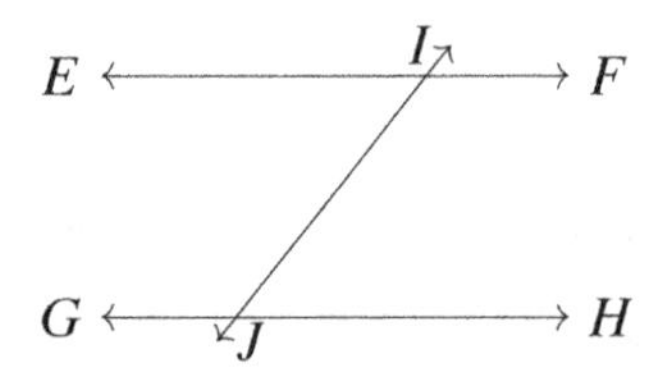

**Solution:** We shall prove this by employing a two-column proof as follows:

| Statements | Reasons |
|---|---|
| 1. Lines $EF$ and $GH$ are parallel. | 1. Given |
| 2. $\angle EIJ$ and $\angle HJI$ are alternate interior angles. | 2. Definition of alternate interior angles |
| 3. Alternate interior angles are congruent when lines are parallel. | 3. Alternate Interior Angles Theorem |
| 4. $\angle EIJ \cong \angle HJI$ | 4. Application of the Alternate Interior Angles Theorem |

## 2.10 Proving Angles Congruent

In geometry, to prove that angles are equal, we often use specific postulates, properties, or theorems. Here are some common methods and concepts to demonstrate that angles are congruent:

**Angle Congruence Postulate:** Angles with identical measures are congruent.

**Vertical Angles Theorem:** Angles opposite each other at intersecting lines are always congruent.

**Adjacent Angles:** Angles that share a vertex and a side but not interior points. They might or might not be congruent, but sometimes we can use their relationship to other angles to prove congruence.

**Complementary and Supplementary Angles:** Angles that are complementary or supplementary to the same angle (or to congruent angles) are congruent themselves.

**Alternate Interior and Exterior Angles:** For parallel lines intersected by a transversal, both alternate interior and exterior angle pairs are congruent.

**Corresponding Angles:** For parallel lines intersected by a transversal, corresponding angle pairs are congruent.

**Angles Formed by Perpendicular Lines:** Intersecting lines at right angles create four congruent right angles.

Consistency in reasoning is crucial in geometric proofs. Congruence of angles can often be inferred from their relationships and positions relative to each other and other shapes.

Line $l_1$ is parallel to line $l_2$, and both are cut by a transversal $t$. Prove: $\angle 1$ is congruent to $\angle 5$.

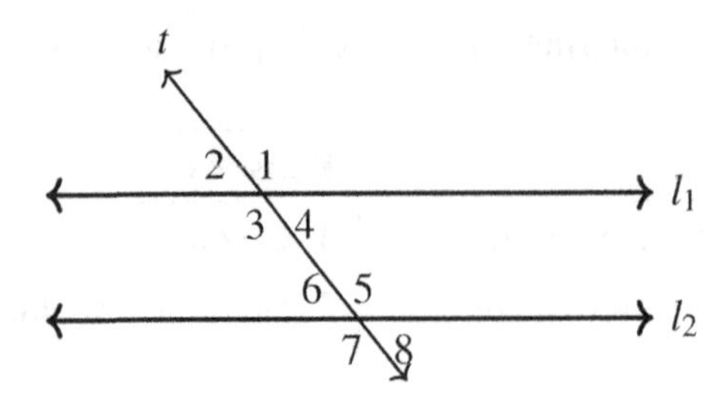


**Solution:** Two-Column Proof:

| **Statements** | **Reasons** |
|---|---|
| 1. Line $l_1 \parallel l_2$ | 1. Given |
| 2. $t$ is a transversal of $l_1$ and $l_2$ | 2. By definition of a transversal |
| 3. $\angle 1$ and $\angle 5$ are corresponding angles | 3. Definition of corresponding angles |
| 4. $\angle 1 \cong \angle 5$ | 4. Corresponding angles are congruent if the lines are parallel |

## 2.11 Practices

**1)** Determine if the following conjectures are true or identify counterexamples where applicable.

1-1) The square of any integer is always positive.

1-2) The sum of two negative integers is always negative.

1-3) The product of two negative numbers is positive.

1-4) All numbers divisible by 6 are also divisible by 3.

1-5) Every triangle has at least two acute angles.

1-6) If $n$ is an integer, then $n^2 > n$.

1-7) All numbers that end in 1 are prime numbers.

1-8) Any three points that are coplanar are also collinear.

1-9) All girls like ice cream.

1-10) All high school students are in choir.

1-11) For any angle, there exists a complementary angle.

1-12) All teenagers can drive.

**2)** Draw conclusions using inductive reasoning.

2-1) The first few terms of a sequence are $5, 10, 15, 20, \ldots$

2-2) In a game of dice, rolling a 6 has occurred more frequently than any other number in the last 10 rolls.

**3)** Using Inductive Reasoning.

3-1) A dot pattern is shown below.

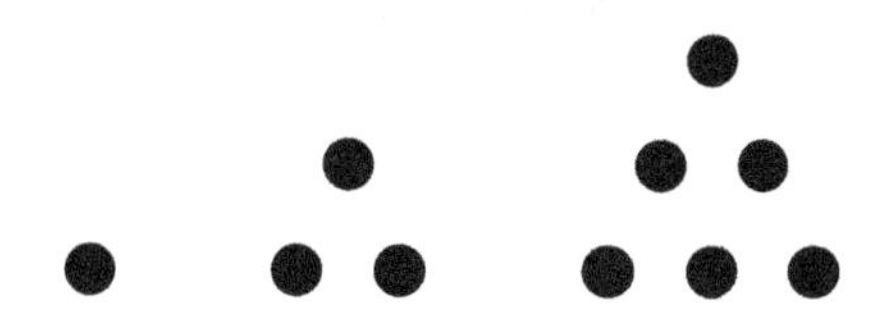

How many dots would there be in the 4-th figure? Draw the 6-th figure.

3-2) How many circles would be in the 9-th figure?

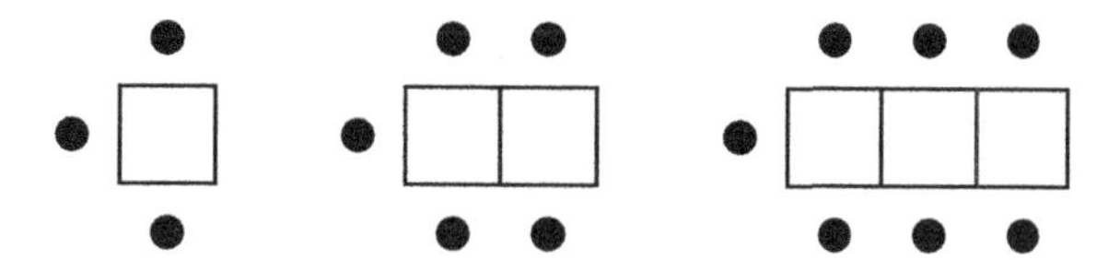

3-3) Look at the pattern $7, 12, 17, 22, 27, \ldots$ What is the 19-th term in the pattern?

3-4) Look at the pattern: $-1, 3, -9, 27, -81, \ldots$ What is the next term in the pattern? The 10-th term? Make a rule for the $n$-th term.

**4)** Identify the hypothesis and conclusion of each conditional.

4-1) If you can see the sun, then it is day.

Hypothesis:

Conclusion:

4-2) If two lines are perpendicular, then they form right angle.

Hypothesis:

Conclusion:

4-3) If $\angle A = 90^\circ$, then $\angle A$ is a right angle.

Hypothesis:

Conclusion:

4-4) If $4x - 2 = x + 1$, then: $x = 1$.

Hypothesis:

Conclusion:

**5)** Write a conditional statement from each of the following.

5-1) A tricycle has three wheels.

5-2) All birds have two wings.

**6)** For each conditional, write the converse and a biconditional statement.

6-1) Conditional: If it is the second Sunday in May, then it is Mother's day.

Converse:

Biconditional:

6-2) Conditional: If a figure has 8 sides, then it is an octagon.

Converse:

Biconditional:

**7)** Write a truth table for the following variables.

7-1) $(p \wedge q) \vee r$

7-2) $(p \vee q) \vee \neg r$

7-3) $p \vee (\neg q \vee r)$

7-4) $(\neg p \vee \neg q) \wedge r$

7-5) $p \wedge (q \vee \neg r)$

7-6) $(\neg p \wedge \neg q) \wedge r$

**8)** Is the following a valid argument? If so, what law is being used? (Hint: Statements could be out of order).

8-1) $p \rightarrow q,\ r \rightarrow p,\ \therefore r \rightarrow q$

8-2) $p \rightarrow r,\ \neg r,\ \therefore \neg p$

8-3) $p \rightarrow q,\ q \rightarrow r,\ \therefore p \rightarrow r$

8-4) $r \rightarrow q,\ p \rightarrow r,\ \therefore q \rightarrow p$

**9)** Determine its converse, inverse, and contrapositive, and if the statements are true or false. Then, if they are false, find a counterexample.

9-1) If a shape is a square, then it has four equal sides.

9-2) If I am at Disneyland, then I am in California.

9-3) If two lines intersect, then they intersect at exactly one point.

9-4) If two sides of a triangle are congruent, then the angles opposite those sides are also congruent.

**10)** Determine the two true conditional statements from the given biconditional statements.

10-1) A whole number is prime if and only if it has exactly two distinct factors.

10-2) Points are collinear if and only if there is a line that contains the points.

10-3) $5x = 15$ if and only if $x = 3$.

**11)** Solve.

11-1) Write the biconditional for the following statement: A shape is a rectangle if it has four right angles.

11-2) Identify the two conditional statements embedded in this biconditional: A polygon is regular if and only if all its sides and angles are congruent.

11-3) Write a definition using a biconditional for the term "odd number."

**12)** What are the relationships between $AD$ and $FC$, if $FE = AB$ and $ED = BC$.

**13)** Mark "$\surd$" in the grid as each added piece of information helps you match the numbers with the people they describe. There are four pencil cases on a table. One pencil case has 12 pencils, one pencil case has 16 pencils, one pencil case has 6 pencils and the other pencil case has 32 pencils.

- Jack has less pencils than Sarah.
- Jacob has more pencils than Sarah.
- David has the most number of pencils.

| | 6 | 12 | 16 | 32 |
|---|---|---|---|---|
| Sarah | | | | |
| Jack | | | | |
| Jacob | | | | |
| David | | | | |

**14)** Use the given property or properties of equality to fill in the blank. $a$, $b$, and $c$ are real numbers.

14-1) Symmetric: If $a+b=b+c$, then $c+b=$ ______.

14-2) Transitive: If $XY=6$ and $XY=ZW$, then $ZW=$ ______.

14-3) Substitution: If $a=b-5$ and $a=c+3$, then $b-5=$ ______.

14-4) Distributive: If $4(2a+3)=b$, then $b=$ ________.

**15)** Solve.

15-1) Given points $O$, $P$, and $Q$ and $OP=7$, $PQ=7$, and $OQ=12$. Are the three points collinear? Is $P$ the midpoint?

15-2) If $\angle RST=60^\circ$ and $\angle RST+\angle UVW=180^\circ$, explain how $\angle UVW$ must be an obtuse angle.

15-3) Identify which property is being used: "If $2x+3=7$ and $2x=4$, then $4+3=7$".

15-4) Given $\triangle ABC$ and $\triangle DEF$ such that $AB\cong DE$ and $BC\cong EF$, if $AC\cong DF$, which property helps us conclude $\triangle ABC\cong\triangle DEF$?

15-5) If $x=5$ and $5=y$, which property of equality allows us to deduce $x=y$?

**16)** Prove it.

16-1) Given vertices $A$, $B$ and $C$ form a triangle. Write a proof to show that $A$, $B$ and $C$ determine a plane.

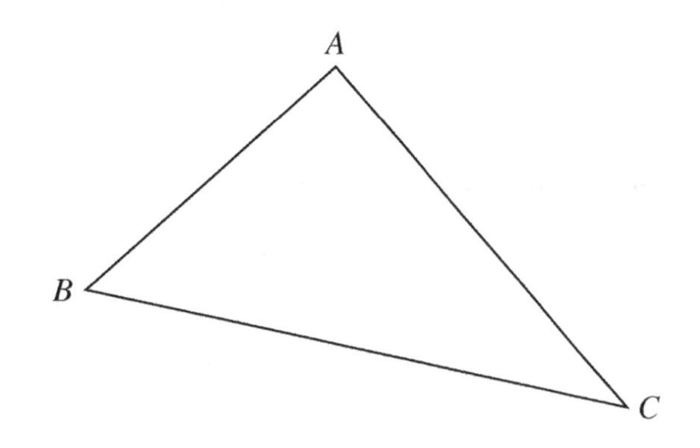

16-2) Given that $D$ is the midpoint of $BC$ and $\angle L \cong \angle B$. Write a proof to show that $LD \cong DC$.

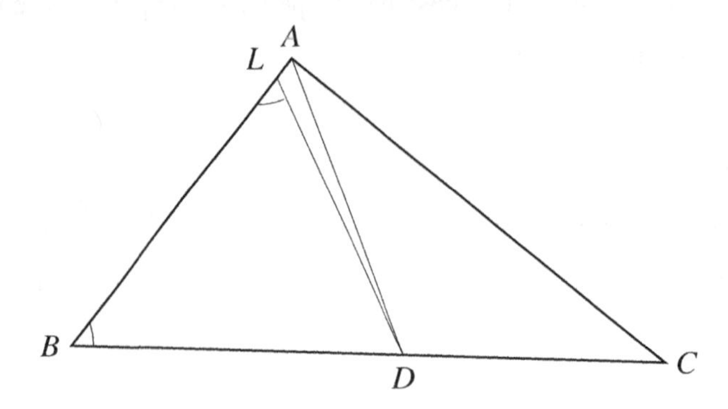


16-3) Given that $B$ is the midpoint of $AC$, $D$ is the midpoint of $CE$, and $AB = DE$. Write a proof to show that $AE = 4AB$.

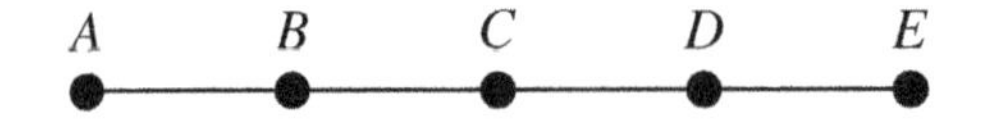


**17)** Solve.

17-1) If $A$, $B$, $C$, $D$, and $E$ are points on a line, in the given order, and $AB = DE$, prove $AD = BE$.

17-2) Given that $u \perp v$, $\angle\alpha \cong \angle\beta$, and $\angle\theta \cong \angle\lambda$. Prove that $\angle\theta = \angle\gamma + \angle\delta$.

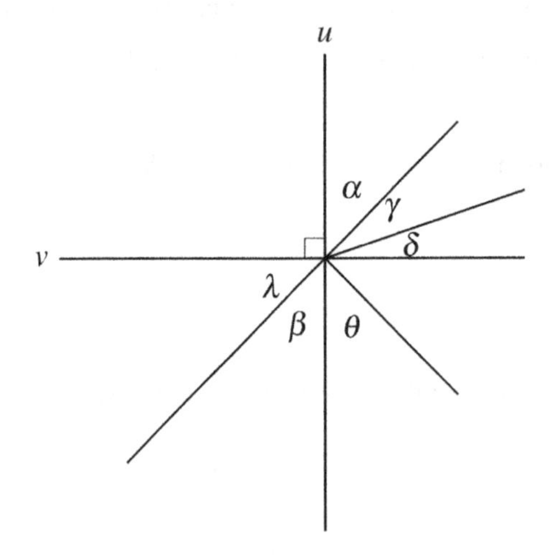


17-3) Given: $\angle a$ and $\angle b$ are right angles, and $\angle c = 40°$. Prove: $\angle a + \angle d + \angle c = 270°$.

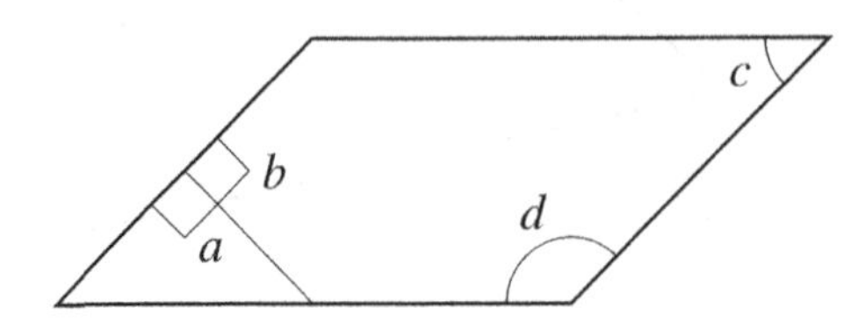

17-4) Given: $\angle BOD = 90^\circ$, $\angle COD = 35^\circ$, and $\angle AOB = 40^\circ$. Prove: $\angle FOE + \angle BOC = 95^\circ$.

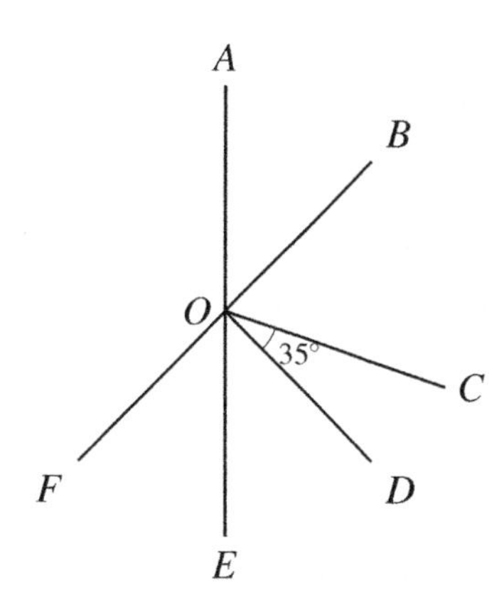

## 2.12 Answers

**1)**

1-1) **Counterexample:** The square of 0 is $0^2 = 0$, which is not positive.

1-2) **True.** The sum of two negative integers always results in a negative sum.

1-3) **True.** The product of two negative numbers is always positive.

1-4) **True.** Any number divisible by 6 is also divisible by 3, due to the prime factors of 6.

1-5) **True.** Every triangle has at least two acute angles, which are angles less than 90°.

1-6) **Counterexample:** For $n = 1$, $1^2 = 1$, hence $n^2$ is not greater than $n$.

1-7) **Counterexample:** The number 91 ends in 1 but is not prime, as it is divisible by 7 and 13.

1-8) **Counterexample:** Three vertices of a triangle are coplanar but not collinear.

1-9) **Counterexample:** There may be girls who dislike ice cream.

1-10) **Counterexample:** Alex is a high school student who is only involved in the chess club and not in the choir.

1-11) **Counterexample:** A 120° angle does not have a complementary angle, as complementary angles sum to 90°.

1-12) **Counterexample:** In many countries, the legal driving age is 18, so for example, a 15-year-old teenager in those countries cannot legally drive.

**2)**

2-1) **Inductive Reasoning Conclusion:** The sequence seems to be increasing by 5 each time. Based on this observed pattern, the next term is likely to be 25.

2-2) **Inductive Reasoning Conclusion:** While dice are designed to give random results, you might infer that there seems to be a trend of rolling 6*s* more frequently in this specific set of rolls. However, this doesn't necessarily mean a 6 will come up in the next roll; statistically, each roll should be independent and have an equal chance of landing on any face.

**3)**

3-1) 10 and 

3-2) 19

3-3) 97

3-4) 243, 19683, and $-1(-3)^{(n-1)}$

**4)**

4-1) Hypothesis: You can see the sun
Conclusion: It is day

4-2) Hypothesis: Two lines are perpendicular
Conclusion: They form right angle

4-3) Hypothesis: $\angle A = 90°$
Conclusion: $\angle A$ is a right angle

4-4) Hypothesis: $4x - 2 = x + 1$
Conclusion: $x = 1$

**5)**

5-1) If it is a tricycle, then it has three wheels.

5-1) If they are birds, then they have two wings.

**6)**

6-1) Converse: If it is Mother's day, then it is the second Sunday in May.
Biconditional: It is the second Sunday in May if and only if it is Mother's day.

6-2) Converse: If a figure is octagon, then it has 8 sides.
Biconditional: A figure has 8 sides if and only if it is an octagon.

**7)**

7-1) $(p \wedge q) \vee r$

| $p$ | $q$ | $r$ | $p \wedge q$ | $(p \wedge q) \vee r$ |
|---|---|---|---|---|
| T | T | T | T | T |
| T | T | F | T | T |
| T | F | T | F | T |
| T | F | F | F | F |
| F | T | T | F | T |
| F | T | F | F | F |
| F | F | T | F | T |
| F | F | F | F | F |

7-2) $(p \vee q) \vee \neg r$

| $p$ | $q$ | $r$ | $\neg r$ | $p \vee q$ | $(p \vee q) \vee \neg r$ |
|---|---|---|---|---|---|
| T | T | T | F | T | T |
| T | T | F | T | T | T |
| T | F | T | F | T | T |
| T | F | F | T | T | T |
| F | T | T | F | T | T |
| F | T | F | T | T | T |
| F | F | T | F | F | F |
| F | F | F | T | F | T |

7-3) $p \vee (\neg q \vee r)$

| $p$ | $q$ | $r$ | $\neg q$ | $\neg q \vee r$ | $p \vee (\neg q \vee r)$ |
|---|---|---|---|---|---|
| T | T | T | F | T | T |
| T | T | F | F | F | T |
| T | F | T | T | T | T |
| T | F | F | T | T | T |
| F | T | T | F | T | T |
| F | T | F | F | F | F |
| F | F | T | T | T | T |
| F | F | F | T | T | T |

7-4) $(\neg p \vee \neg q) \wedge r$

| $p$ | $q$ | $r$ | $\neg p$ | $\neg q$ | $\neg p \vee \neg q$ | $(\neg p \vee \neg q) \wedge r$ |
|---|---|---|---|---|---|---|
| T | T | T | F | F | F | F |
| T | T | F | F | F | F | F |
| T | F | T | F | T | T | T |
| T | F | F | F | T | T | F |
| F | T | T | T | F | T | T |
| F | T | F | T | F | T | F |
| F | F | T | T | T | T | T |
| F | F | F | T | T | T | F |

7-5) $p \wedge (q \vee \neg r)$

| $p$ | $q$ | $r$ | $\neg r$ | $q \vee \neg r$ | $p \wedge (q \vee \neg r)$ |
|---|---|---|---|---|---|
| T | T | T | F | T | T |
| T | T | F | T | T | T |
| T | F | T | F | F | F |
| T | F | F | T | T | T |
| F | T | T | F | T | F |
| F | T | F | T | T | F |
| F | F | T | F | F | F |
| F | F | F | T | T | F |

7-6) $(\neg p \wedge \neg q) \wedge r$

| $p$ | $q$ | $r$ | $\neg p$ | $\neg q$ | $\neg p \wedge \neg q$ | $(\neg p \wedge \neg q) \wedge r$ |
|---|---|---|---|---|---|---|
| T | T | T | F | F | F | F |
| T | T | F | F | F | F | F |
| T | F | T | F | T | F | F |
| T | F | F | F | T | F | F |
| F | T | T | T | F | F | F |
| F | T | F | T | F | F | F |
| F | F | T | T | T | T | T |
| F | F | F | T | T | T | F |

**8)**

8-1) This is a valid argument. If $p$ implies $q$ and $r$ implies $p$, then $r$ implies $q$.

8-2) This is a valid argument. If $p$ implies $r$ and $r$ is not true, then $p$ must not be true either.

8-3) This is a valid argument. If $p$ implies $q$ and $q$ implies $r$, then $p$ implies $r$.

8-4) This argument is not valid. The given premises do not support the conclusion. The conclusion $q \rightarrow p$ is not necessarily true based on the premises provided.

**9)**

9-1) This statement is true.

Converse: If a shape has four equal sides, then it is a square. (False: A rhombus has four equal sides but is not necessarily a square)

Inverse: If a shape is not a square, then it does not have four equal sides. (False: A rhombus is not a square but it has four equal sides).

Contrapositive: If a shape does not have four equal sides, then it is not a square. (True).

9-2) This statement Not necessarily true. There is a Disneyland in Paris, Tokyo, Hong Kong, and Shanghai

Converse: If I am in California, then I am at Disneyland. (False: You can be in California without being at Disneyland).

Inverse: If I am not at Disneyland, then I am not in California. (False: You can be elsewhere in California).

Contrapositive: If I am not in California, then I am not at Disneyland. (False: If I am in Tokyo or Paris, I can be in Disneyland).

9-3) This statement is true.

Converse: If two lines intersect at exactly one point, then they intersect. (True).

Inverse: If two lines do not intersect, then they do not intersect at exactly one point. (True).

Contrapositive: If two lines do not intersect at exactly one point, then they do not intersect. (True).

9-4) This statement is true.

Converse: If the angles opposite two sides of a triangle are congruent, then the two sides are congruent. (True).

Inverse: If two sides of a triangle are not congruent, then the angles opposite those sides are not congruent. (True).

Contrapositive: If the angles opposite two sides of a triangle are not congruent, then the two sides are not congruent. (True).

**10)**

10-1) True.

Conditional Statement 1: If a whole number is prime, then it has exactly two distinct factors.

Conditional Statement 2: If a whole number has exactly two distinct factors, then it is prime.

10-2) True.

Conditional Statement 1: If points are collinear, then there is a line that contains the points.

Conditional Statement 2: If there is a line that contains the points, then the points are collinear.

10-3) True.

Conditional Statement 1: If $5x = 15$, then $x = 3$.

Conditional Statement 2: If $x = 3$, then $5x = 15$.

**11)**

11-1) Biconditional: A shape is a rectangle if and only if it has four right angles.

11-2) Conditional Statement 1: If a polygon is regular, then all its sides and angles are congruent. Conditional Statement 2: If all the sides and angles of a polygon are congruent, then the polygon is regular.

11-3) Biconditional: A number is odd if and only if it cannot be divided evenly by 2.

**12)** $AD = AB + BC + CD$ and $FC = FE + ED + DC$, So

$$AB + BC = FE + ED \Rightarrow AD = FC.$$

**13)**

| | 6 | 12 | 16 | 32 |
|---|---|---|---|---|
| Sarah | | √ | | |
| Jack | √ | | | |
| Jacob | | | √ | |
| David | | | | √ |

**14)**

14-1) $a + b$

14-2) 6

14-3) $c + 3$

14-4) $8a + 12$

**15)**

15-1) Using the Segment Addition Postulate: if $O$, $P$, and $Q$ are collinear, then $OP + PQ = OQ$. $7 + 7 = 14$, which is not equal to 12. Therefore, the three points are not collinear. Since $OQ \neq OP + PQ$, $P$ is not the midpoint of segment $OQ$.

15-2) To find $\angle UVW$, subtract $\angle RST$ from $180^\circ$: $180^\circ - 60^\circ = 120^\circ$. Since $\angle UVW = 120^\circ$ and it is greater than $90^\circ$, $\angle UVW$ is an obtuse angle.

15-3) This is using the Substitution Property of Equality.

15-4) If two triangles have all three pairs of corresponding sides congruent, then the triangles are congruent by the Side-Side-Side (SSS) Postulate.

15-5) The Transitive Property of Equality states that if $a = b$, then $b = a$. Using this property, we can deduce that if $x = 5$ and $5 = y$, then $x = y$.

**16)**

16-1) Consider a triangle $ABC$ where line $AB$ and line $AC$ intersect. Point $A$ is common to both lines, whereas points $B$ and $C$ lie uniquely on $AB$ and $AC$, respectively. Given that points $B$ and $C$ do not share the same line, points $A$, $B$, and $C$ are noncollinear. It is a fundamental principle that three noncollinear points define a plane. Therefore, points $A$, $B$, and $C$ determine a plane.

16-2) We know that $BD = DC$ (definition of midpoint) and we know that triangle $LDB$ is isosceles, because $\angle L \cong \angle B$. In an isosceles triangle, both legs are equal. Therefore, $LD = DB$ and $LD = DC$, and finally, $LD \cong DC$.

16-3) We know that $AB = BC$ and $CD = DE$ (definition of midpoint) and $AB = DE$ (given). By using the Segment Addition Postulate, we have: $AC = AB + BC$, $CE = CD + DE$, and $AE = AC + CE$. By substitution, we have: $AE = AB + BC + CD + DE$ and since $AB = BC = CD = DE$, we get: $AE = AB + AB + AB + AB$. Finally, by simplifying, we conclude: $AE = 4AB$.

**17)** Two-Column Proof:

17-1)

| Statements | Reasons |
|---|---|
| 1) $A$, $B$, $C$, $D$, and $E$ are collinear, in that order. | 1) Given |
| 2) $AB = DE$ | 2) Given |
| 3) $BD = BD$ | 3) Reflexive |
| 4) $AB + BD = DE + BD$ | 4) Addition |
| 5) $AB + BD = AD$ and $DE + BD = BE$ | 5) Segment Addition Postulate |
| 6) $AD = BE$ | 6) Substitution or Transitive |

17-2)

| Statement | Reason |
|---|---|
| 1) $\angle\alpha \cong \angle\beta$ | 1) Given |
| 2) $u \perp v$ | 2) Given |
| 3) $\angle\theta \cong \angle\lambda$ | 3) Given |
| 4) $\angle\alpha = \angle\beta$ | 4) $\cong$ angles have $=$ measures |
| 5) $\angle\lambda \cong \angle\gamma + \angle\delta$ | 5) Vertical Angles Theorem |
| 6) $\angle\lambda = \angle\gamma + \angle\delta$ | 6) $\cong$ angles have $=$ measures |
| 7) $\angle\theta = \angle\lambda$ | 7) $\cong$ angles have $=$ measures |
| 8) $\angle\theta = \angle\gamma + \angle\delta$ | 8) Transitive |

17-3)

| Statement | Reason |
|---|---|
| 1) $\angle a$ and $\angle b$ are right angles | 1) Given |
| 2) $\angle c = 40°$ | 2) Given |
| 3) $\angle a = 90°$ and $\angle b = 90°$ | 3) Definition of a right angle |
| 4) $\angle d = 140°$ | 4) Definition of supplementary angles |
| 5) $90° + 140° + 40° = 270°$ | 5) Addition of real numbers |
| 6) $\angle a + \angle d + \angle c = 270°$ | 6) Substitution |

17-4)

| Statement | Reason |
|---|---|
| 1) $\angle BOD = 90°$ | 1) Given |
| 2) $\angle COD = 35°$ | 2) Given |
| 3) $\angle AOB = 40°$ | 3) Given |
| 4) $\angle BOD = \angle BOC + \angle COD = 90°$ | 4) Definition of complementary angles |
| 5) $\angle BOC + 35° = 90°$ | 5) Substitution |
| 6) $\angle BOC = 55°$ | 6) Subtraction |
| 7) $\angle AOB \cong \angle FOE$ | 7) Vertical Angles Theorem |
| 8) $\angle AOB = \angle FOE$ | 8) $\cong$ angles have = measures |
| 9) $40° + 55° = 95°$ | 9) Addition of real numbers |
| 10) $\angle FOE + \angle BOC = 95°$ | 10) Substitution |

# 3. Coordinate Geometry

## 3.1 Finding Slope

The concept of slope is fundamental in coordinate geometry. It describes how steep a line is and gives you an indication of the direction in which the line moves across the coordinate plane.

Key Point

The slope of a line, given by two points $A(x_1,y_1)$ and $B(x_2,y_2)$, is calculated using the formula:

$$m = \frac{y_2 - y_1}{x_2 - x_1} = \frac{\text{rise}}{\text{run}}.$$

This represents the ratio of the vertical change to the horizontal change between these two points.

A line's slope, indicated by $m$ in the equation $y = mx + b$, reveals if the line ascends, descends, or remains horizontal as one moves from left to right along the line. Here, $b$ represents the point where the line intersects the $y$-axis, also known as the $y$-intercept.

Find the slope of the line passing through the points $A(1,-2)$ and $B(3,2)$.

**Solution:** The order of points does not affect the slope, so it does not matter which point you choose for $(x_1,y_1)$ and $(x_2,y_2)$. Applying the formula, we get:

$$\text{slope} = \frac{2-(-2)}{3-1} = \frac{4}{2} = 2.$$

The slope of the line through these points is 2, which indicates that for every unit of horizontal change (to the right), there is a 2-unit vertical change (upwards).

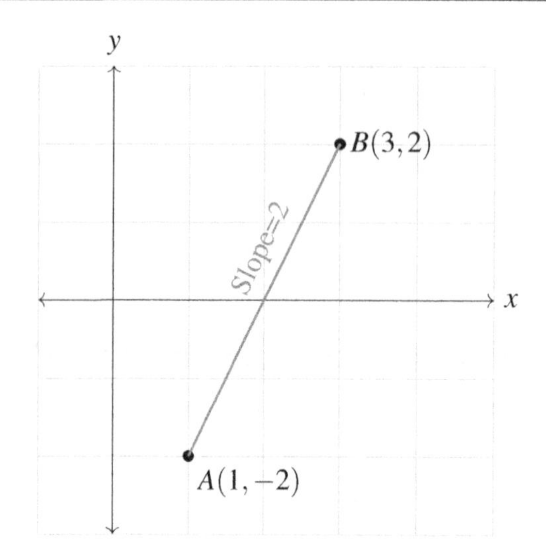


**Example** Determine the slope of the line given by the equation $y = -3x + 4$.

**Solution:** The slope of a line can be identified when the equation is in the format $y = mx + b$, where $m$ represents the slope. For the line $y = -3x + 4$, the slope is the coefficient of $x$, which in this case is $-3$.

## 3.2 Writing Linear Equations

The slope-intercept form of a linear equation is one of the most useful formats in coordinate geometry.

The equation of a line in slope-intercept form is $y = mx + b$, where $m$ represents the slope of the line, and $b$ represents the $y$-intercept of the line.

**Key Point**

To write an equation of a line, you need to identify two things: the slope ($m$) and the $y$-intercept ($b$). Recall that the $y$-intercept is the point where the line crosses the $y$-axis, so its $x$-coordinate is always 0.

If you are given a slope and a point through which the line passes, you can find the $y$-intercept by plugging the slope and the point's coordinates into the slope-intercept form and solving for $b$.

**Example** What is the equation of the line that passes through the point $(1, 2)$ and has a slope of $\frac{1}{2}$?

**Solution:** We start with the slope-intercept form: $y = mx + b$. Here, $m = \frac{1}{2}$, and we need to find $b$. Using the point $(1, 2)$, we substitute $x = 1$ and $y = 2$ into the equation: $2 = \left(\frac{1}{2}\right)(1) + b$. Now solve for $b$:

$$2 = \frac{1}{2} + b \Rightarrow b = 2 - \frac{1}{2} \Rightarrow b = \frac{4}{2} - \frac{1}{2} \Rightarrow b = \frac{3}{2}.$$

The equation of the line is therefore: $y = \frac{1}{2}x + \frac{3}{2}$.

 Write the equation of the line with a slope of 3 that goes through the point $(-2,-1)$.

**Solution:** Given the point $(-2,-1)$ and $m=3$, we substitute into the slope-intercept form:

$$-1=(3)(-2)+b \Rightarrow -1=-6+b \Rightarrow b=-1+6 \Rightarrow b=5.$$

The equation of the line is $y=3x+5$.

## 3.3 Finding Midpoint

The midpoint of a line segment is the point that is equidistant from both endpoints of the segment. To find this point in a Cartesian coordinate system where the endpoints are given as $A(x_1,y_1)$ and $B(x_2,y_2)$, we can use the following formula: $M\left(\frac{x_1+x_2}{2},\frac{y_1+y_2}{2}\right)$.

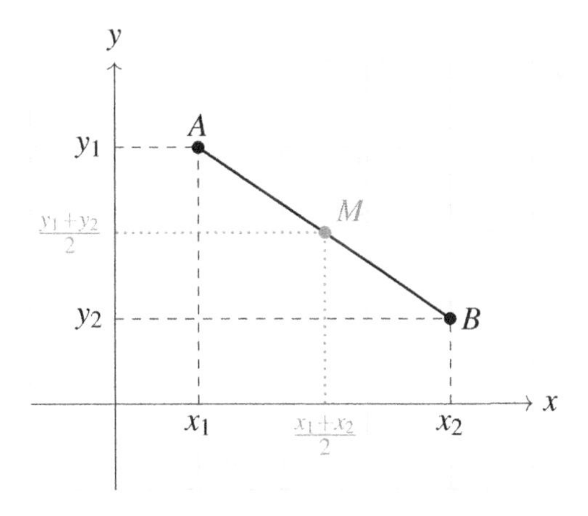


**Key Point**

The midpoint formula is derived by averaging the $x$ and $y$ coordinates of the two endpoints.

The practicality of the midpoint formula can be appreciated by working through examples. Let us examine how this works with an actual pair of endpoints.

 **Example** Find the midpoint of the line segment with endpoints $A(2,-4)$ and $B(6,8)$.

**Solution:** Using the midpoint formula:

$$M=\left(\frac{x_1+x_2}{2},\frac{y_1+y_2}{2}\right),$$

with $(x_1,y_1)=(2,-4)$ and $(x_2,y_2)=(6,8)$, we get:

$$M=\left(\frac{2+6}{2},\frac{-4+8}{2}\right)=\left(\frac{8}{2},\frac{4}{2}\right)=(4,2).$$

The midpoint of the line segment with the given endpoints is $M(4,2)$.

Find the midpoint of the line segment with endpoints $A(0,-4)$ and $B(4,-2)$.

**Solution:** Using the midpoint formula:

$$M = \left(\frac{x_1+x_2}{2}, \frac{y_1+y_2}{2}\right),$$

with $(x_1,y_1) = (0,-4)$ and $(x_2,y_2) = (4,-2)$, we get:

$$M = \left(\frac{0+4}{2}, \frac{-4+(-2)}{2}\right) = \left(\frac{4}{2}, \frac{-6}{2}\right) = (2,-3).$$

The midpoint of the line segment with the given endpoints is $M(2,-3)$.

## 3.4 Finding Distance of Two Points

The distance between two points can be determined using the Pythagorean theorem. If we have points $A(x_1,y_1)$ and $B(x_2,y_2)$, we can think of the distance between them as the hypotenuse of a right triangle where the other two sides are parallel to the coordinate axes.

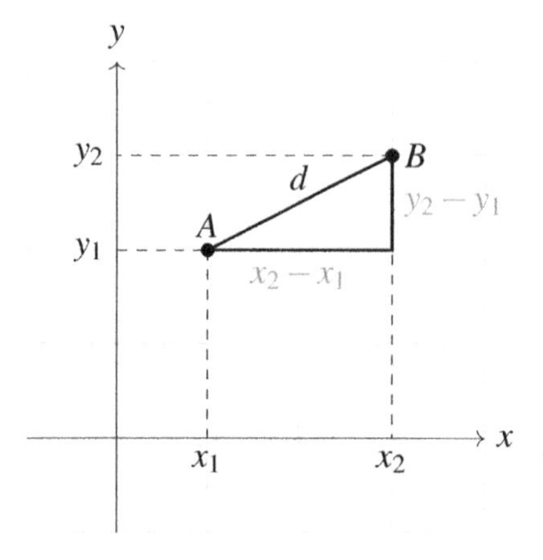


Key Point

The distance $d$ between points $A(x_1,y_1)$ and $B(x_2,y_2)$ is given by the formula:

$$d = \sqrt{(x_2-x_1)^2+(y_2-y_1)^2}.$$

To calculate the distance, just subtract the corresponding coordinates of $A$ and $B$, square the results, add them together, and then take the square root of that sum.

 Find the distance between the points $C(3,-2)$ and $D(7,4)$ on the coordinate plane.

**Solution:** Using the distance formula, we have: $d=\sqrt{(x_2-x_1)^2+(y_2-y_1)^2}$. Substitute $x_1=3$, $y_1=-2$, $x_2=7$, $y_2=4$:

$$d=\sqrt{(7-3)^2+(4-(-2))^2}=\sqrt{4^2+6^2}=\sqrt{16+36}=\sqrt{52}\approx 7.21.$$

So, the distance between points $C$ and $D$ is approximately 7.21 units.

 Find the distance between $(4,2)$ and $(-5,-10)$ on the coordinate plane.

**Solution:** Let point $A$ be $(4,2)$ and point $B$ be $(-5,-10)$. Using the distance formula:

$$d=\sqrt{(-5-4)^2+(-10-2)^2}=\sqrt{(-9)^2+(-12)^2}=\sqrt{81+144}=\sqrt{225}=15.$$

Thus, the distance between the points is 15 units.

## 3.5 Finding a Graph's Slope

The concept of slope is crucial in understanding the behavior of lines in the coordinate plane. In the context of coordinate geometry, the slope provides us with a measure of the steepness or inclination of a line.

**Key Point**

The slope of a line in the coordinate plane is the ratio of the vertical change (*rise*) to the horizontal change (*run*) between two points on the line.

To calculate the slope, often denoted by $m$, you can use the formula:

$$m=\frac{\Delta y}{\Delta x}=\frac{y_2-y_1}{x_2-x_1},$$

where $(x_1,y_1)$ and $(x_2,y_2)$ are any two distinct points on the line.

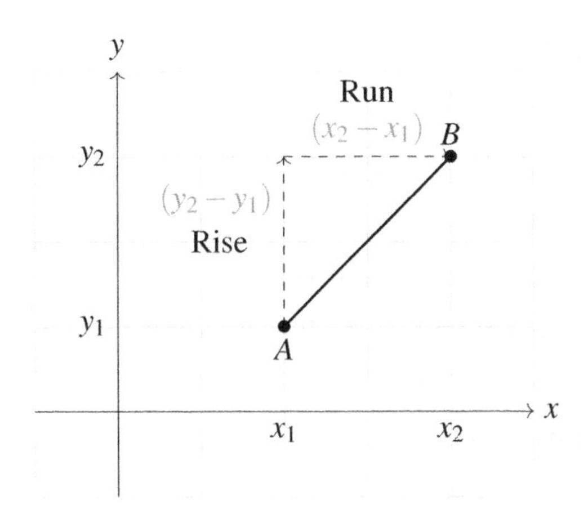

The rise indicates how high or low you go from Point $A$ to Point $B$. The run is how far you go from Point $A$ to Point $B$.

Find the slope of the line passing through the points $A(1,2)$ and $B(4,6)$.

**Solution:** Applying the slope formula:

$$m = \frac{6-2}{4-1} = \frac{4}{3}.$$

So, the slope of the line is $\frac{4}{3}$, which means for every 3 units the line runs horizontally, it rises 4 units vertically.

Determine the slope of the line shown in the graph below.

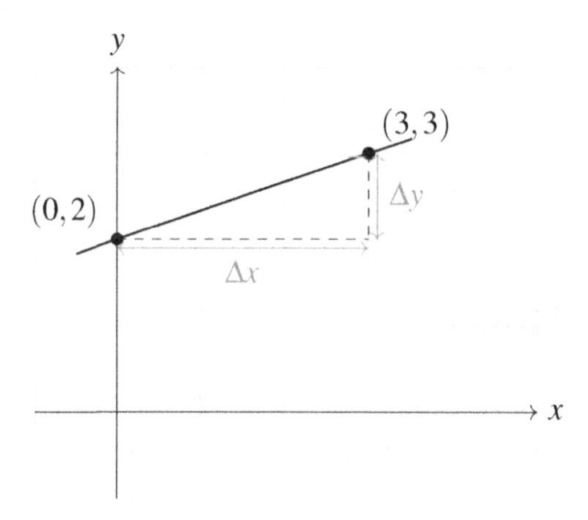


**Solution:** From the graph, we select two points on the line, namely $(0,2)$ and $(3,3)$. Then, we calculate the rise and run:

$$m = \frac{\Delta y}{\Delta x} = \frac{3-2}{3-0} = \frac{1}{3}.$$

Hence, the slope of the line is $\frac{1}{3}$, indicating a gentle incline.

## 3.6 Graphing Lines Using Slope–Intercept Form

The slope-intercept form is a popular way of expressing the equation of a line. This form is particularly useful for quickly sketching graphs of lines and understanding their characteristics.

The slope-intercept form of a line with slope $m$ and $y$-intercept $b$ is given by the equation $y = mx + b$.

To graph a line using this method, follow these simple steps:

1. Start by plotting the $y$-intercept, the point where the line crosses the $y$-axis. This is given by $(0,b)$.
2. Use the slope $m$ to determine the rise over run. From the $y$-intercept, move up or down by the rise and right or left by the run to find a second point on the line.
3. Draw a straight line through the two points.

Remember, if the slope $m$ is positive, the line slopes upwards to the right, and if $m$ is negative, the line slopes downwards to the right.

Sketch the graph of $y = \frac{3}{2}x + 2$.

**Solution:** Following our steps:

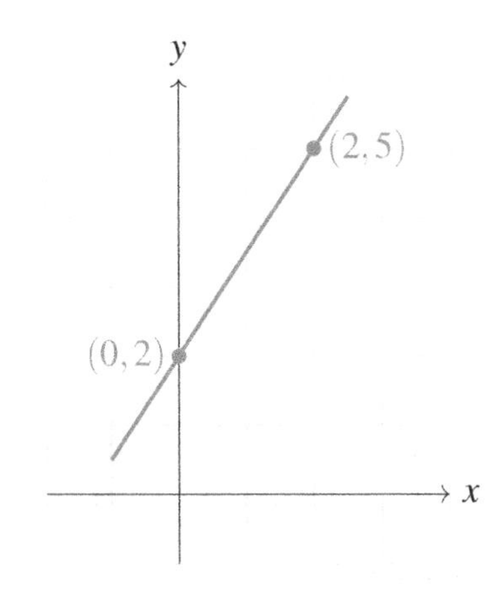


1. Plot the $y$-intercept $(0, 2)$.
2. With a slope of $\frac{3}{2}$, from the point $(0, 2)$, move up 3 units (rise) and to the right 2 units (run) to find a second point on the line, which is $(2, 5)$.
3. Draw the line through both points.

Determine the equation of a line that passes through the point $(4, -2)$ and has a slope of $-\frac{1}{2}$ and sketch its graph.

**Solution:** First, to get the $y$-intercept $b$, we can substitute the given point into the slope-intercept equation and solve for $b$:

$$y = mx + b \Rightarrow -2 = -\frac{1}{2}(4) + b \Rightarrow -2 = -2 + b \Rightarrow b = 0.$$

So, the equation is $y = -\frac{1}{2}x$. To graph the line:

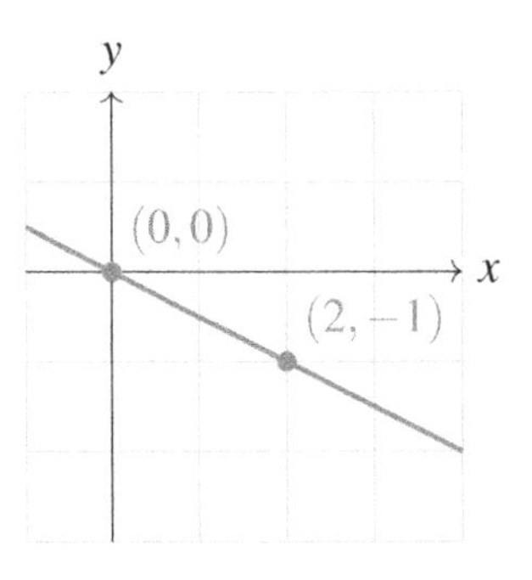


1. Plot the $y$-intercept $(0, 0)$, which is also the origin.
2. With a slope of $-\frac{1}{2}$, we move down 1 unit (fall) and to the right 2 units (run) from the origin to locate another point $(2, -1)$.
3. Connect the points and extend the line through in both directions.

In this case, we used a known point and the slope to graph the line rather than calculating two points from the equation directly.

## 3.7 Writing Linear Equations from Graphs

We will explore how to use the slope-intercept form to write the equations of lines from their graphical representations on an $xy$-coordinate plane. This process involves interpreting the graph to determine the line's slope ($m$) and y-intercept ($b$), and then converting these graphical insights into the mathematical equation:

$$y = mx + b.$$

Let us now look at some examples.

**Example** The graph below shows a straight line. Write the equation of this line.

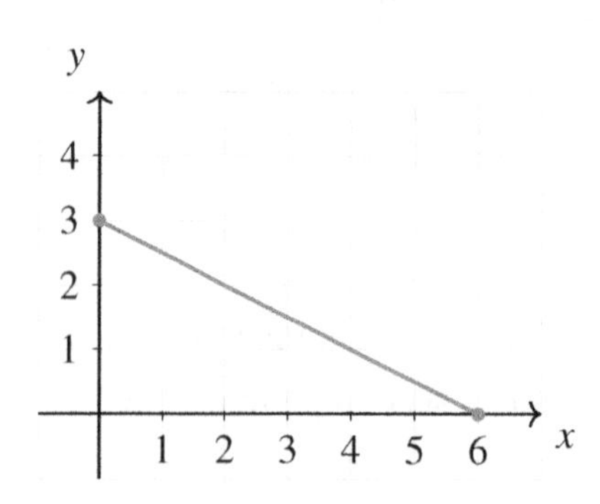


**Solution:** We begin by determining the $y$-intercept, which from the graph is at $(0,3)$. So, $b = 3$. To find the slope $m$, we choose two points on the line. One point is the $y$-intercept, $(0,3)$, and another could be the $x$-intercept, $(6,0)$. We plug these into the slope formula: $m = \frac{3-0}{0-6} = -\frac{1}{2}$. Now that we have both $m$ and $b$, we write the equation of the line: $y = -\frac{1}{2}x + 3$.

**Example** The graph below shows a straight line. Write the equation of this line using two points on the graph.

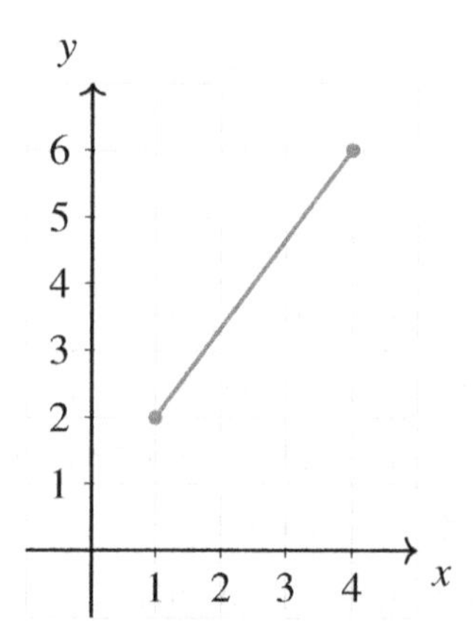


**Solution:** To find the equation of the line, we select two points on the graph. Let's choose the points $(1,2)$ and $(4,6)$. We calculate the slope $m$ using the formula $m = \frac{y_2 - y_1}{x_2 - x_1}$. Substituting the values, $m = \frac{6-2}{4-1} = \frac{4}{3}$. Now, we can use one of the points and the slope to find the $y$-intercept $b$. Using the point $(1,2)$ and the equation $y = mx + b$, we get $2 = \frac{4}{3} \cdot 1 + b$, which simplifies to $b = 2 - \frac{4}{3} = \frac{2}{3}$. Therefore, the equation of the line is $y = \frac{4}{3}x + \frac{2}{3}$.

## 3.8 Converting Between Standard and Slope-Intercept Forms

In coordinate geometry, equations of lines can be expressed in various forms. Two of the most common forms are the standard form and the slope-intercept form. The standard form of a linear equation is:

$$Ax + By = C,$$

where $A$, $B$, and $C$ are constants. The slope-intercept form is written as:

$$y = mx + b,$$

where $m$ represents the slope of the line and $b$ is the $y$-intercept. To convert from standard form to slope-intercept form, follow these steps:

1. Isolate $y$ by moving $Ax$ to the other side of the equation.
2. Divide every term by the coefficient of $y$, which is $B$.

**Key Point**

The slope-intercept form makes it easy to identify the slope and $y$-intercept of a line, which is critical for graphing.

Let us see how this works through examples.

Convert the following standard form equation to slope-intercept form: $2x + 3y = 6$.

**Solution:** We want to isolate $y$, so, first, we subtract $2x$ from both sides: $3y = -2x + 6$. Next, we divide every term by the coefficient of $y$, which is 3: $y = -\frac{2}{3}x + 2$. The equation is now in slope-intercept form with a slope of $-\frac{2}{3}$ and a $y$-intercept of 2.

Convert from standard form to slope-intercept form: $6x - 9y = 18$.

**Solution:** Starting by isolating $y$, we add $6x$ to both sides: $-9y = -6x + 18$. We then divide each term by $-9$, the coefficient of $y$: $y = \frac{6}{9}x - 2$, simplifying the fraction gives us: $y = \frac{2}{3}x - 2$. This equation is now in slope-intercept form, with the slope $m = \frac{2}{3}$ and $y$-intercept $b = -2$.

## 3.9 Slope-intercept Form and Point-slope Form

There are two commonly used forms of linear equations: slope-intercept form and point-slope form.

**Key Point**

The *slope-intercept form* of a line is given by $y = mx + b$, where $m$ is the slope of the line and $b$ is the $y$-intercept, the point where the line crosses the $y$-axis.

**Key Point**

The *point-slope form* of a line is $y - y_1 = m(x - x_1)$ where $(x_1, y_1)$ is a point on the line and $m$ is the slope.

The point-slope form is especially useful when you know one point on the line and the slope. To convert it to the slope-intercept form, simplify it to solve for $y$.

**Example** Find the equation of a line that passes through the point $(-2, 4)$ and has a slope of 2.

**Solution:** First, write the equation using the point-slope form: $y - y_1 = m(x - x_1)$. Substituting the given point and slope, we get: $y - 4 = 2(x + 2)$. Now we will convert it to the slope-intercept form by expanding and simplifying:

$$y - 4 = 2x + 4 \Rightarrow y = 2x + 8.$$

Therefore, the slope-intercept form of the line is $y = 2x + 8$.

**Example** Find the equation of a line that passes through the points $(1, 3)$ and $(4, 11)$.

**Solution:** First, calculate the slope $m$ using the formula: $m = \frac{y_2 - y_1}{x_2 - x_1}$. With our points, this becomes: $m = \frac{11-3}{4-1} = \frac{8}{3}$. Now, use one of the points and the slope in the point-slope form:

$$y - y_1 = m(x - x_1),$$

substituting the point $(1, 3)$ and slope $\frac{8}{3}$, we get: $y - 3 = \frac{8}{3}(x - 1)$. To convert this into slope-intercept form, expand and simplify:

$$y - 3 = \frac{8}{3}x - \frac{8}{3} \Rightarrow y = \frac{8}{3}x + \frac{1}{3}.$$

Therefore, the slope-intercept form of the line is $y = \frac{8}{3}x + \frac{1}{3}$.

## 3.10 Write a Point-slope Form Equation from a Graph

The point-slope form is particularly useful when we know the slope of a line and a single point through which it passes. With this information, we can quickly write down the line's equation.

**Key Point**

The point-slope form of the equation of a line is given by $y - y_1 = m(x - x_1)$, where $m$ represents the slope of the line and $(x_1, y_1)$ are the coordinates of a known point on the line.

The procedure for determining a line's equation from a graph can be summarized in three steps:

1. Determine the slope of the line, denoted by $m$. To find it, select two points on the line and use the slope formula $m = \frac{\Delta y}{\Delta x} = \frac{y_2 - y_1}{x_2 - x_1}$.
2. Plug the slope and the coordinates of one of the points into the point-slope form equation.
3. Simplify, if necessary, to arrive at the point-slope form of the line's equation.

Let us apply these steps through an example.

Given the graph below, determine the equation of the line in point-slope form.

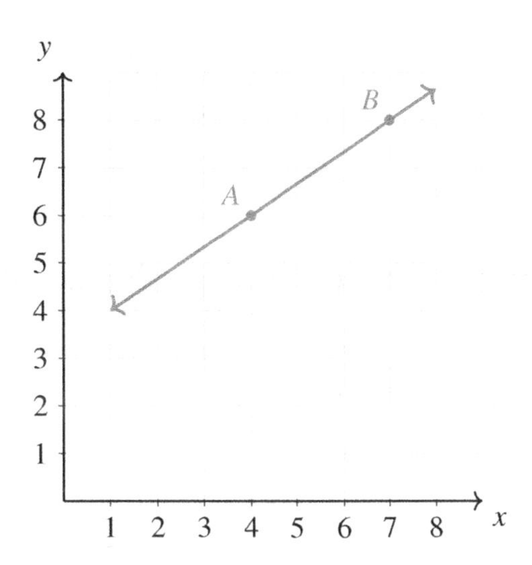


**Solution:** First, we identify two points on the line. Let us use: Point $A$ with coordinates $(4,6)$, marked on the graph, and another point with coordinates $(7,8)$. Now, we calculate the slope $(m)$ of the straight line: $m = \frac{8-6}{7-4} = \frac{2}{3}$. With the slope known, and using the coordinates of point $A$, we write the equation in point-slope form:

$$y - y_1 = m(x - x_1) \Rightarrow y - 6 = \frac{2}{3}(x - 4),$$

thus, the equation of the line in point-slope form is: $y = \frac{2}{3}x + \frac{10}{3}$.

## 3.11 Writing Linear Equations From y-Intercept and a Slope

Here, we will discuss how to write linear equations using the $y$-intercept and the slope. We know that the general formula for a line in slope-intercept form is:

$$y = mx + b,$$

where $m$ represents the slope, and $b$ is the $y$-intercept. The slope $m$ measures the steepness of the line, with positive slopes indicating an upward tilt and negative slopes a downward tilt. The $y$-intercept $b$ indicates the point where the line crosses the $y$-axis.

**Key Point**

The $y$-intercept provides a starting value for $y$ when $x = 0$. This point is useful for sketching the graph of the linear function.

To write an equation with a given $y$-intercept $b$ and a slope $m$, simply plug these values into the slope-intercept formula:

1. Input the slope $m$ into your slope-intercept formula as the coefficient of $x$.
2. Input the $y$-intercept $b$ as the constant term in your equation.

Let us consider an example to consolidate our understanding.

**Example** Write the equation of the line with a $y$-intercept of $-8$ and a slope of 6.

**Solution:** Using the slope-intercept form $y = mx + b$, we substitute $m = 6$ and $b = -8$. Thus, the equation will be: $y = 6x - 8$.

## 3.12 Comparison of Linear Functions: Equations and Graphs

When discussing linear functions in a coordinate plane, the standard form used is $f(x) = mx + b$, where $m$ represents the slope and $b$ represents the $y$-intercept. These functions can be compared by looking at their equations, plotting their values in tables, or graphing them in a coordinate plane.

**Key Point**

The slope of a linear function indicates how fast $y$ changes with respect to $x$. A linear function's slope is also referred to as a rate of change. A steeper slope means a greater rate of change.

**Example** If you have the linear functions $f(x) = 2x + 3$ and $g(x) = 5x - 1$, which one has a greater slope?

**Solution:** The slope of $f(x)$ is 2 and the slope of $g(x)$ is 5. Therefore, $g(x)$ has a greater slope, meaning it is steeper than $f(x)$.

**Example** Which of the following functions is steeper: $h(x) = \frac{1}{4}x - 2$ or $k(x) = \frac{1}{3}x + 1$?

**Solution:** The slope of $h(x)$ is $\frac{1}{4}$ and the slope of $k(x)$ is $\frac{1}{3}$. Since $\frac{1}{3}$ is greater than $\frac{1}{4}$, function $k(x)$ is steeper than $h(x)$.

**Example** Consider the function $y = \frac{3}{2}x + 1$ and an unknown function $y = mx + b$, defined

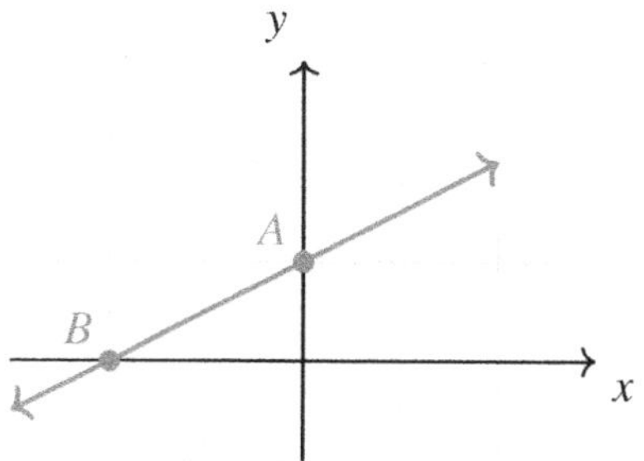


by two points on its graph, $A(0,1)$ and $B(-2,0)$. Which function has a greater slope: $y=\frac{3}{2}x+1$ or the one represented by points $A$ and $B$?

**Solution:** For $y=\frac{3}{2}x+1$, the slope is clearly $\frac{3}{2}$. For the second function, we use the points to find the slope: $m=\frac{1-0}{0-(-2)}=\frac{1}{2}$. Thus, the slope of the first function $\frac{3}{2}$ is greater than the slope of the second function $\frac{1}{2}$.

## 3.13 Equations of Horizontal and Vertical lines

Horizontal lines have a slope of 0. Therefore, the equation of a horizontal line can be simplified to $y=b$, where $b$ is a constant and represents the $y$-coordinate of all points on the line.

Vertical lines have undefined slope and can only be represented by the equation $x=a$, where $a$ is a constant and represents the $x$-coordinate of every point on the line.

**Key Point**

Horizontal lines are characterized by a slope of 0 and are described by the equation $y=b$, while vertical lines have an undefined slope and are represented by $x=a$, where $b$ and $a$ are constants corresponding to their respective $y$ and $x$ coordinates.

Write an equation for the horizontal line that passes through the point $(6,2)$.

**Solution:** As a horizontal line maintains a constant $y$-coordinate for all points on it. For the point provided $(6,2)$, the $y$-coordinate is 2. Therefore, the equation for this horizontal line is $y=2$.

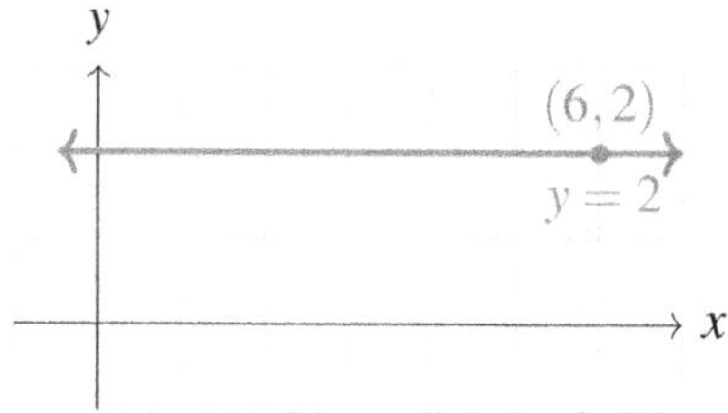


Notice that for the horizontal line passing through $(6,2)$, the fact that it runs parallel to the $x$-axis ensures that $y$ will always equal 2, regardless of the value of $x$.

**Example**

Find the equation for a vertical line passing through the point $(-3,1)$.

**Solution:** For a vertical line, the $x$-coordinate remains constant.

Given the point $(-3, 1)$, the $x$-coordinate is $-3$. Hence, the equation of this vertical line is $x = -3$.

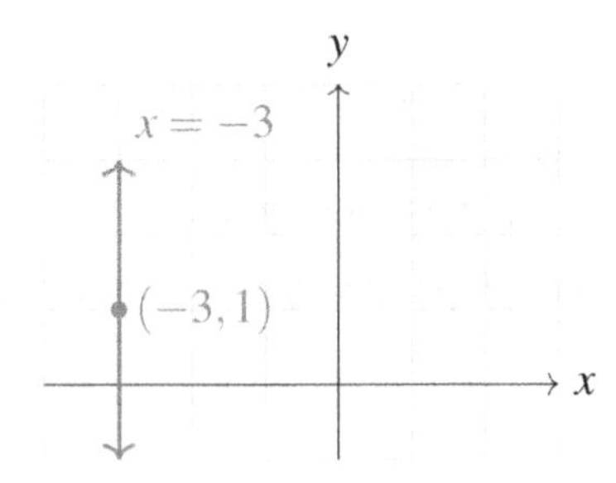


## 3.14 Graph of a Horizontal or Vertical Line

Graphing a line on the coordinate plane generally requires a point and the slope of the line. Yet, horizontal and vertical lines are exceptions to this rule due to their unique properties.

A *horizontal line* is a line that extends left to right, parallel to the $x$-axis. Its notable characteristic is a slope of 0, which arises from the fact that the slope, calculated as 'rise over run', is zero due to the absence of vertical change.

**Key Point**

The mathematical representation of a horizontal line is always: $y = a$, where $a$ is a constant value representing the $y$-coordinate of all points on the line.

Conversely, a *vertical line* extends vertically, parallel to the $y$-axis. Its slope is considered undefined, stemming from the fact that the horizontal change, or 'run', is zero, and division by zero is not defined in mathematics.

**Key Point**

The equation of a vertical line is given by: $x = a$, where $a$ is a constant value indicating the $x$-coordinate of every point on the line.

**Example** Graph the equation: $y = -2$.

**Solution:** The equation $y = -2$ represents a horizontal line where every point on the line has a $y$-coordinate of $-2$. To graph the horizontal line, select a few points such as $(-1, -2)$, $(0, -2)$, and $(1, -2)$, and draw a straight line that passes through these points.

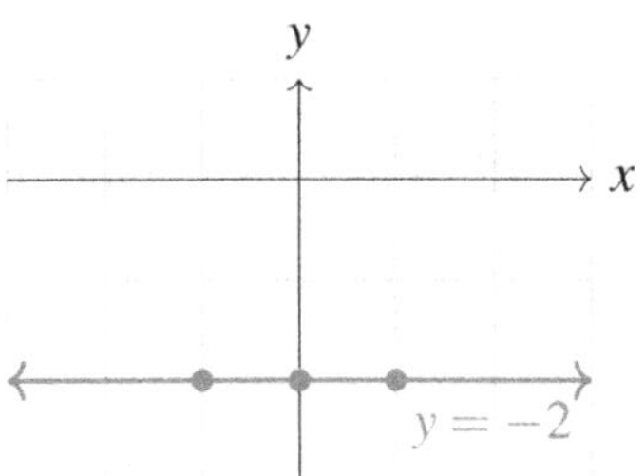

**Example** Graph the equation: $x = 3$.

**Solution:** The equation $x = 3$ corresponds to a vertical line. All points on the line have an $x$-coordinate of 3. To graph this line, take points such as $(3,-1)$, $(3,0)$, and $(3,1)$, and draw a line passing through them that extends infinitely in both the up and down directions.

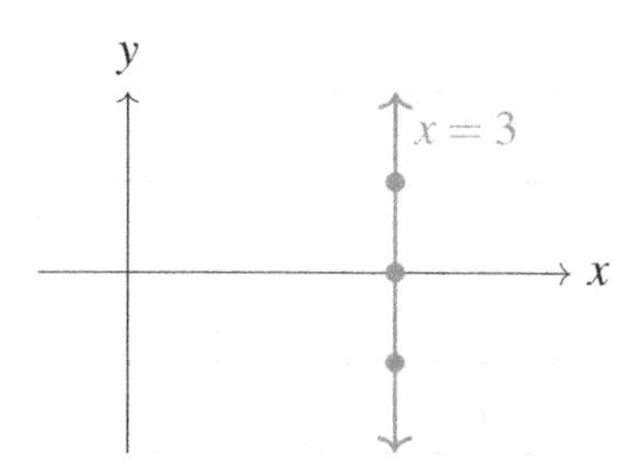


## 3.15 Equation of Parallel and Perpendicular Lines

Knowing whether lines are parallel or perpendicular to each other helps determine their position and behavior in the Cartesian plane.

A key characteristic of *parallel lines* is that they never intersect, no matter how far they are extended. This can only be the case if the two lines rise and run at the same rate. If we have a line $y = mx + b$ and we want to find the equation of another line parallel to it that passes through a point $(x_1, y_1)$, we use the slope $m$ with the point-slope form of a line equation: $y - y_1 = m(x - x_1)$.

**Key Point**

Parallel lines, marked by their congruent slopes, invariably maintain identical slope values.

*Perpendicular lines* intersect each other at a 90° angle. We use a fundamental property of their slopes to determine if two lines are perpendicular. We use the opposite reciprocal of the given slope and point-slope form to find the equation of a line perpendicular to a given line and passing through a specific point.

**Key Point**

Two lines are perpendicular if and only if their slopes are opposite reciprocals of each other. If the slope of one line is $m$, then the slope of a line perpendicular to it is $-\frac{1}{m}$. This relationship is characterized by the product of their slopes being $-1$.

**Example** Find the equation of a line that is parallel to $y = \frac{3}{4}x + 5$ and passes through the point $(-2, 1)$.

**Solution:** Since parallel lines have identical slopes, the slope of our new line is $\frac{3}{4}$. Using the point-slope form, we plug in our slope and our point: $y - 1 = \frac{3}{4}(x - (-2))$. Simplifying, we get: $y - 1 = \frac{3}{4}x + \frac{3}{2}$. Bringing the 1 to the other side: $y=\frac{3}{4}x+\frac{5}{2}$, this is the equation of the line we are looking for.

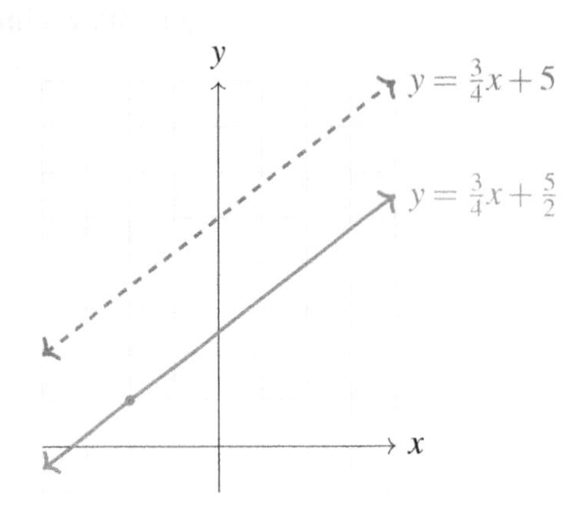


**Example** Find the equation of a line that is perpendicular to $y=-2x+3$ and passes through the point $(4,-1)$.

**Solution:** The slope of the given line is $-2$. The slope of the perpendicular line will be the opposite reciprocal: $\frac{1}{2}$. Using the point-slope form of a line: $y-(-1)=\frac{1}{2}(x-4)$. Simplifying the equation: $y+1=\frac{1}{2}x-2$. And then bringing the 1 to the other side: $y=\frac{1}{2}x-3$.

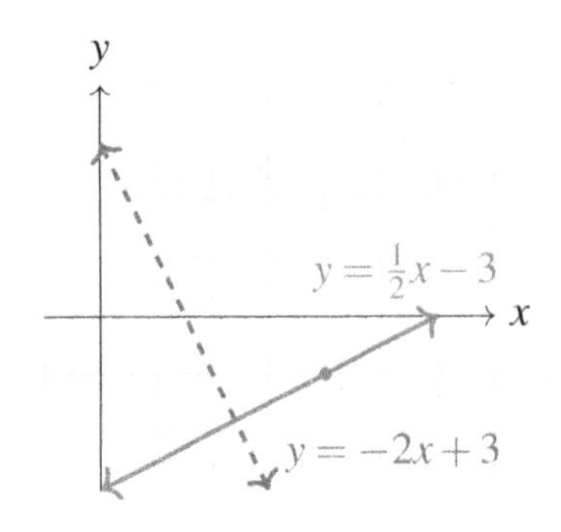


## 3.16 Practices

**1)** Find the slope of each line.

1-1) $y=x-5$

1-2) $y=2x+6$

1-3) $y=-5x-8$

1-4) Line through $(2,6)$ and $(5,0)$

1-5) Line through $(8,0)$ and $(-4,3)$

1-6) Line through $(-2,-4)$ and $(-4,8)$

**2)** Solve.

2-1) What is the equation of a line with slope 4 and intercept 16?

2-2) What is the equation of a line with slope 3 and passes through point $(1,5)$?

2-3) What is the equation of a line with slope $-5$ and passes through point $(-2,7)$?

2-4) The slope of a line is $-4$ and it passes through $(-6,2)$. What is the equation of the line?

2-5) The slope of a line is $-3$ and it passes through $(-3,-6)$. What is the equation of the line?

**3)** Find the midpoint of the line segment with the given endpoints.

3-1) $(5,0)$, $(1,4)$

3-2) $(2,3)$, $(4,7)$

3-3) $(8,1)$, $(2,5)$

3-4) $(5,10)$, $(3,6)$

3-5) $(4,-1)$, $(-2,7)$

3-6) $(2,-5)$, $(4,1)$

3-7) $(7,6)$, $(-5,2)$

3-8) $(-2,8)$, $(4,-6)$

**4)** Find the distance between each pair of points.

4-1) $(-2,8)$, $(-6,8)$

4-2) $(4,-4)$, $(14,20)$

4-3) $(-1,9)$, $(-5,6)$

4-4) $(0,3)$, $(4,3)$

4-5) $(0,-2)$, $(5,10)$

4-6) $(4,3)$, $(7,-1)$

4-7) $(2,6)$, $(10,-9)$

4-8) $(3,3)$, $(6,-1)$

4-9) $(-2,-12)$, $(14,18)$

4-10) $(2,-2)$, $(12,22)$

**5)** Look at the graphs and find the slope of the line.

5-1)

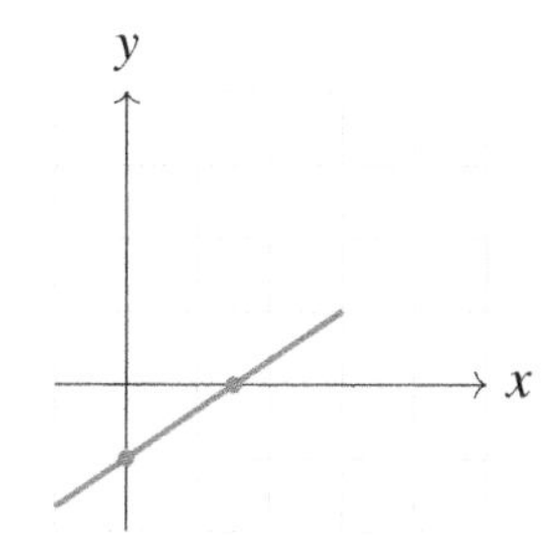


5-2)

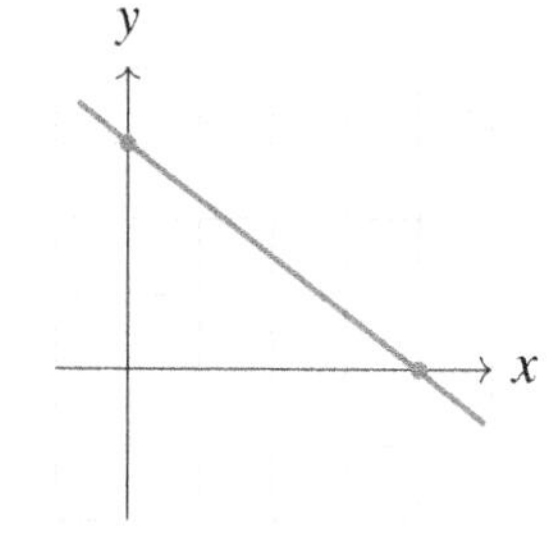

**6)** Sketch the graph of each line. (Using Slope–Intercept Form)

6-1)

$y = x + 4$

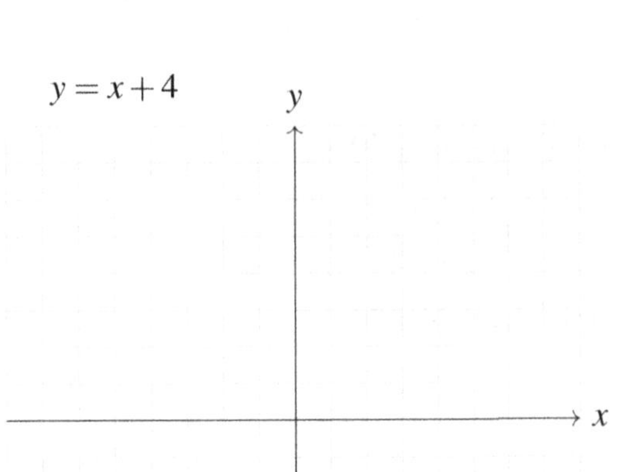


6-2)

$y = 2x - 5$

**7)** Look at the graph below. Write the equation of this line in slope-intercept form.

7-1)

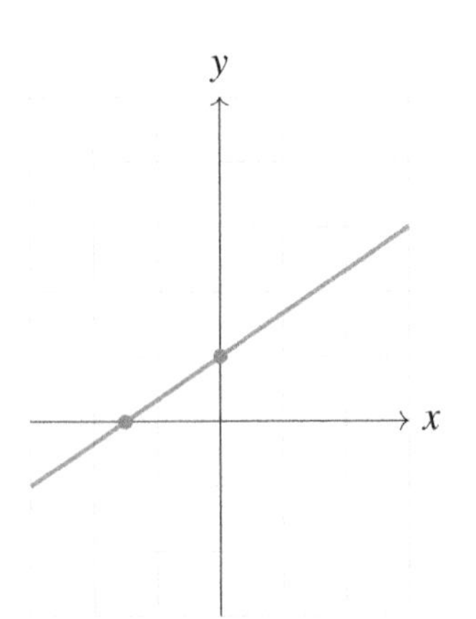


7-2)

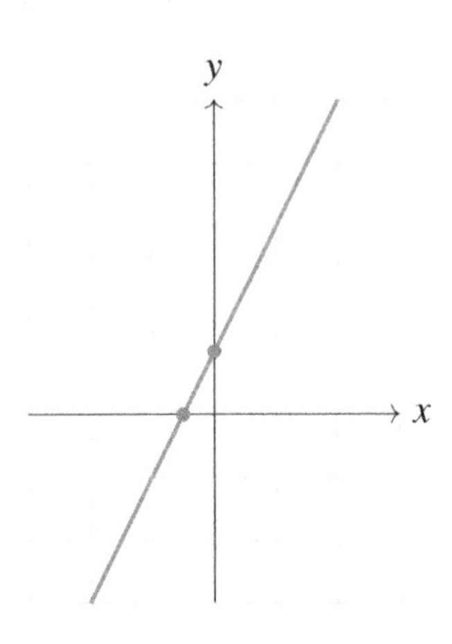


**8)** Convert the equation of the following line from standard form to Slope-Intercept form.

8-1) $6x + 13y = 12$

8-2) $7x + 14y = 21$

8-3) $3x - y = -17$

8-4) $-5x + 4y = 20$

**9)** Write the equation of the graphs below in slope-intercept form.

9-1)

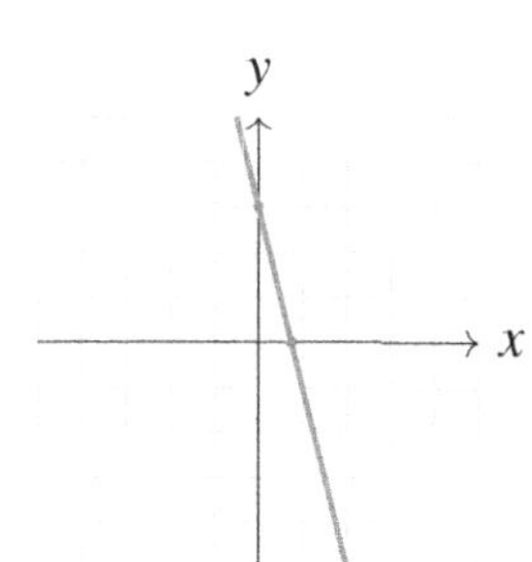


9-2)

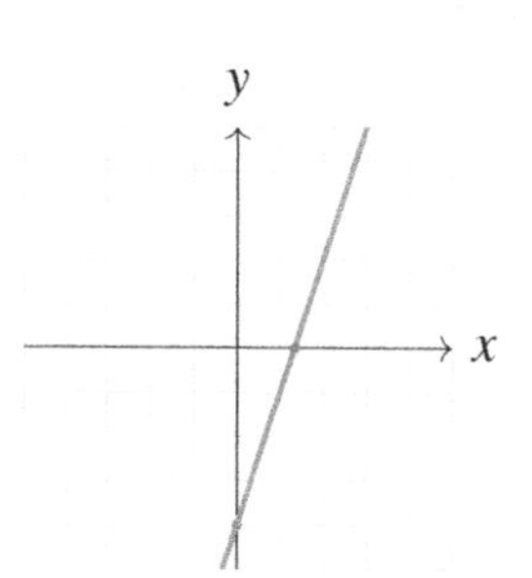


**10)** Find the equation of each line.

10-1) Through: $(6,-6)$, slope $= -2$

Point-slope form: ___________

Slope-intercept form: ___________

10-2) Through: $(-7,7)$, slope $= 4$

Point-slope form: ___________

Slope-intercept form: ___________

**11)** Write an equation of each of the following lines in point-slope form.

11-1)

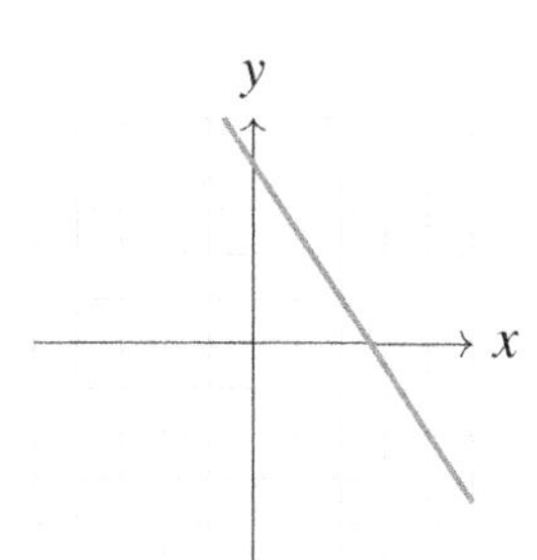


11-2)

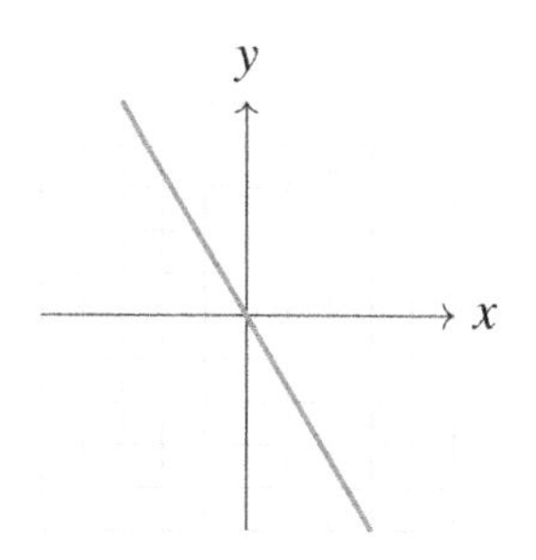


**12)** Solve.

12-1) Write the equation of the line with a $y$-intercept of $\frac{9}{11}$ and a slope of $-\frac{1}{7}$.

12-2) Write the equation of the line with a $y$-intercept of 0 and a slope of 10.

12-3) Write the equation of the line with a $y$-intercept of $-\frac{3}{10}$ and a slope of $-4$.

12-4) Write the equation of the line with a $y$-intercept of 17 and a slope of 1.

12-5) Write the equation of the line with a $y$-intercept of $-6$ and a slope of $-1$.

12-6) Write the equation of the line with a $y$-intercept of $\frac{7}{3}$ and a slope of $-\frac{5}{6}$.

**13)** Compare the slope of the function $A$ and function $B$.

13-1) Function A:

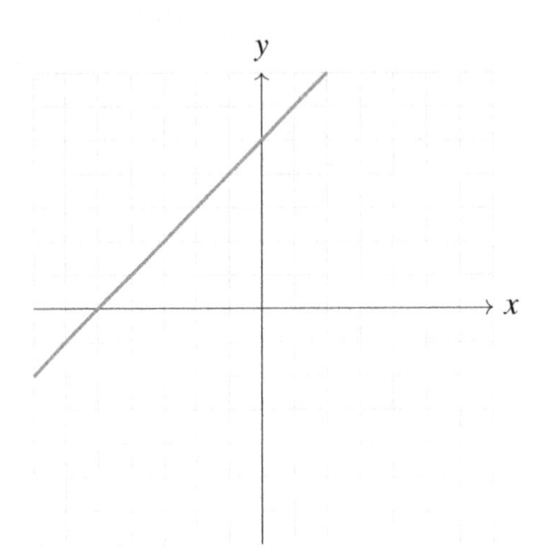


Function B: $y = 6x - 3$

13-2) Function A:

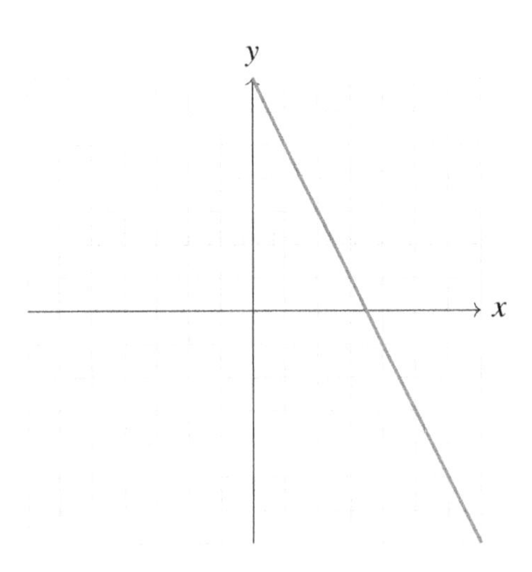


Function B: $y = -2x - 1$

13-3) Function A:

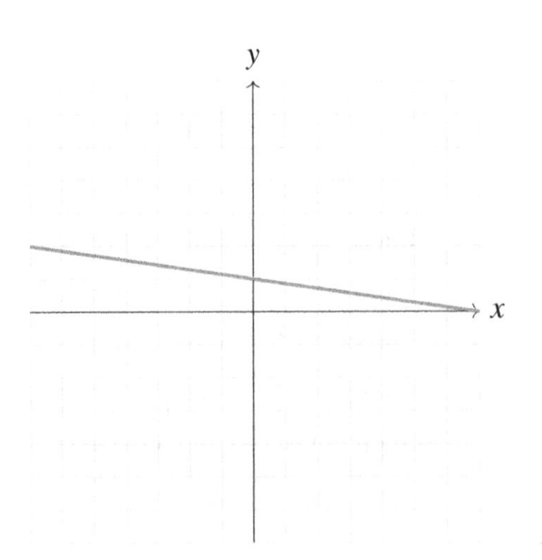

Function B: $y = 7x - 1$

**14)** Sketch the graph of each line.

14-1)

Vertical line that passes through $(2,6)$

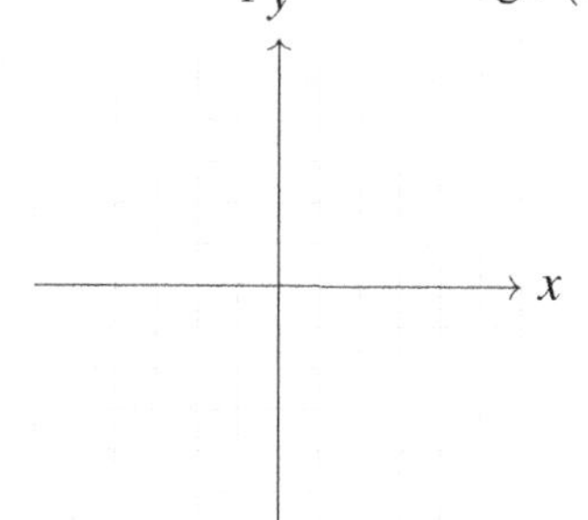


14-2)

Horizontal line that passes through $(5,3)$

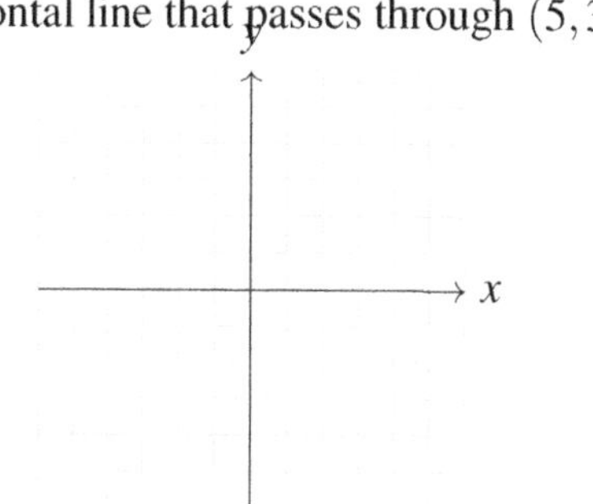


**15)** Find the equation of each line with the given information.

15-1) Through: $(4,4)$

Parallel to $y = -6x + 5$

Equation: ____________

15-2) Through: $(7,1)$

perpendicular to $y = -\frac{1}{2}x - 4$

Equation: ____________

15-3) Through: $(2,0)$

Parallel to $y = x$

Equation: ____________

15-4) Through: $(0,-4)$

perpendicular to $y = 2x + 3$

Equation: ____________

15-5) Through: $(-1,1)$

Parallel to $y = 2$

Equation: ____________

15-6) Through: $(3,4)$

perpendicular to $y = -x$

Equation: ____________

## 3.17 Answers

**1)**

1-1) 1

1-2) 2

1-3) $-5$

1-4) $-2$

1-5) $-\frac{1}{4}$

1-6) $-6$

**2)**

2-1) $y = 4x + 16$

2-2) $y = 3x + 2$

2-3) $y = -5x - 3$

2-4) $y = -4x - 22$

2-5) $y = -3x - 15$

**3)**

3-1) $(3,2)$

3-2) $(3,5)$

3-3) $(5,3)$

3-4) $(4,8)$

3-5) $(1,3)$

3-6) $(3,-2)$

3-7) $(1,4)$

3-8) $(1,1)$

**4)**

4-1) 4

4-2) 26

4-3) 5

4-4) 4

4-5) 13

4-6) 5

4-7) 17

4-8) 5

4-9) 34

4-10) 26

**5)**

5-1) $\frac{2}{3}$

5-2) $-\frac{3}{4} = -0.75$

**6)**

6-1) $y = x + 4$

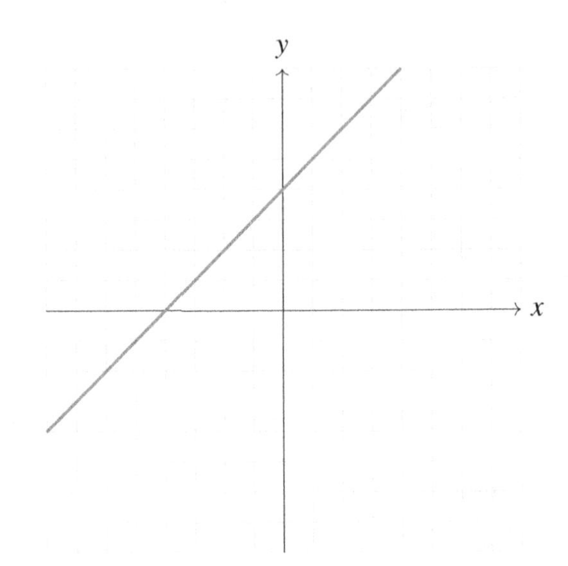


6-2) $y = 2x - 5$

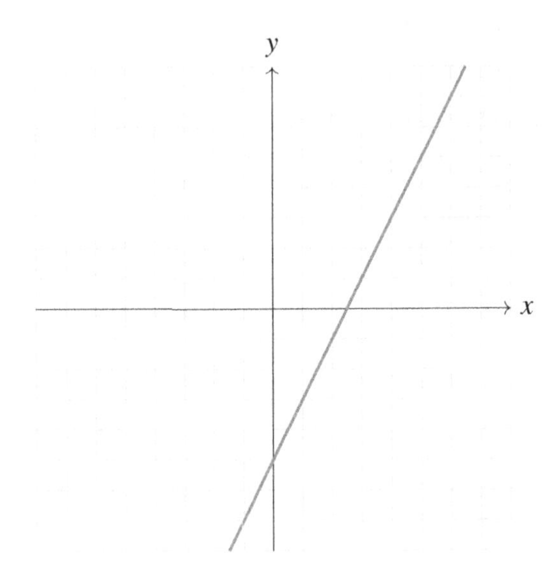


**7)**

7-1) $y = \frac{2}{3}x + 1$

7-2) $y = 2x + 1$

**8)**

8-1) $y = -\frac{6}{13}x + \frac{12}{13}$

8-2) $y=-\frac{1}{2}x+\frac{3}{2}$

8-3) $y=3x+17$

8-4) $y=\frac{5}{4}x+5$

**9)**

9-1) $y=-4x+3$

9-2) $y=3x-4$

**10)**

10-1) Point-slope form: $y+6=-2(x-6)$
Slope-intercept form: $y=-2x+6$

10-2) Point-slope form: $y-7=4(x+7)$
Slope-intercept form: $y=4x+35$

**11)**

11-1) $y-1=-\frac{3}{2}(x-2)$

11-2) $y=-\frac{5}{3}x$

**12)**

12-1) $y=-\frac{1}{7}x+\frac{9}{11}$

12-2) $y=10x$

12-3) $y=-4x-\frac{3}{10}$

12-4) $y=x+17$

12-5) $y=-x-6$

12-6) $y=-\frac{5}{6}x+\frac{7}{3}$

**13)**

13-1) The slope of function $A$ is 1 and is lower than the slope of function $B$ (6).

13-2) The slopes of both functions are $-2$. Two functions are parallel.

13-3) The slope of function $A$ is $-\frac{1}{7}$ and the slope of function $B$ is 7. Two functions are perpendicular.

**14)**

14-1)

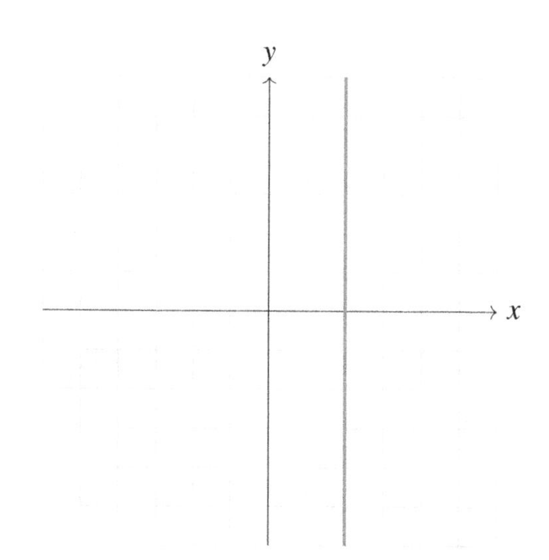


14-2)

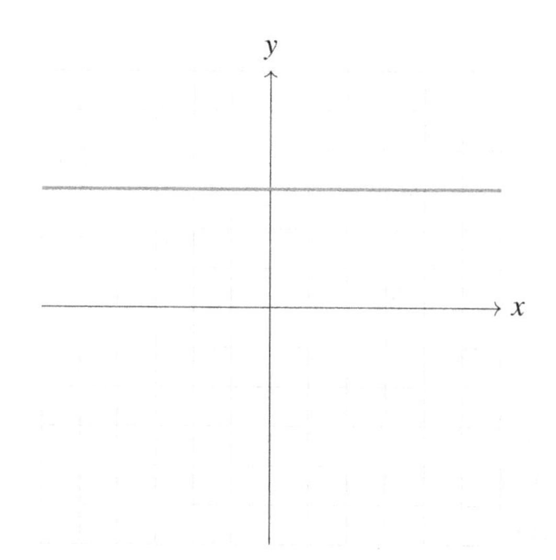


**15)**

15-1) $y=-6x+28$

15-2) $y=2x-13$

15-3) $y=x-2$

15-4) $y=-\frac{1}{2}x-4$

15-5) $y=1$

15-6) $y=x+1$

# 4. Transformations, Rigid Motions, and Congruence

## 4.1 Transformations on the Coordinate Plane

In geometry, transformations refer to the movement of figures on the coordinate plane. These transformations can be classified into several types, each having a unique set of properties and effects on the figures involved. The most common types of transformations are translations, reflections, rotations, and dilations.

**Key Point**

A transformation in geometry means changing a figure's position, size, or orientation on the coordinate plane. This can be achieved through various means, including shifting (translation), flipping (reflection), turning (rotation), and resizing (dilation).

**Overview of Transformations**

**Translation:** Moves a figure in a straight line from one position to another, without changing its size, shape, or orientation.

**Reflection:** Flips a figure over a line, creating a mirror image of the original figure.

**Rotation:** Turns a figure around a fixed point, known as the center of rotation, through a specified angle.

**Dilation:** Changes the size of a figure, either enlarging or reducing it, while maintaining its shape.

Each of these transformations maintains certain properties of the figures, such as their shape or size, while altering others. The subsequent sections of this book will delve into each of these transformations in detail, exploring their properties, the rules governing them, and their applications in geometry.

Determine whether the given picture represents a reflection, rotation, or translation.

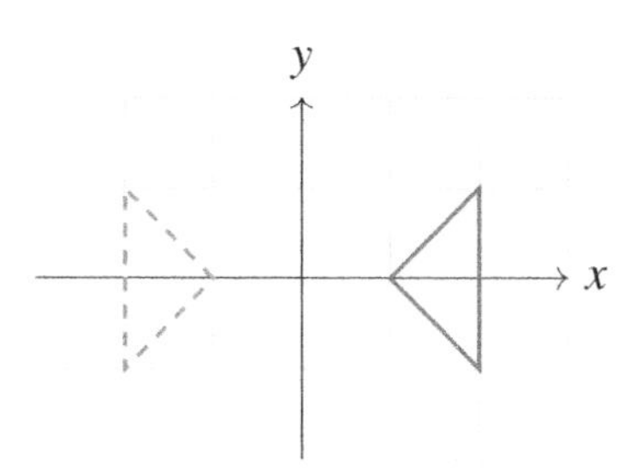


**Solution:** The given picture shows the figure flipped over a line. Thus, it represents a reflection.

## 4.2 Translations on the Coordinate Plane

A translation is a type of transformation that moves every point of a figure the same distance in the same direction on the coordinate plane. This motion slides a shape or point to a new position, but its size, orientation, and shape remain unchanged.

**Key Point**

A translation does not rotate, reflect, or resize a figure; it strictly moves the figure from one location to another on the coordinate plane.

For a given point, a translation is defined by a pair of numbers, denoted as $(a,b)$, where '$a$' represents the horizontal shift, and '$b$' represents the vertical shift. These two numbers are known as the components of a translation vector. The general rule to perform a translation is given by the notation:

$$(x,y) \to (x+a, y+b),$$

this implies that a point $(x,y)$ when translated by $(a,b)$, will move to a new point $(x+a, y+b)$.

In the diagram below, triangle $BGT$ is translated 5 units to the right and 3 units up.

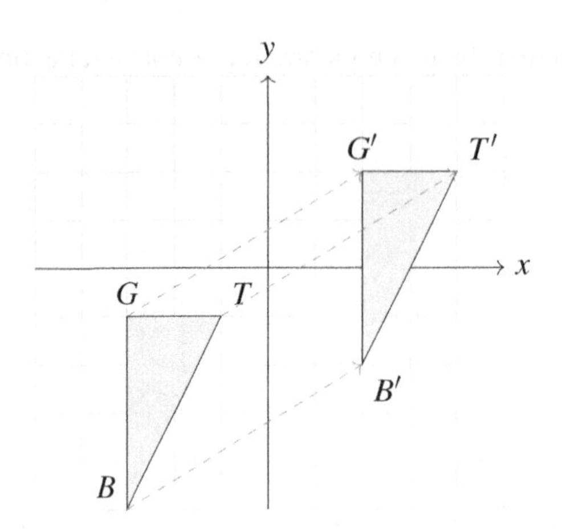

 Translate the point $P = (2, -4)$ by 7 units to the left and 2 units down.

**Solution:** To translate the point $P$ left and down, we use a negative '$a$' for left and a negative '$b$' for down. Here, $a = -7$ and $b = -2$. According to our rule, $(x, y) \to (x + a, y + b)$, we get:

$$P' = (2 + (-7), -4 + (-2)) = (-5, -6).$$

 Translate the rectangle with vertices $A = (1, 2)$, $B = (1, 5)$, $C = (4, 5)$, and $D = (4, 2)$ by 3 units to the right and 4 units down.

**Solution:** Using the translation rule, for each point of the rectangle we add 3 to the $x$-coordinate and subtract 4 from the $y$-coordinate:

$$A' = (1 + 3, 2 - 4) = (4, -2)$$
$$B' = (1 + 3, 5 - 4) = (4, 1)$$
$$C' = (4 + 3, 5 - 4) = (7, 1)$$
$$D' = (4 + 3, 2 - 4) = (7, -2)$$

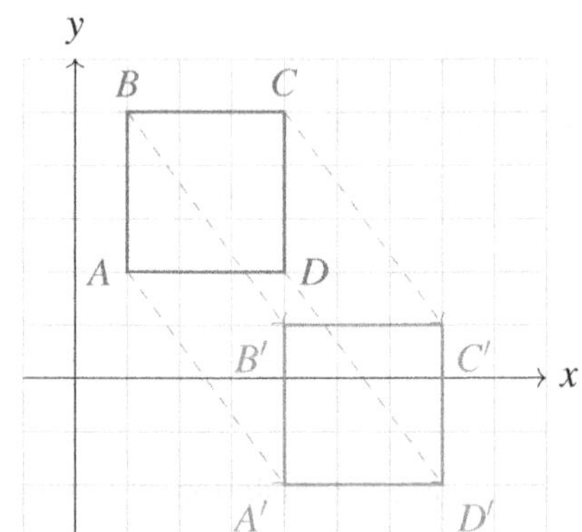


## 4.3 Reflections on the Coordinate Plane

Reflection flips an object across a line, known as the line of reflection, without changing its size or shape. Reflections on the coordinate plane can occur over the $x$-axis, $y$-axis, the line $y = x$, the line $y = -x$, or even the origin.

**Key Point**

Reflection is flipping an object across a line without changing its size or shape.

Reflections change the coordinates of the points that form shapes in predictable ways:

- The reflection of the point $(x, y)$ across the $x$-axis is the point $(x, -y)$.
- The reflection of the point $(x, y)$ across the $y$-axis is the point $(-x, y)$.
- The reflection of the point $(x, y)$ across the line $y = x$ is the point $(y, x)$.
- The reflection of the point $(x, y)$ across the line $y = -x$ is the point $(-y, -x)$.
- The reflection of the point $(x, y)$ at the origin results in the point $(-x, -y)$.

These rules allow us to determine the new coordinates of a figure after reflection.

Graph the image of triangle $ABC$ under reflection across the $x$-axis. Given coordinates:

$$A = (-3, 4),\ B = (-3, 1),\ \text{and } C = (0, 1).$$

**Solution:** By applying the rule for reflections over the $x$-axis, we find the coordinates of the reflected image:

$$A' = (-3, -4),\ B' = (-3, -1),\ \text{and } C' = (0, -1).$$

The image of triangle $ABC$ after reflection is triangle $A'B'C'$.

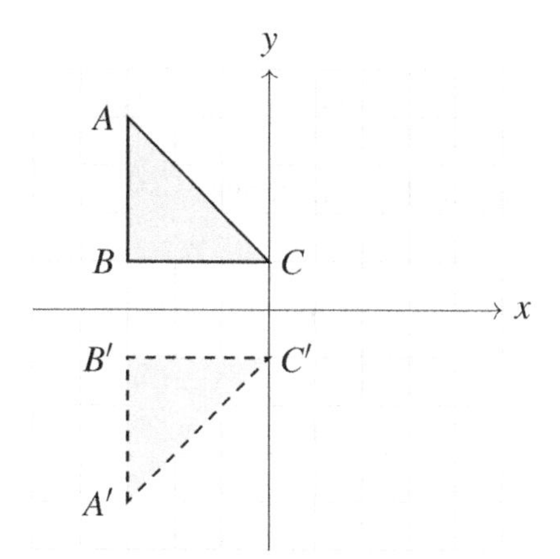


## 4.4 Rotations on the Coordinate Plane

A rotation spins an object about a fixed point called the center of rotation. In a two-dimensional coordinate plane, this center is often the origin, denoted by $(0, 0)$. There are two possible directions for a rotation: clockwise and counterclockwise. In this section, we will consider counterclockwise rotations as the positive direction, which is a common convention in mathematics.

When rotating figures on the coordinate plane, there are specifically three common angles that are widely used in geometric problems: 90°, 180°, and 270°. Each of these angles has a corresponding rule that dictates how the coordinates of a point change after the rotation.

**Key Point**

Rotating a point around the origin by 90° counterclockwise changes the position of the point from $(x,y)$ to $(-y,x)$.

**Key Point**

A 180° rotation around the origin maps the point $(x,y)$ to $(-x,-y)$, essentially flipping the point across the origin.

**Key Point**

Finally, rotating a point by 270° counterclockwise results in the point moving from $(x,y)$ to $(y,-x)$.

**Example** Suppose we have a triangle $\Delta ABC$ with vertices $A=(4,5)$, $B=(5,1)$, and $C=(1,1)$. Graph the original triangle $\Delta ABC$ and its image after rotating it 90° counterclockwise about the origin.

**Solution:** Applying the rotation rule for 90° counterclockwise, we find the new coordinates for each vertex.

$$A=(4,5)\to A'=(-5,4)$$
$$B=(5,1)\to B'=(-1,5)$$
$$C=(1,1)\to C'=(-1,1)$$

Now, let us graph both the original triangle and the rotated one.

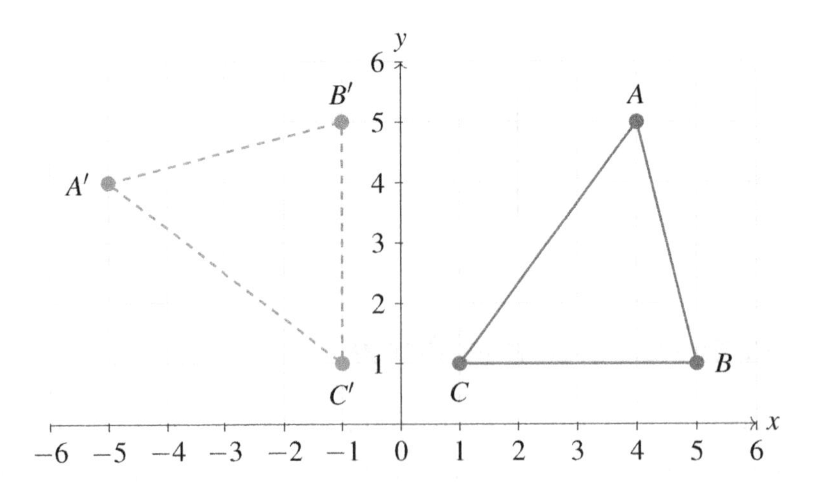


Notice how each vertex of the triangle has moved according to the rotation rule and the new triangle $\Delta A'B'C'$ is in the correct position after a 90° counterclockwise rotation.

## 4.5 Dilation on the Coordinate Plane

Dilation is a transformation resulting in a figure that is similar to the original figure, meaning that the shape is the same, but the size can be different.

**Key Point**

A dilation either enlarges or reduces a figure by a scale factor, relative to a point called the center of dilation.

In a dilation on the coordinate plane, each point $P(x,y)$ of the figure is moved along a straight line that is drawn from a fixed point called the center of dilation. The new position of the point $P'$, after dilation, is determined based on the scale factor.

The scale factor $k$ can be defined as follows:

$$k = \frac{\text{image length}}{\text{original length}} = \frac{\text{distance of image from center of dilation}}{\text{distance of object from center of dilation}}.$$

A key distinction in dilation is:

- If $k > 1$, the image is an enlargement (the figure becomes larger).
- If $0 < k < 1$, the image is a reduction (the figure becomes smaller).

Dilate the image of rectangle $ABCD$ with vertices at $A(-2,1)$, $B(2,1)$, $C(-2,-1)$, and $D(2,-1)$ by a scale factor of 2, using the origin as the center of dilation.

**Solution:** To perform the dilation, we multiply each coordinate by the scale factor of 2:

$$A' = (-4,2),\ B'(4,2),\ C' = (-4,-2),\ \text{and } D'(4,-2).$$

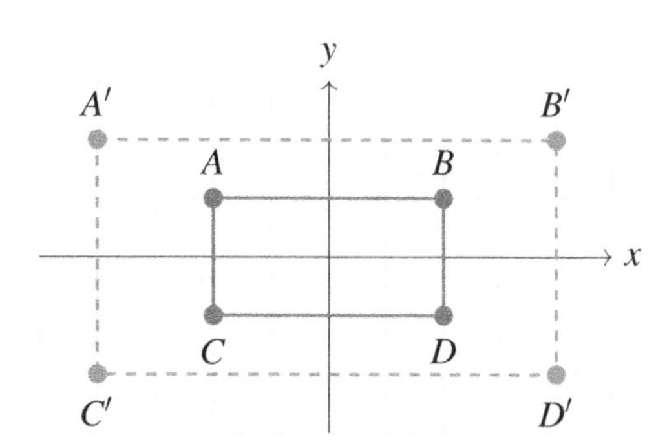


The resulting figure will be an image twice the size of the original rectangle $ABCD$, which means we expect an enlargement. The figure will have the same shape but will be larger.

## 4.6 Dilations: Scale Factor and Center

The scale factor is an essential concept in dilation because it tells us how much we should expand or contract the figure.

**Key Point**

The scale factor is the ratio of the side lengths of the image to the corresponding side lengths of the pre-image. If the scale factor is greater than 1, the figure enlarges; if it is between 0 and 1, the figure shrinks.

When we have two corresponding points between the pre-image and the image, $(x_1, y_1)$ and $(x_2, y_2)$, we can find the center of dilation $(x_o, y_o)$ by solving the following equations:

$$(x_o, y_o) = \left(\frac{kx_1 - x_2}{k-1}, \frac{ky_1 - y_2}{k-1}\right),$$

where $k$ is the scale factor of the dilation.

**Example** Consider a dilation with a scale factor of 2 and two corresponding points $A(2,3)$ in the pre-image and $A'(6,8)$ in the image. Find the center of dilation.

**Solution:** Using the given points $A(2,3)$ and $A'(6,8)$ with the scale factor $k = 2$, we apply the formula:

$$(x_o, y_o) = \left(\frac{(2\times 2) - 6}{2-1}, \frac{(2\times 3) - 8}{2-1}\right) = (-2, -2).$$

Therefore, the center of dilation is at $(-2, -2)$.

**Example** In a dilation, triangle $ABC$ with sides $AB = 5$, $BC = 7$, and $AC = 6$ becomes triangle $A'B'C'$ with sides $A'B' = 10$, $B'C' = 14$, and $A'C' = 12$. What is the scale factor of the dilation?

**Solution:** Use sides $A'B'$ and $AB$:

$$K = \frac{A'B'}{AB} = \frac{10}{5} = 2.$$

The scale factor is 2.

## 4.7 Dilations: Finding a Coordinate

When we perform a dilation, each point of the figure moves along a straight line away from or towards a fixed point, known as the center of dilation.

**Key Point**

Dilations are transformations that produce an image that is the same shape as the original, but a different size. The scale factor, $k$, determines how much the figure is enlarged or reduced.

To find a coordinate on the dilated image, we use the center of dilation, $(x_o, y_o)$, the original coordinate $(x_1, y_1)$, and the scale factor $k$:

$$(x_2, y_2) = (k(x_1 - x_o) + x_o, k(y_1 - y_o) + y_o).$$

If the center of dilation is the origin, which means $x_o = 0$ and $y_o = 0$, then the formula simplifies to:

$$(x_2, y_2) = (kx_1, ky_1).$$

For a scale factor $k > 1$, the image is enlarged; for $0 < k < 1$, the image is reduced.

**Example** A triangle with vertices at $A(2,7)$, $B(2,1)$, and $C(6,1)$ is dilated with a scale factor of 2 and the origin as the center of dilation. Find the coordinates of the dilated triangle's vertices.

**Solution:** Using the simplified dilation formula since the center of dilation is the origin:

$$A'(x_2, y_2) = 2(2,7) = (4,14),$$
$$B'(x_2, y_2) = 2(2,1) = (4,2),$$
$$C'(x_2, y_2) = 2(6,1) = (12,2).$$

So, the dilated triangle's vertices are $A'(4,14)$, $B'(4,2)$, and $C'(12,2)$.

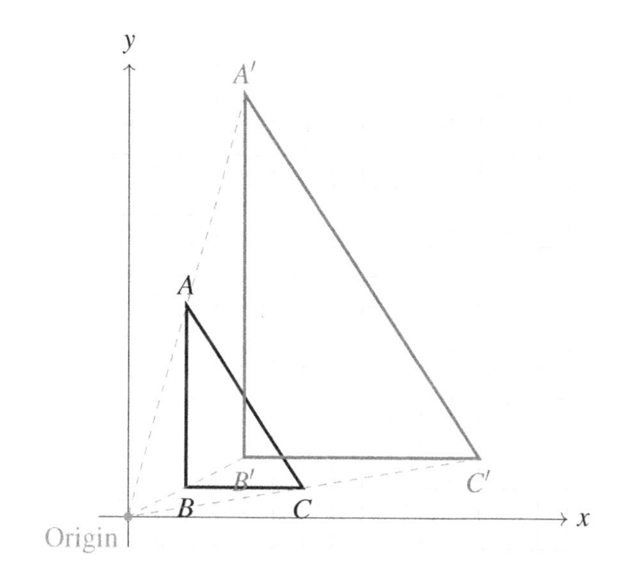


## 4.8 Congruence and Rigid Motions

As we delve into the topic of congruence and rigid motions, let us remember that in geometry, congruence signifies that two figures have the same size and the same shape.

**Translation** Translation is the simplest rigid motion, where every point of a shape is moved the same distance in the same direction. Imagine sliding a book across a table; it is the same book just in a different location.

**Rotation** A rotation takes place around a fixed point known as the center of rotation. Just like the hands of a clock move around its center, our shapes in geometry rotate around this pivot point, staying true to their original form.

**Reflection** A reflection creates a mirror image of a shape over a line of reflection. Think of looking at your reflection in a lake. The mirrored image maintains the exact proportions of the original shape, only flipped across that imaginary line.

**Glide Reflection** Glide reflection is a two-step process, where a shape is first translated and then reflected. The Moonwalk dance move can be thought of as a real-life example of glide reflection.

**Key Point**

Rigid motions, also known as isometries, are movements that preserve the distances and angles between points in a shape. If a shape can be transformed into another shape using only rigid motions and both figures coincide perfectly, they are congruent.

**Example** Are triangles $ABC$ and $DEF$ congruent on the coordinate plane?

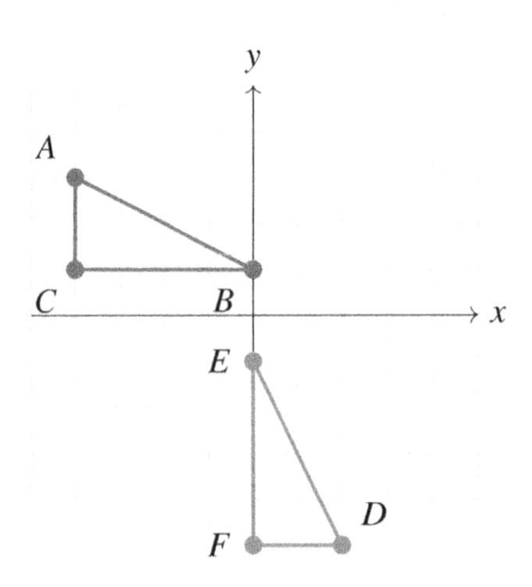


**Solution:** To determine if triangles $ABC$ and $DEF$ are congruent, we perform the following transformations on triangle $ABC$:

**Translation:** Translate triangle $ABC$ 2 units downwards. The new coordinates of the vertices will be $A'(-4,1)$, $B'(0,-1)$, and $C'(-4,-1)$.

**Rotation:** Rotate triangle $A'B'C'$ $90°$ counterclockwise about point $B'$. This will place $A'$ at $A''(-2,-5)$, $B'$ at $B''(0,-1)$, and $C'$ at $C''(0,-5)$.

**Reflection:** Reflect triangle $A''B''C''$ about the $y$-axis. The vertices will now align with $D$, $E$, and $F$ of triangle $DEF$, showing that the triangles are congruent.

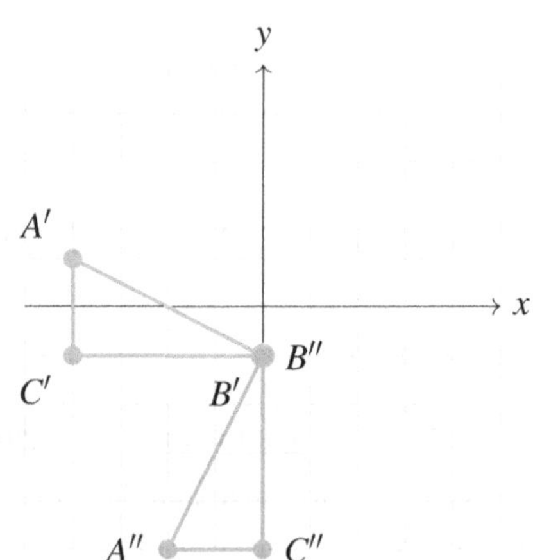


## 4.9 Symmetries of a Figure

Symmetry in geometry implies that a figure remains unchanged under certain transformations, maintaining its size and shape as in rigid motions. The primary types of symmetry are reflectional and rotational, both fundamental in assessing the balance and harmony of geometric figures.

**Key Point**

Reflectional symmetry (Line Symmetry) exists when a figure can be folded over a line such that its two halves coincide perfectly. This axis of fold is known as the "axis of symmetry" or "mirror line."

**Key Point**

Rotational symmetry is when a figure can be rotated around a central point by a particular angle and still look unchanged. The minimum angle at which this occurs is called the "angle of rotational symmetry."

Now, let us consider some examples to illustrate these concepts.

**Example** A regular pentagon is placed on a plane. How many axes of symmetry does it have, and what kind of symmetry is it?

**Solution:** A regular pentagon has reflectional symmetry with 5 axes of symmetry. These lines pass through each vertex and bisect the opposite side. Each line divides the pentagon into two congruent halves.

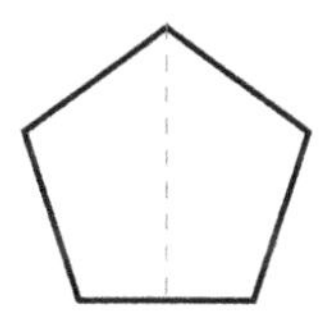

**Example** What kind of symmetry does a figure have if it possesses two lines of symmetry that intersect at right angles, and it also has rotational symmetry of 180°?

**Solution:** This figure likely resembles a rectangle (excluding a square which has four lines of symmetry). A rectangle has two lines of symmetry, one running parallel to the shorter sides and one running parallel to the longer sides, and rotational symmetry of 180°.

180°

## 4.10 Practices

**1)** Determine whether the given picture represents a reflection, rotation, or translation.

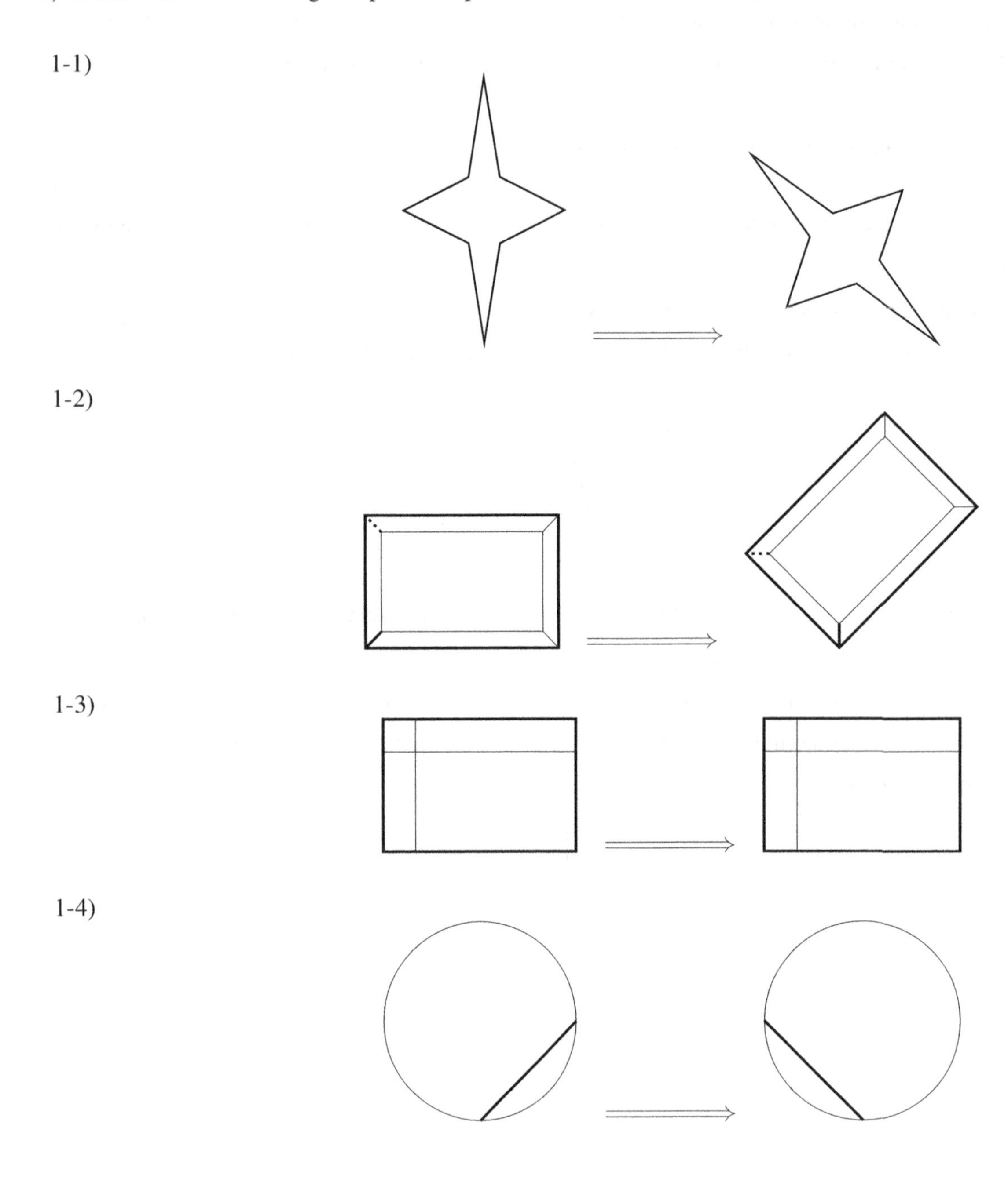


**2)** Evaluate.

2-1) Translation: 4 units left and 1 unit down

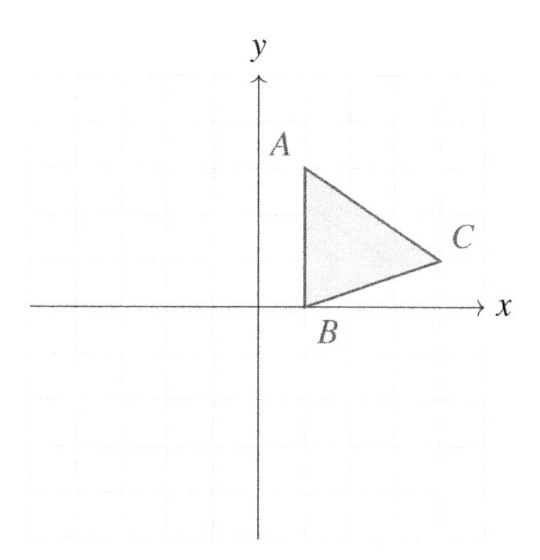


2-2) Rotation: $90^{\circ}$ (counter-clockwise)

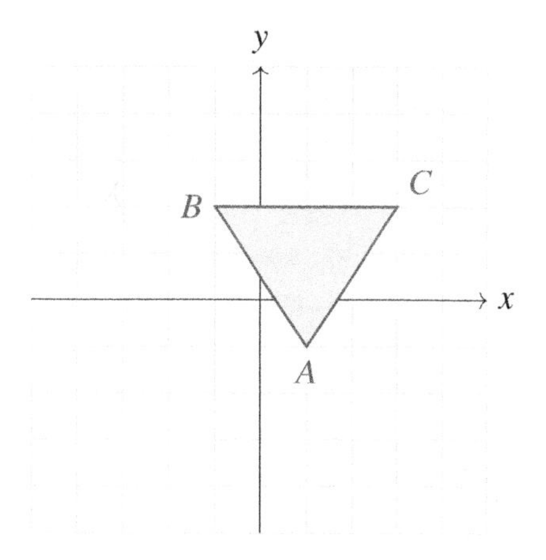


2-3) Reflection across line: $y = x$

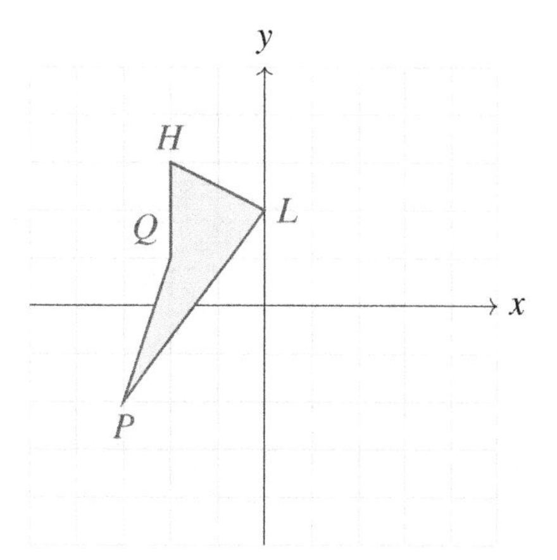

2-4) Dilation of 0.5

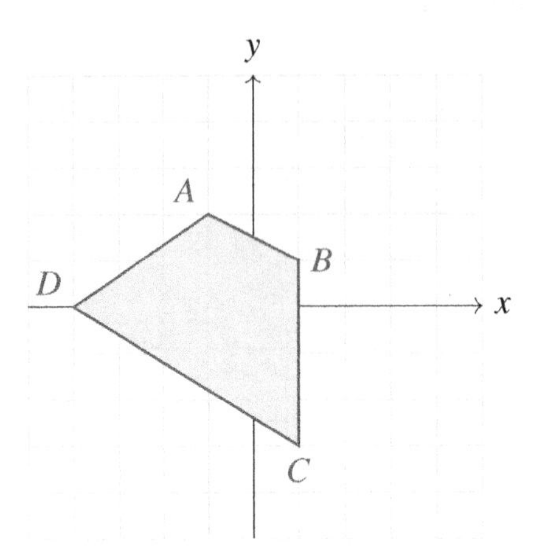


2-5) Reflection across line: $y = -x$

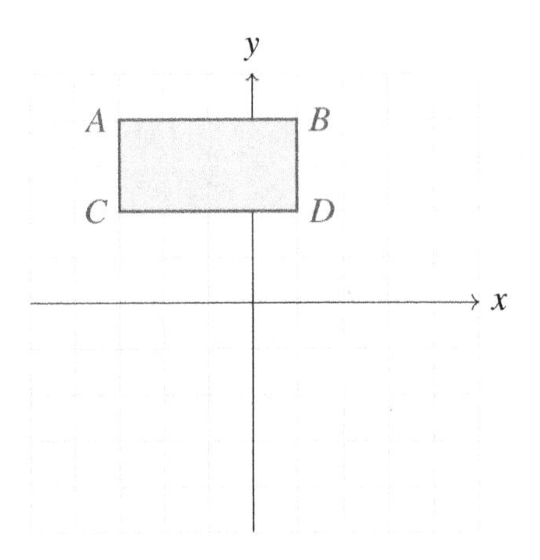


2-6) Rotation: $270^{\circ}$ (counter-clockwise)

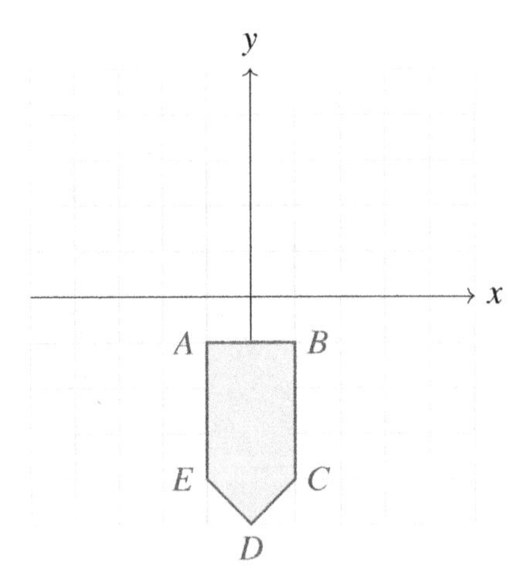


**3)** What is the dilation scale factor in the following figures?

3-1)

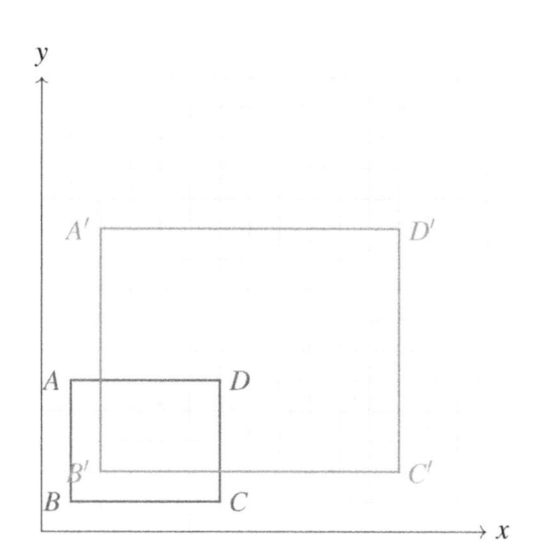


3-2)

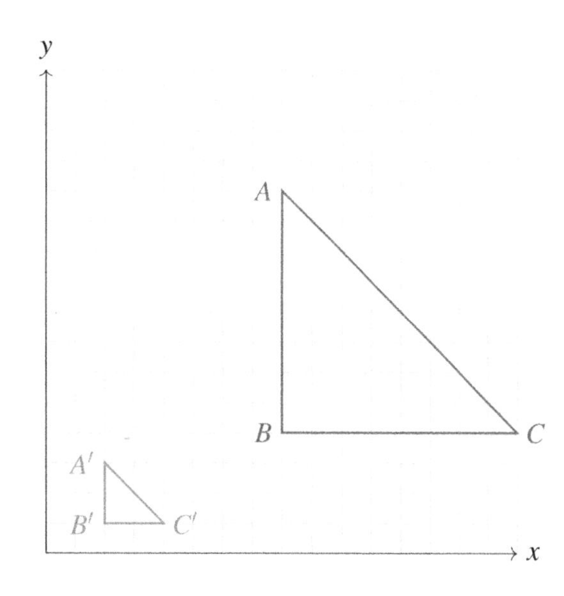


**4)** Get the coordinates of the vertices of the following figures after dilation. The center of dilation is the origin and k is the scale factor.

4-1) $k = 3$

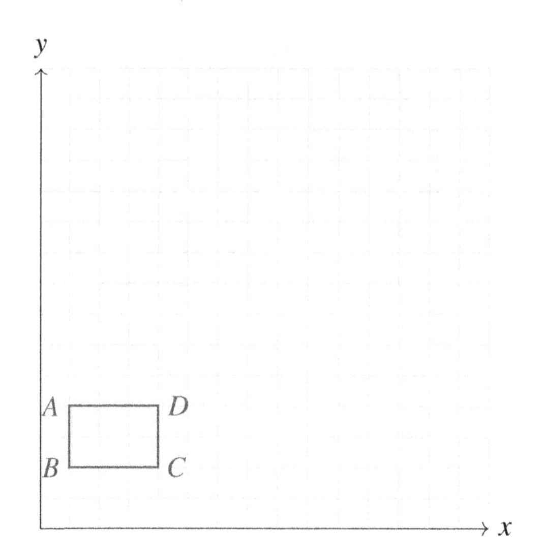


4-2) $k = \frac{1}{3}$

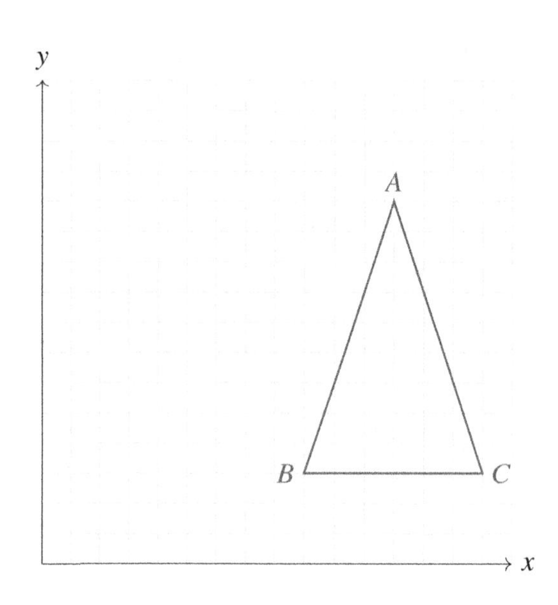


**5)** Graph the image of the figure using the transformation given.

5-1) Translation: 4 units right and 1 unit down

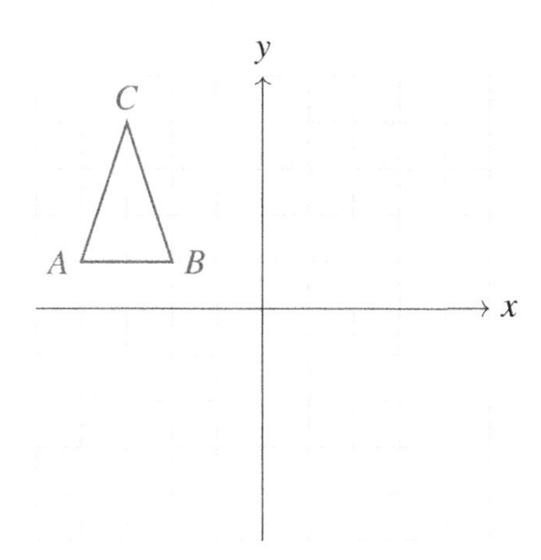


5-2) Translation: 4 units right and 2 units up

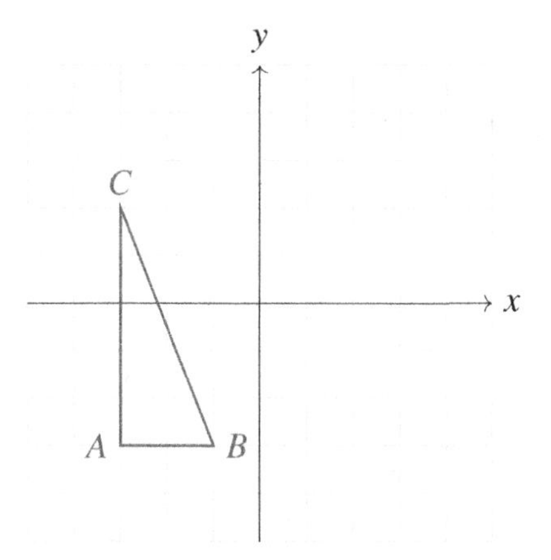


**6)** Write a rule to describe each transformation.

6-1)

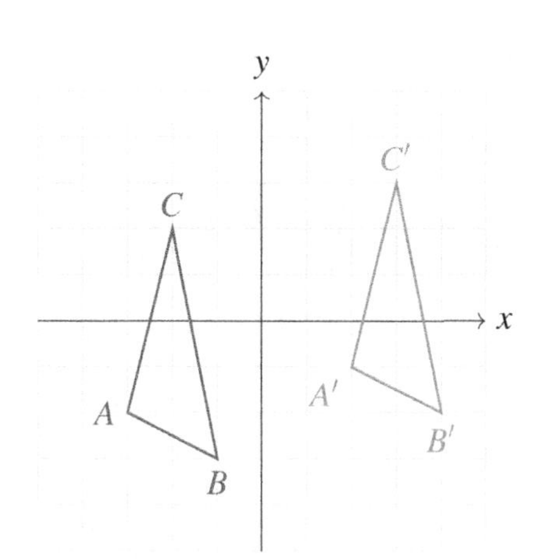

6-2)

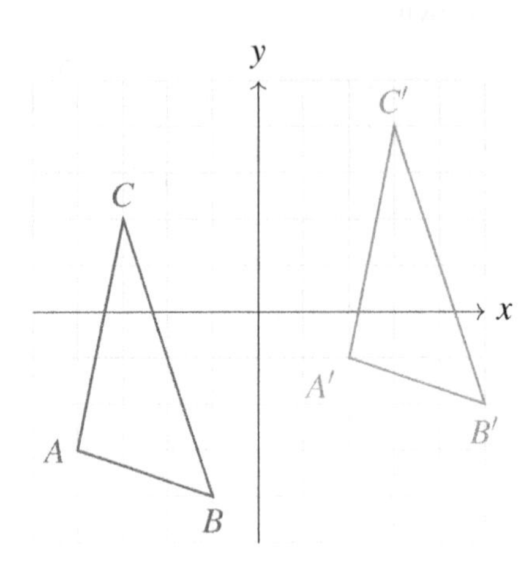


**7)** Are the figures congruent on the coordinate planes?

7-1)

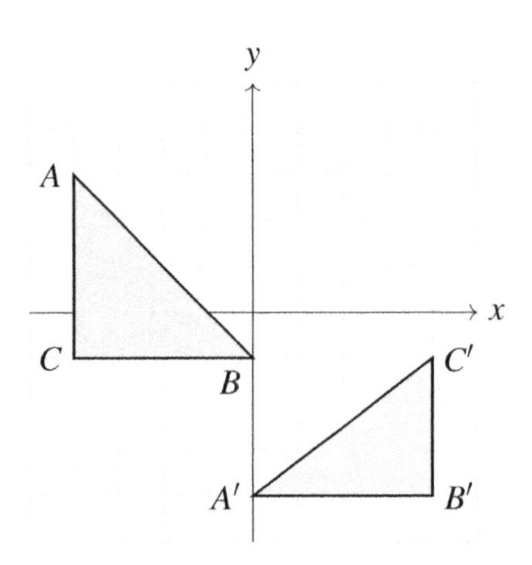


7-2)

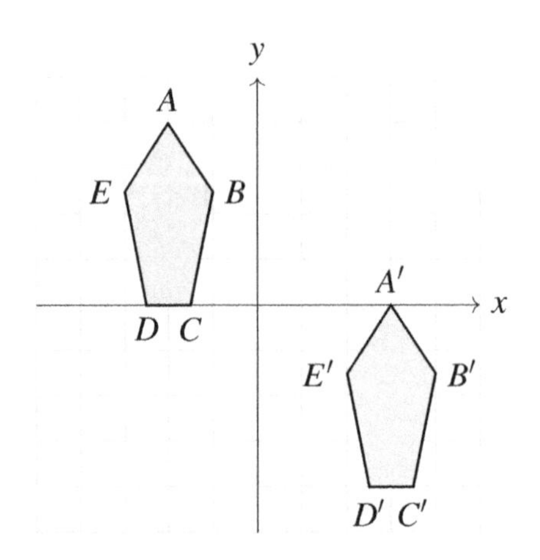

7-3)

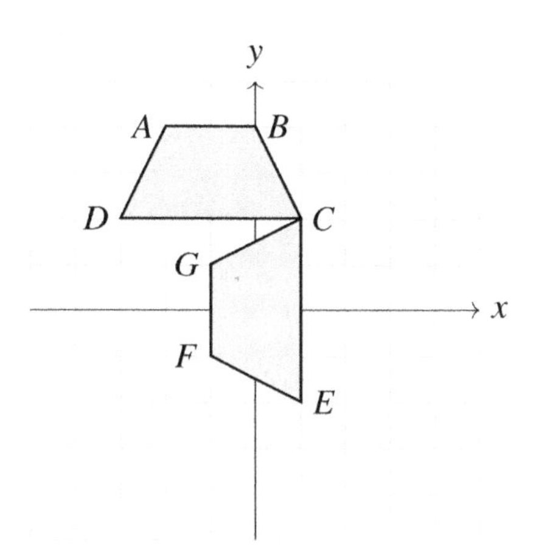


7-4)

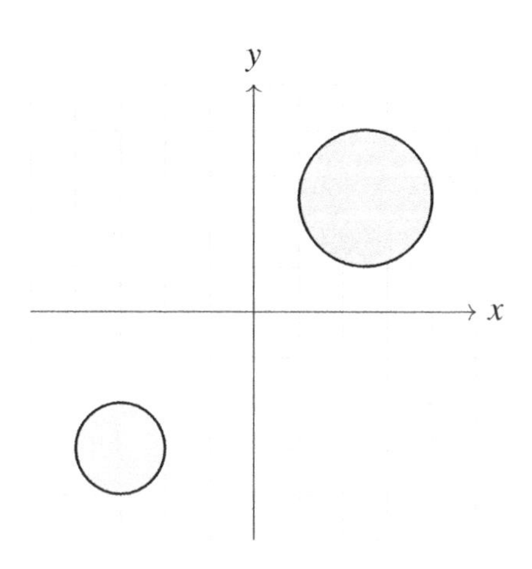


**8)** Solve.

8-1) Given an isosceles triangle ABC with base BC and vertex $A$, if you draw the altitude $AD$ from $A$ to $BC$, is $AD$ the axis of symmetry?

8-2) Given a parallelogram $ABCD$, if you draw the altitude $HL$ from $AB$ to $CD$, is $HL$ the axis of symmetry?

8-3) Does the circle have reflection symmetry?

8-4) How many reflection symmetrical does a regular hexagon have? and how many rotations symmetrical does it have?

## 4.11 Answers

**1)**

1-1) Rotation

1-2) Rotation

1-3) Translation

1-4) Reflection

**2)**

2-1)

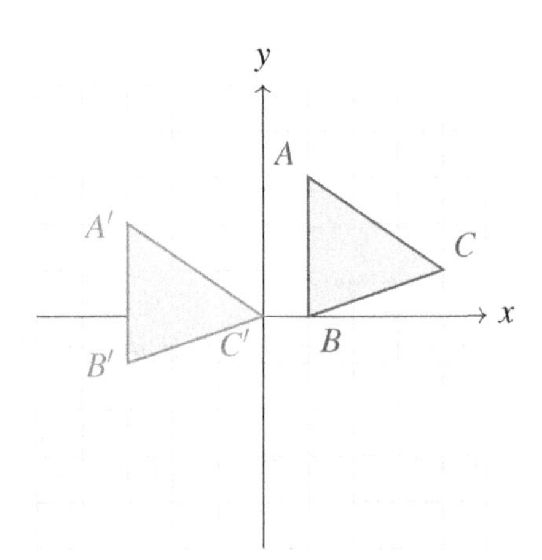


2-2)

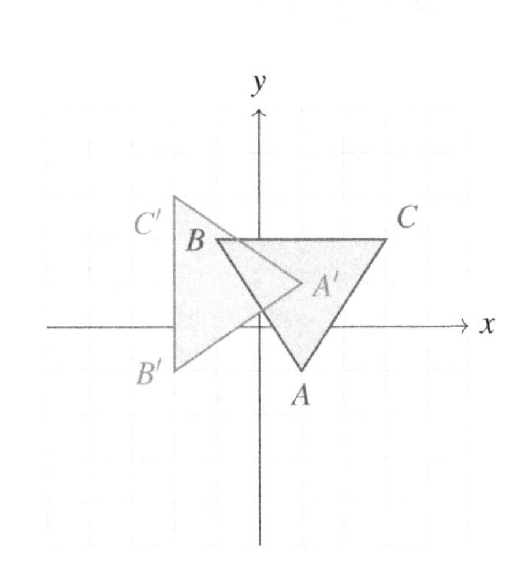


2-3)

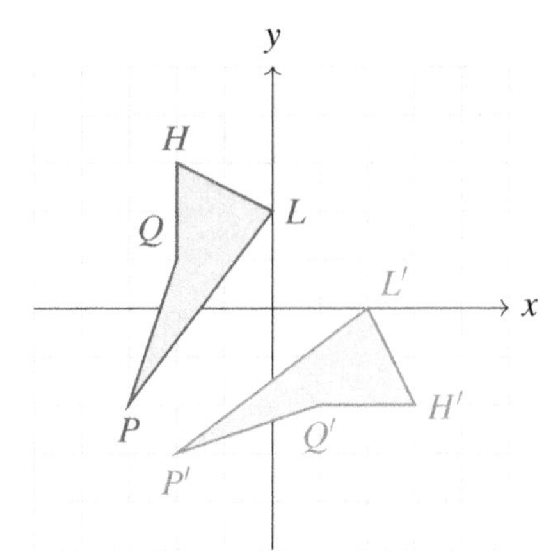


2-4)

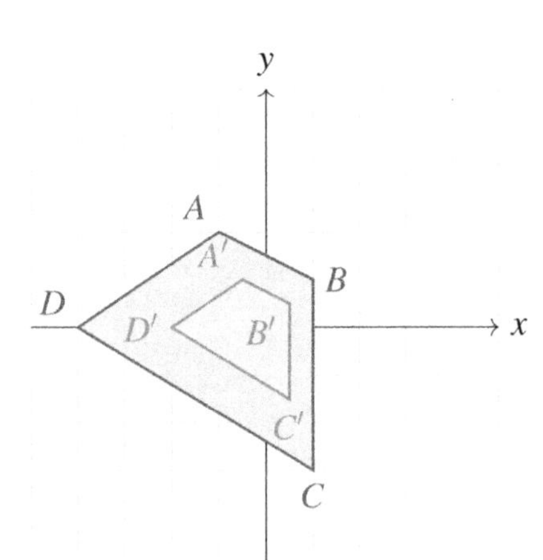


2-5)

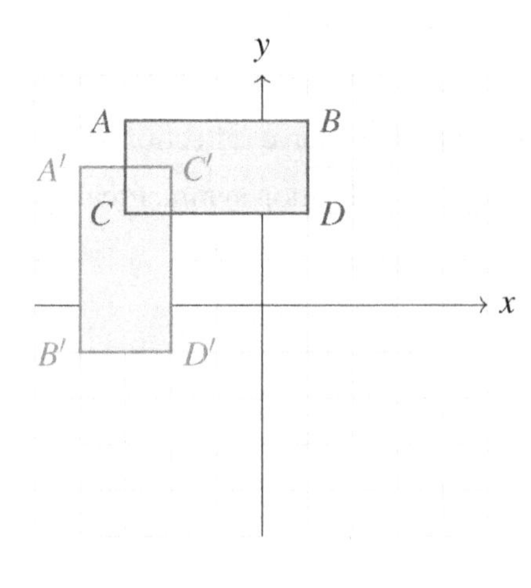

2-6)

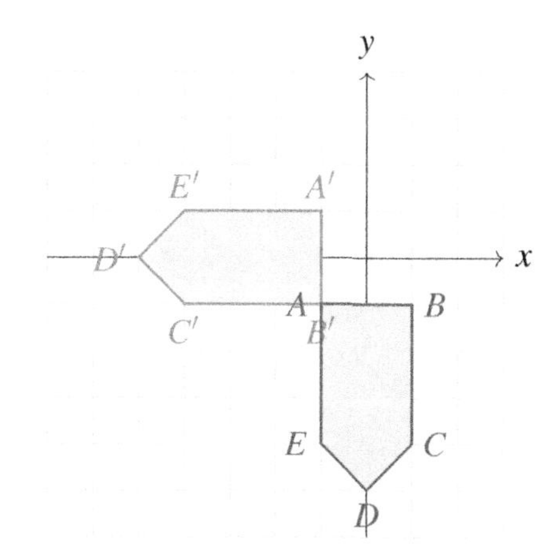


**3)**

3-1) 2

3-2) $\frac{1}{4}$

**4)**

4-1)

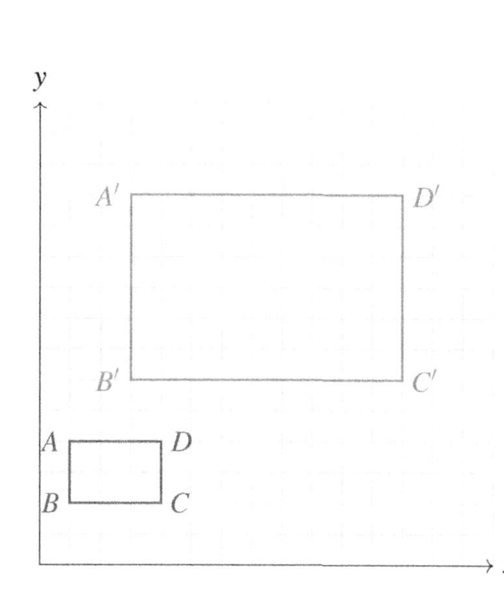


4-2)

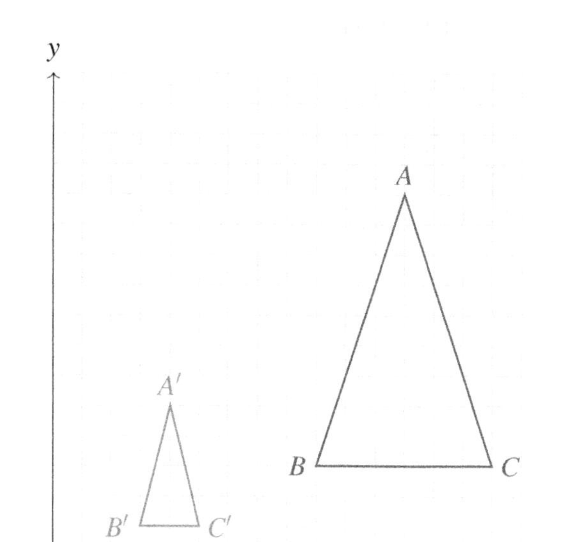


**5)**

5-1)

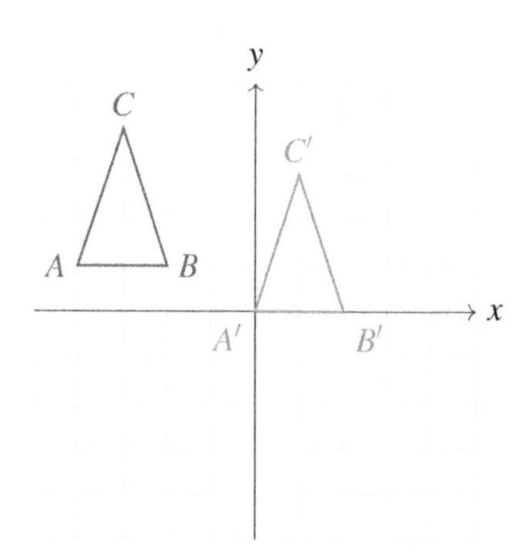


5-2)

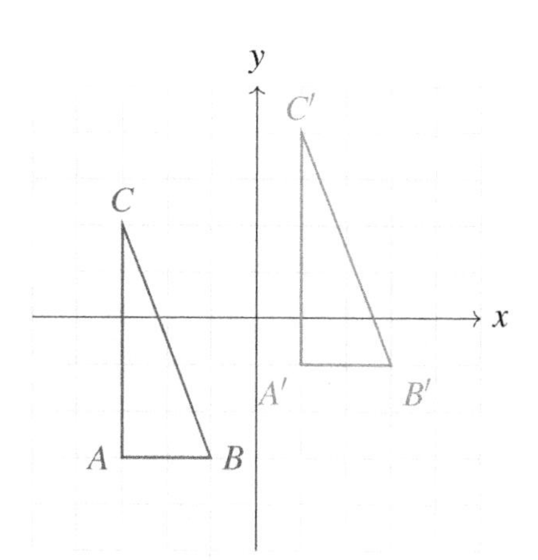


**6)**

6-1) Translation: 5 units right and 1 unit up

6-2) Translation: 6 units right and 2 unit up

**7)**

7-1) No

7-2) Yes

7-3) Yes

7-4) No

**8)**

8-1) Yes

8-2) No, it does not have reflection Symmetry.

8-3) Yes

8-4) 6 and 6

# 5. Quadrilaterals and Polygons

## 5.1 Classifying Polygons

Polygons are closed 2-dimensional shapes comprised of straight line segments. These segments are called sides. The classification of polygons can be done in several ways based on their properties such as the number of sides, types of internal angles, lengths of sides, and their symmetry. The following key points address these properties in detail.

**Key Point**

[By Number of Sides] The number of sides of a polygon determines its most basic classification, such as triangle, quadrilateral, pentagon, hexagon, heptagon, octagon, nonagon, and decagon, which have 3, 4, 5, 6, 7, 8, 9, and 10 sides, respectively. Polygons with more than 10 sides are usually referred to by their number as 11-gon, 12-gon, and so on.

**Key Point**

[By Internal Angles] Convex polygons have internal angles all less than 180°, while concave polygons have at least one internal angle greater than 180°.

**Key Point**

[By Length of Sides] Regular polygons have all sides and angles equal, whereas irregular polygons do not have equal sides or angles.

**Key Point**

[By Symmetry] Symmetrical polygons can be folded along an axis or axes so that the halves match up exactly, while asymmetrical polygons lack this kind of symmetry.

Classify the polygon that has 8 sides, each with an internal angle of 135°.

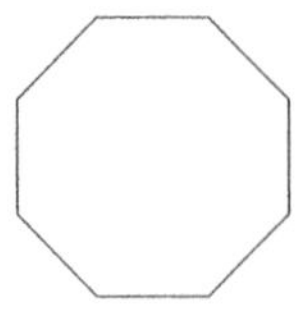

**Solution:** The given shape is an octagon, characterized by its 8 sides. It qualifies as a regular polygon due to the equality of all its internal angles. Additionally, this polygon is convex as each of its internal angles is less than 180°.

Identify the polygon that has 6 sides with varying lengths.

**Solution:** This polygon is classified as a hexagon, distinguished by its 6 sides. Given the unequal lengths of its sides, it is described as an irregular polygon.

## 5.2 Angles in Quadrilaterals

A quadrilateral, by definition, is a polygon with four edges (or sides) and four vertices or corners. Understanding the properties of their angles is essential in Geometry.

Take any quadrilateral *ABCD*.

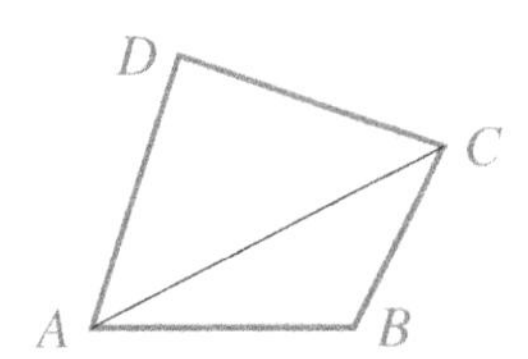


When you draw a diagonal, such as *AC*, you effectively divide the quadrilateral into two triangles: *ABC* and *ADC*. Remember, the sum of the angles in any triangle is 180°. Therefore, the combined angle sum for both triangles *ABC* and *ADC* is $2 \times 180° = 360°$. This establishes the rule for the sum of the angles in the quadrilateral *ABCD*.

**Key Point**

The sum of the interior angles of any quadrilateral is 360°.

Various types of quadrilaterals exhibit unique properties concerning their angles:

- **Rectangle:** All angles are right angles (90°).
- **Square:** A rectangle with equal sides—thus all angles are 90°.
- **Rhombus:** Equilateral quadrilateral but not necessarily equiangular; opposite angles are congruent, and adjacent angles are supplementary.
- **Parallelogram:** Opposite angles are congruent, and adjacent angles are supplementary.
- **Trapezoid:** Only one pair of opposite sides is parallel; angle sums are still 360°, but without the regularity seen in the types above.

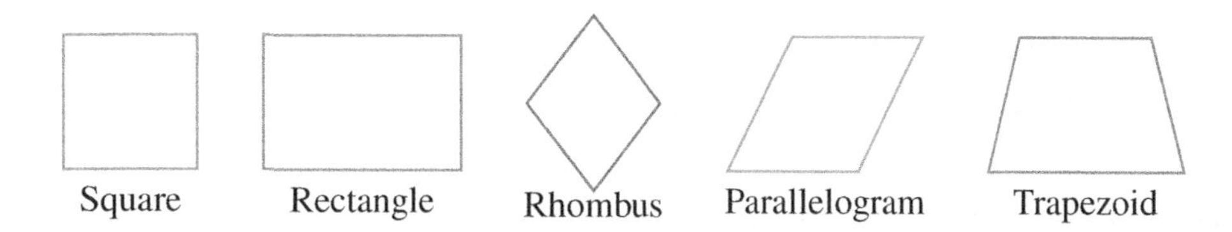


Let us delve into some examples to cement our understanding.

In parallelogram $ABCD$ with $\angle A = 50^\circ$, determine $\angle B$, $\angle C$, and $\angle D$.

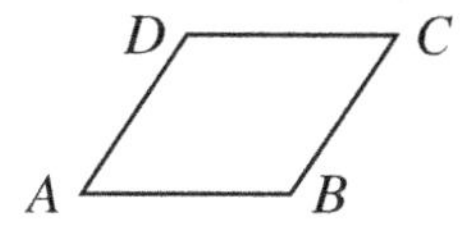


**Solution:** In parallelogram $ABCD$, $\angle A \cong \angle C = 50^\circ$, and $\angle B \cong \angle D = 180^\circ - 50^\circ = 130^\circ$.

Find all angles of a rhombus given that one angle is $60^\circ$.

**Solution:** In a rhombus, opposite angles are congruent, and adjacent angles are supplementary. Therefore, the angles are $60^\circ$, $120^\circ$ (calculated as $180^\circ - 60^\circ$), $60^\circ$, and $120^\circ$.

## 5.3 Properties of Trapezoids

A trapezoid, also known as a trapezium in some countries, is a four-sided figure characterized primarily by a pair of parallel sides.

**Bases and Legs:** The two parallel sides of a trapezoid are known as the *bases*. In the diagram, they are denoted as $AB$ and $CD$. The other two sides, which are not parallel, are called the *legs* of the trapezoid. In this example, they are represented as $AD$ and $BC$.

**Height:** The *height* (or altitude) of a trapezoid is the perpendicular distance measured between the two bases. In the below diagram, the height is depicted as a dashed line from point $C$ to point $E$ on base $AB$. This measurement is crucial for calculating the area of trapezoids.

**Base Angles:** The angles adjacent to each base are termed the *base angles*. In this diagram, on base $AB$, we have angles $\angle A$ and $\angle B$, and on base $CD$, we have angles $\angle C$ and $\angle D$.

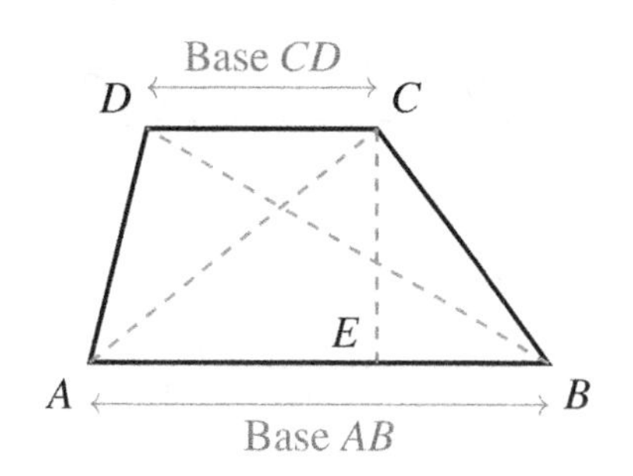

**Key Point**

The bases are the foundational sides of a trapezoid and are by definition, parallel.

**Isosceles Trapezoid:** An isosceles trapezoid is a trapezoid with legs of equal length (see the next plot). This special type of trapezoid has unique properties that distinguish it from a regular trapezoid.

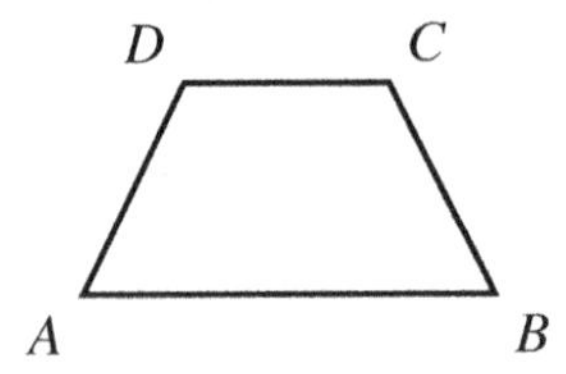


**Key Point**

In an isosceles trapezoid, base angles are congruent, and the diagonals are of equal lengths.

**Midsegment:** A trapezoid's *midsegment* is a line segment connecting the midpoints of the legs. It is also sometimes called the median. The length of the midsegment is the average of the lengths of the two bases:

$$Midsegment = \frac{\text{length of base } AB + \text{length of base } CD}{2}.$$

**Example** Given an isosceles trapezoid with $\overline{AB}$ measuring 27 units, $\overline{CD}$ measuring 9 units, and diagonal $\overline{DB}$ equal to 18 units, find the length of the other diagonal and the midsegment.

**Solution:** Since the trapezoid is isosceles, the diagonals are congruent. Therefore, $\overline{AC} = \overline{DB} = 18$ units. To find the length of the midsegment, use the formula:

$$m = \frac{\overline{AB} + \overline{CD}}{2} \Rightarrow m = \frac{27+9}{2} \Rightarrow m = \frac{36}{2} \Rightarrow m = 18 \text{ units}.$$

## 5.4 Properties of Parallelograms

We encounter a special type of quadrilateral with both pairs of opposite sides parallel. Consider the following diagram illustrating a parallelogram:

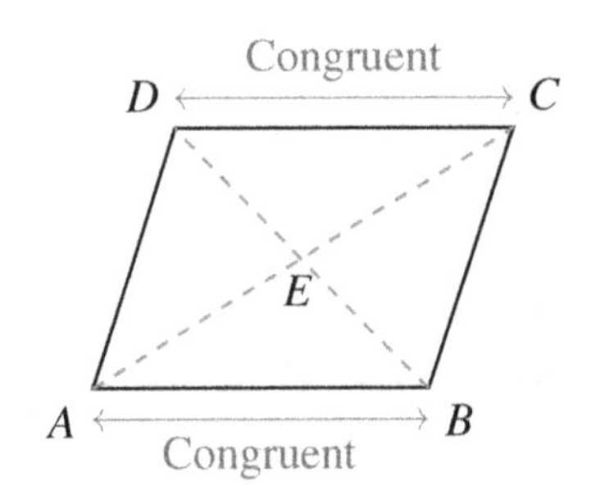

The geometry properties of a parallelogram are as follows:

1. **Parallel Opposite Sides:** Lines $\overline{AB}$ and $\overline{CD}$ are parallel, as well as lines $\overline{AD}$ and $\overline{BC}$.
2. **Congruent Opposite Sides:** $\overline{AB} \cong \overline{CD}$ and $\overline{AD} \cong \overline{BC}$.
3. **Congruent Opposite Angles:** $\angle A \cong \angle C$ and $\angle B \cong \angle D$.
4. **Supplementary Consecutive Angles:** $\angle A + \angle B = 180^\circ$, $\angle B + \angle C = 180^\circ$, $\angle C + \angle D = 180^\circ$, and $\angle D + \angle A = 180^\circ$.
5. **Bisecting Diagonals:** Diagonals $\overline{AC}$ and $\overline{BD}$ bisect each other at the midpoint (point $E$).
6. **Congruent Triangles Formed by Diagonals:** $\triangle ABD \cong \triangle BCD$ and $\triangle ABC \cong \triangle CDA$.
7. **Parallelogram Law:** $AB^2 + BC^2 + CD^2 + DA^2 = AC^2 + BD^2$.

In parallelogram $WXYZ$ with $\overline{WX} = 7$ units and $\overline{WZ} = 9$ units, find $\overline{XY}$ and $\overline{YZ}$.

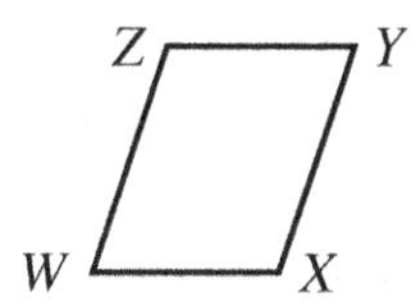


**Solution:** Since opposite sides in a parallelogram are congruent, $\overline{XY} = \overline{WZ} = 9$ units and $\overline{YZ} = \overline{WX} = 7$ units.

For parallelogram $ABCD$ with $\overline{AB} = 11$ units and $\angle A = 80^\circ$, find $\overline{CD}$ and $\angle D$.

**Solution:** In a parallelogram, opposite sides are congruent. Therefore, $\overline{CD} = \overline{AB} = 11$ units. Since consecutive angles are supplementary in a parallelogram, $\angle D = 180^\circ - \angle A = 180^\circ - 80^\circ = 100^\circ$.

## 5.5 Properties of Rectangles

A rectangle is a special type of parallelogram with additional features. Consider the following rectangle diagram:

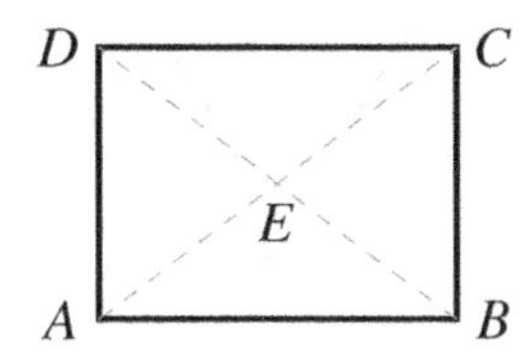


The geometric properties of a rectangle are as follows:

1. **Right Angles:** All interior angles of a rectangle are right angles, each measuring $90^\circ$.

2. **Congruent Opposite Sides:** Opposite sides of a rectangle are congruent; for example, in rectangle $ABCD$, $AB \cong CD$ and $BC \cong AD$.
3. **Parallel Opposite Sides:** The opposite sides are not only equal in length but also parallel to each other: $AB \parallel CD$ and $BC \parallel AD$.
4. **Congruent Diagonals:** Diagonals of a rectangle are congruent, meaning $AC \cong BD$.
5. **Bisecting Diagonals:** Diagonals of a rectangle bisect each other. If the intersection point of the diagonals is denoted as $E$, then $AE \cong CE$ and $BE \cong DE$. Note that while the diagonals bisect each other, they are not perpendicular.

In rectangle $ABCD$, given $AB = 3x - 6$ and $DC = 12$, find $x$.

**Solution:** Because opposite sides of a rectangle are congruent, we equate $AB$ to $DC$: $3x - 6 = 12$. Solving for $x$ gives $x = 6$.

In rectangle $WXYZ$, the length of diagonal $WY$ is 12 cm. Find the lengths of $WE$.

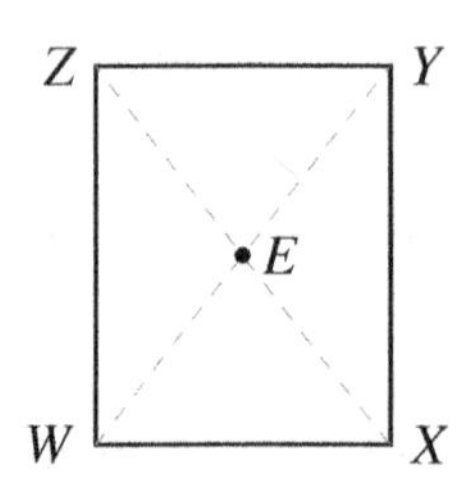


**Solution:** Since diagonals in a rectangle bisect each other, $WY$ is divided into two equal segments at point $E$. Therefore, $WE = \frac{1}{2} \times 12 \text{ cm} = 6 \text{ cm}$.

## 5.6 Properties of The Rhombus

A rhombus is a specific type of parallelogram —a parallelogram with all four sides of equal length. Let us explore the defining characteristics of a rhombus like below:

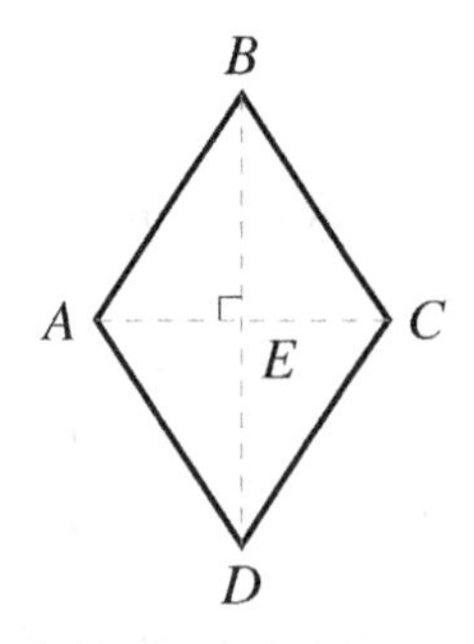


- **Congruent Sides:** All sides of a rhombus are congruent. In rhombus $ABCD$, this means $AB \cong BC \cong CD \cong DA$.
- **Parallel Opposite Sides:** The opposite sides of a rhombus are parallel. Therefore, in rhombus $ABCD$, $AB \parallel CD$ and $BC \parallel AD$.

- **Equal Opposite Angles:** Like all parallelograms, the opposite angles in a rhombus are equal due to the parallel sides creating alternate interior angles that are congruent.
- **Perpendicular Bisectors Diagonals:** The diagonals of a rhombus are perpendicular bisectors of each other. For instance, in rhombus $ABCD$, diagonal $AC$ bisects $BD$ at a right angle, creating four right angles at the intersection.
- **Diagonals Bisect Angles:** Each diagonal in a rhombus bisects a pair of opposite angles, contributing to the symmetry of the shape. For example, diagonal $AC$ in rhombus $ABCD$ bisects angles $A$ and $C$, and diagonal $BD$ bisects angles $B$ and $D$.

Consider a rhombus named $JKLM$.

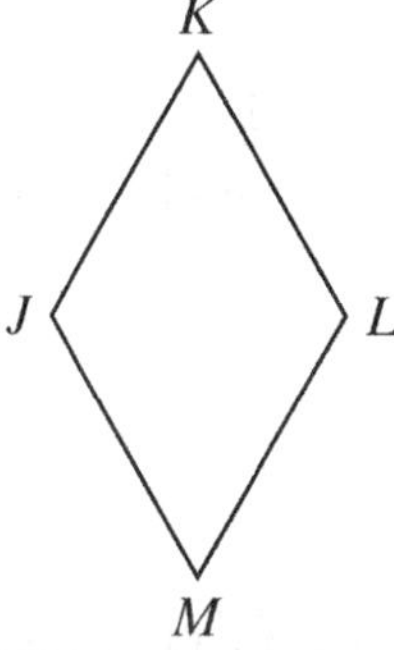


If the angle at vertex $J$ is $120°$, determine the size of the angle at vertex $M$.

**Solution:** In a rhombus, adjacent angles are supplementary, meaning that they add up to $180°$. Since the angle at vertex $J$ is $120°$, the adjacent angles at vertices $K$ and $M$ will both be $180° - 120° = 60°$.
Therefore, the size of the angle at vertex $M$ in rhombus $JKLM$ is $60°$.

## 5.7 Properties of Squares

A square is a quadrilateral with four equal sides and four equal angles. While a rhombus focuses on equal sides, the square brings together both the attributes of a rhombus and a rectangle. Let us explore the defining properties of a square by considering the following square

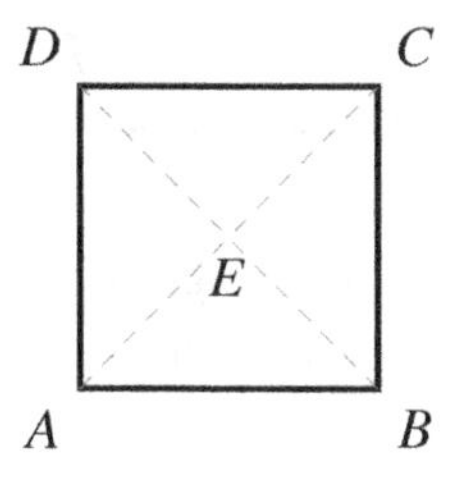


- **Equal Sides:** All sides of a square are equal in length. For square $ABCD$, this means $AB = BC = CD = DA$.
- **Right Angles:** Every angle in a square measures $90°$, making a square a specific type of rectangle.
- **Parallel Opposite Sides:** Like in any parallelogram, the opposite sides of a square are parallel. Thus, in square $ABCD$, $AB \parallel CD$ and $BC \parallel AD$.
- **Congruent Diagonals at Right Angles:** The diagonals of a square are congruent and intersect at right angles. In square $ABCD$, $AC = BD$ and they intersect at $90°$.

- **Angle Bisecting Diagonals:** The diagonals of a square bisect its angles, creating two 45° angles at each vertex.
- **Equal Halves:** The diagonals bisect each other, dividing each other into equal halves at the point of intersection.

Now, let us demonstrate how to apply these properties with a practical example.

Consider square $PQRS$, with diagonal $PR$ measuring $4\sqrt{2}$ units.

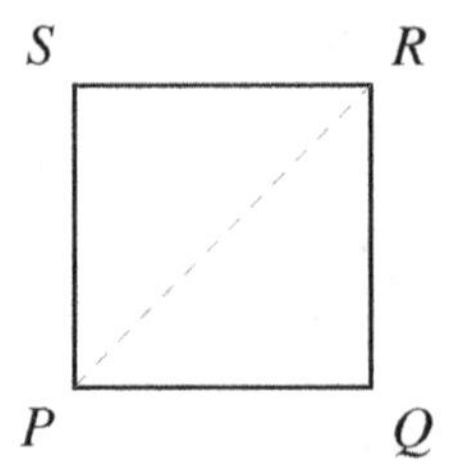


Calculate the length of side $PQ$.

**Solution:** In square $PQRS$, the diagonals are congruent and bisect each other at right angles, forming two $45-45-90$ right triangles. For triangle $\triangle PQR$, using the formula for the hypotenuse of a right-angled triangle:

$$PQ = \frac{PR}{\sqrt{2}} = \frac{4\sqrt{2}}{\sqrt{2}} = 4\ units.$$

## 5.8 Areas of Triangles and Quadrilaterals

Let us explore how to calculate the areas of triangles and quadrilaterals, fundamental shapes in geometry.

- **Triangle:** To find its area, you can multiply half of the base length by the height.
- **Parallelogram:** The area is determined by multiplying the base length by the height.
- **Rectangle:** To calculate its area, multiply the length by the width.
- **Square:** The area is found by squaring the length of one of its sides.
- **Trapezoid:** Its area is half of the sum of the lengths of the two bases, multiplied by the height.
- **Rhombus:** To find its area, multiply half of the lengths of its diagonals together.

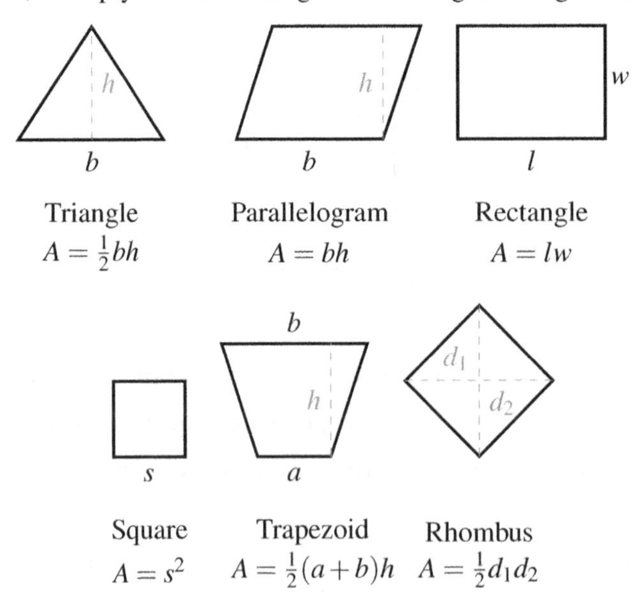

**Example** Find the area of a parallelogram with a base of 8 units and a height of 5 units.

**Solution:** Using the formula for the area of a parallelogram, we get:

$$A = bh = 40 \text{ square units.}$$

**Example** Calculate the area of a trapezoid with bases of 6 units and 10 units, and a height of 4 units.

**Solution:** Applying the area of a trapezoid formula:

$$A = \frac{1}{2}(a+b)h = \frac{1}{2}(6+10)\times 4 = \frac{1}{2}\times 16\times 4 = 32 \text{ square units.}$$

**Example** Determine the area of a rhombus whose diagonals measure 8 and 6 units.

**Solution:** Utilizing the formula for the area of a rhombus with given diagonals:

$$A = \frac{1}{2}d_1 d_2 = \frac{1}{2}\times 8\times 6 = 24 \text{ square units.}$$

## 5.9 Perimeter of Polygons

The perimeter is the total distance around the shape. We measure this distance by adding up the lengths of all the sides. Let us discuss how to calculate the perimeter for several common polygons.

- **Square:** The perimeter of a square is simply four times one of its sides since all sides are equal.
- **Rectangle:** For a rectangle, the perimeter is the sum of twice its width and twice its length.
- **Trapezoid:** A trapezoid has four sides that could be of different lengths. Thus, its perimeter is the sum of these four sides.
- **Regular Polygon:** A regular polygon with $n$ sides has all sides of equal length. Its perimeter is $n$ times the length of one side.
- **Parallelogram:** Finally, the perimeter of a parallelogram is twice the sum of the width and the length.

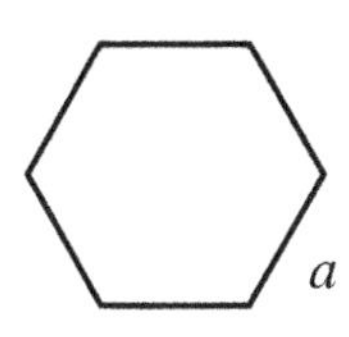


Regular Hexagon
$P = 6a$

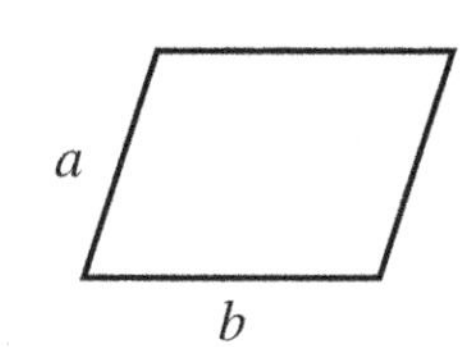


Parallelogram
$P = 2(a+b)$

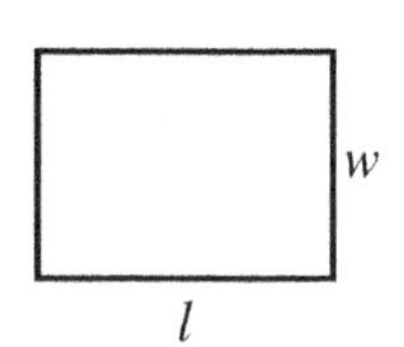


Rectangle
$P = 2(l+w)$

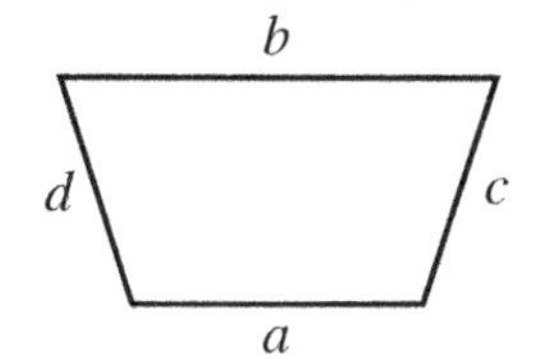


Trapezoid
$P = a+b+c+d$

To calculate the perimeter of any polygon, you add the lengths of all its sides.

**Example** Find the perimeter of a regular hexagon with a side length of 8 m.

**Solution:** Since the hexagon is regular, all sides are equal. Thus the perimeter of the hexagon is

$$P = 6 \times \text{one side} = 6 \times 8\,\text{m} = 48\,\text{m}.$$

**Example** Find the perimeter of a trapezoid with sides measuring 7 ft, 8 ft, 8 ft, and 10 ft.

**Solution:** The perimeter of a trapezoid is the sum of all its side lengths.

$$P = 7\,\text{ft} + 8\,\text{ft} + 8\,\text{ft} + 10\,\text{ft} = 33\,\text{ft}.$$

**Example** Find the perimeter of a parallelogram with width 3 m and length 5 m.

**Solution:** The perimeter of a parallelogram is twice the sum of the width and length.

$$P = 2(l + w) = 2(5\,\text{m} + 3\,\text{m}) = 2(8\,\text{m}) = 16\,\text{m}.$$

## 5.10 Polygons and Angles

A polygon is a closed figure formed by the connection of straight line segments, where the endpoints of these segments are known as vertices. Let's explore the relationship between polygons and their angles.

**Interior Angles of a Polygon**: The interior angles are those found within the boundaries of a polygon. A key concept here is that the sum of the interior angles of a polygon depends only on the number of sides ($n$) the polygon has.

**Key Point**

The sum of the interior angles ($S$) of an $n$-sided polygon is given by

$$S = (n - 2) \times 180^\circ.$$

**Each Interior Angle of a Regular Polygon**: If a polygon is regular, meaning all sides and angles are congruent, we can deduce the measure of each individual interior angle with ease.

**Key Point**

The measure of each interior angle ($A$) of an $n$-sided regular polygon is

$$A = \frac{(n-2) \times 180^\circ}{n}.$$

**Exterior Angles of a Polygon**: An exterior angle is formed when a side of a polygon is extended and we examine the angle between the extended line and the adjacent side. The sum of exterior angles of any polygon, irrespective of the number of sides, is always the same.

**Key Point**

The sum of the exterior angles of any polygon is $360^\circ$, regardless of the number of sides.

For regular polygons, each exterior angle can be evenly distributed.

**Key Point**

In a regular polygon with $n$ sides, each exterior angle ($E$) is $E = \frac{360^\circ}{n}$.

With the foundation established, let us put these concepts to the test with some practical examples.

Calculate the sum of the interior angles of a decagon (a 10-sided polygon).

**Solution:** Using the formula for the sum of interior angles $S = (n-2) \times 180^\circ$, for a decagon $n = 10$, we get

$$S = (10-2) \times 180^\circ = 8 \times 180^\circ = 1440^\circ.$$

So, the sum of the interior angles of a decagon is $1440^\circ$.

Find the measure of each exterior angle of a regular hexagon.

**Solution:** We know that the sum of exterior angles of any polygon is $360^\circ$. For a regular hexagon, there are 6 sides, so each exterior angle $E$ is given by

$$E = \frac{360^\circ}{n} = \frac{360^\circ}{6} = 60^\circ.$$

Hence, each exterior angle of a regular hexagon measures $60^\circ$.

## 5.11 Practices

**1)** State the most specific name for each figure.

1-1)

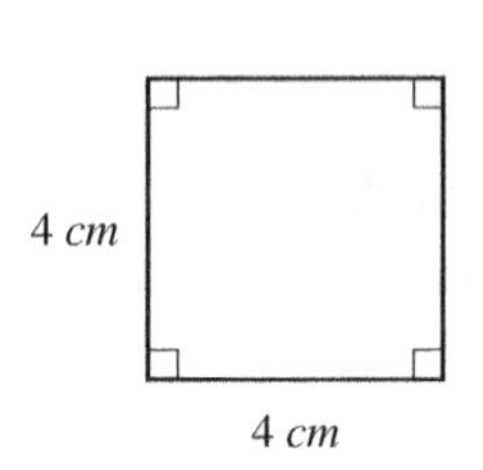


1-2)

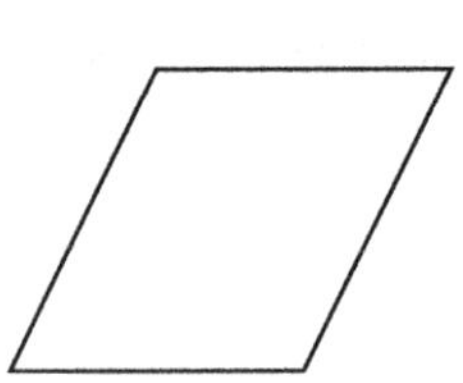

1-3)

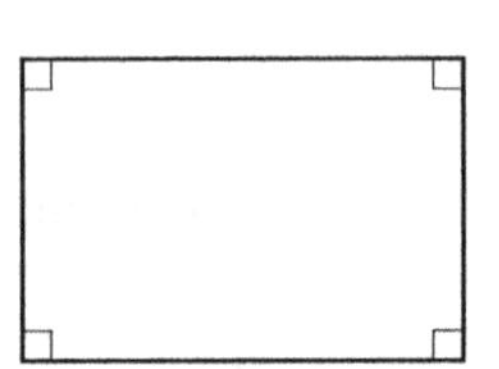

1-4)

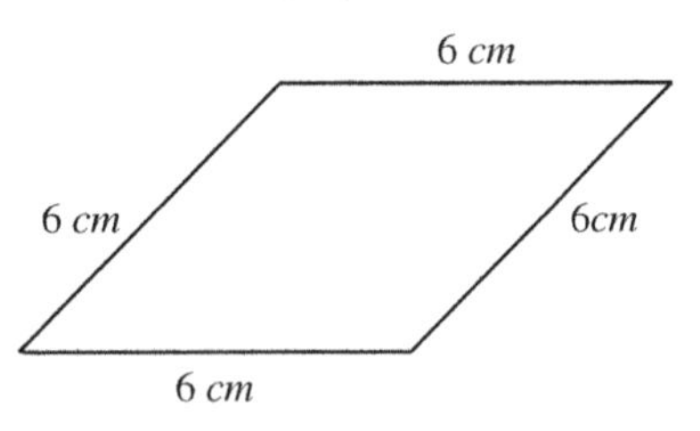


1-5)

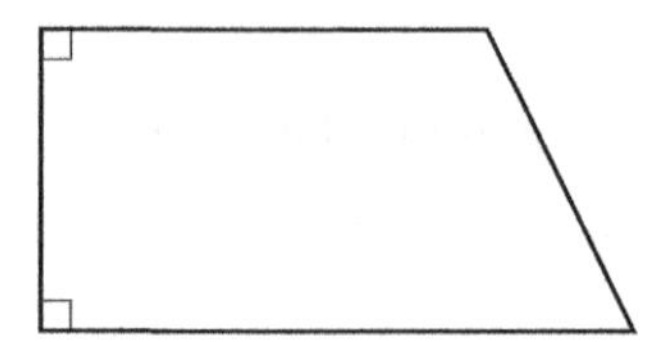

1-6)

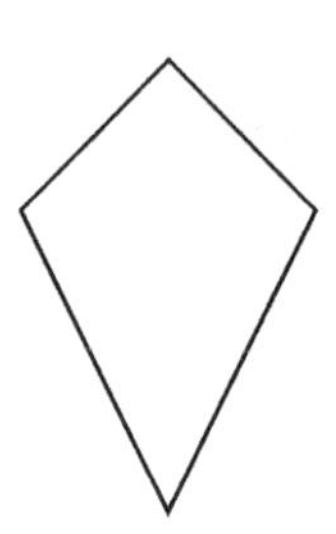

1-7)

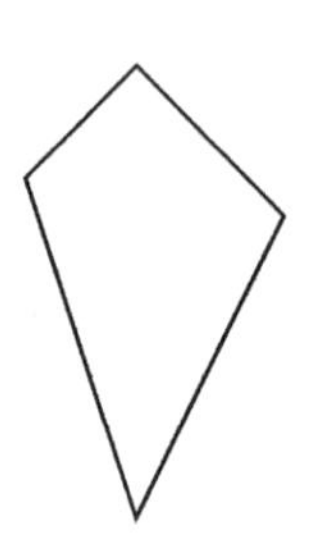

**2)** State all possible name for each figure.

2-1)

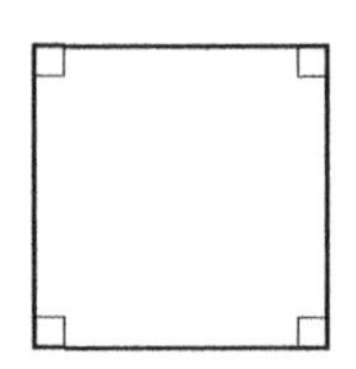

2-2)

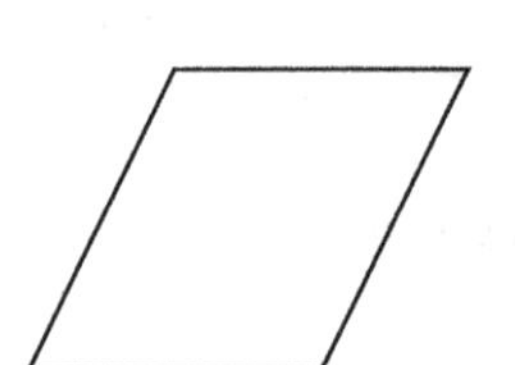

2-3)

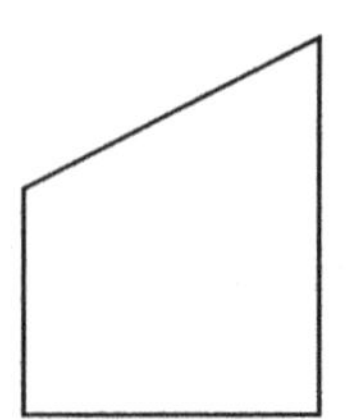

**3)** Find the measure of each angle indicated.

3-1)

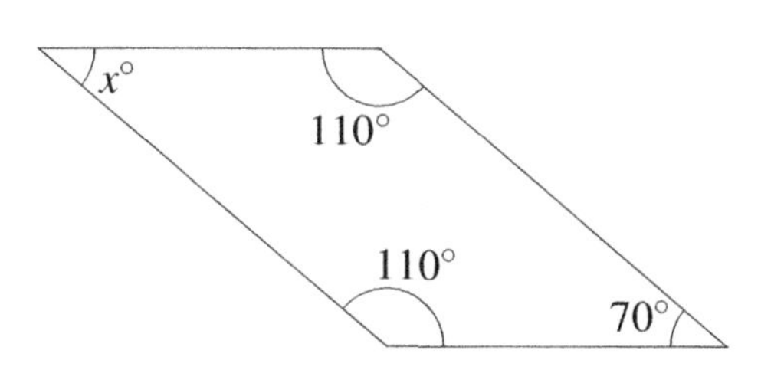


3-2)

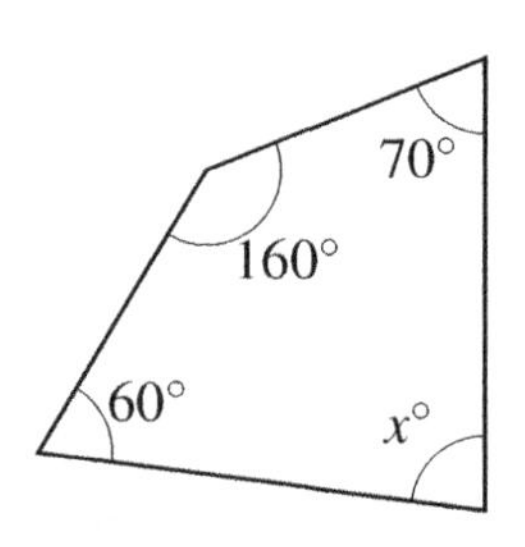


3-3)

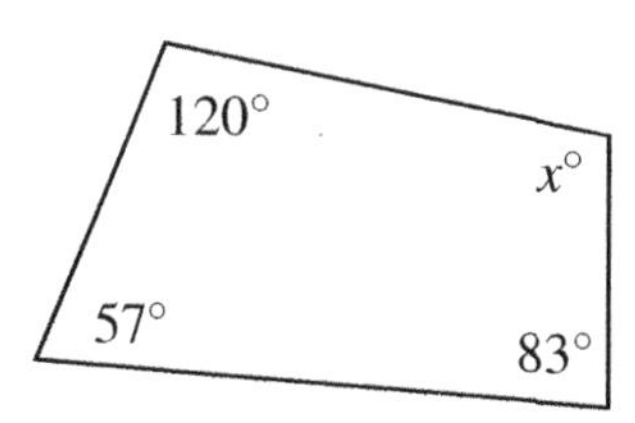


3-4)

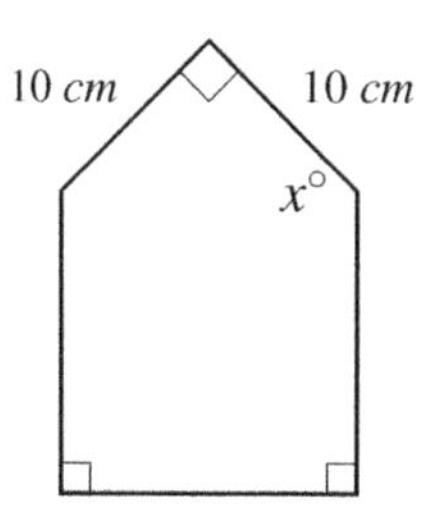


**4)** Solve for $x$.

4-1)

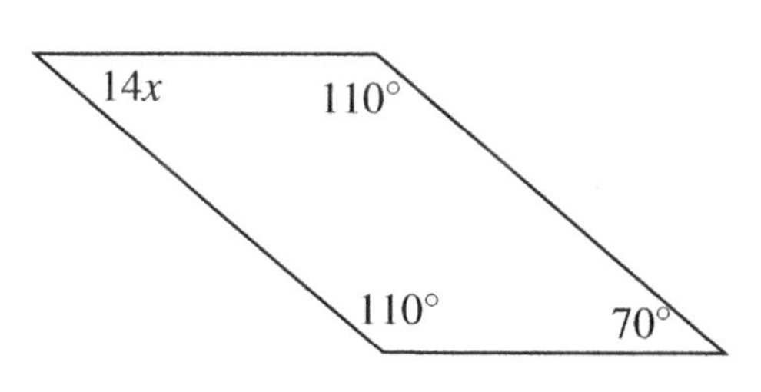


4-2)

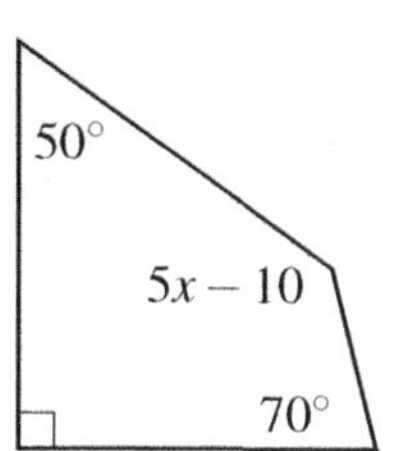


4-3)

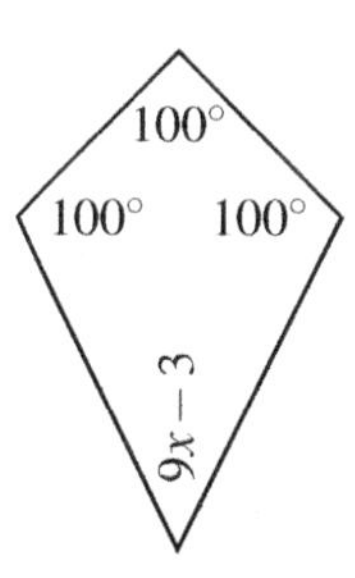


4-4)

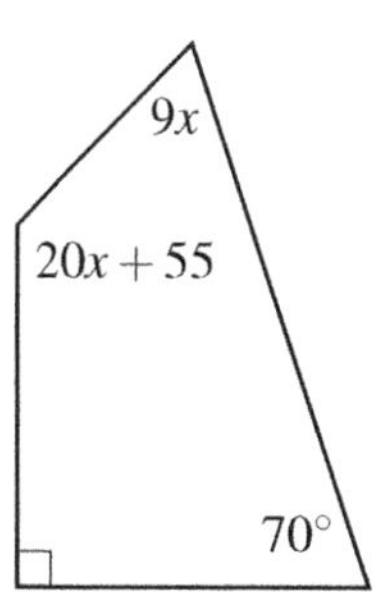


**5)** Find the length of the angle indicated for each trapezoid.

5-1)

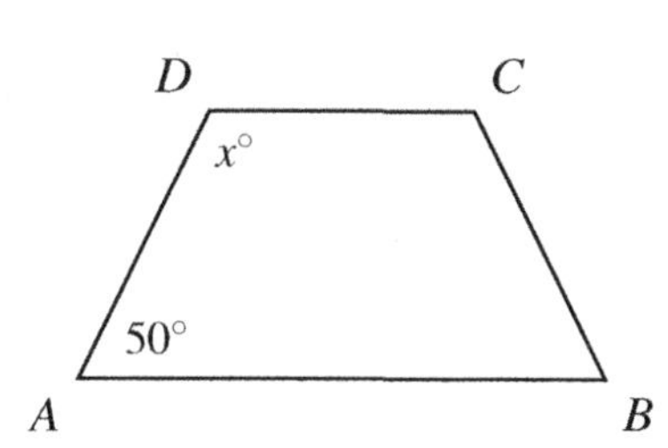

5-2)

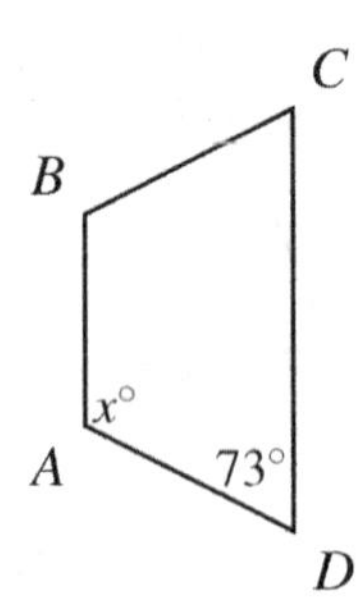


**6)** Find the length of the median of each trapezoid.

6-1)

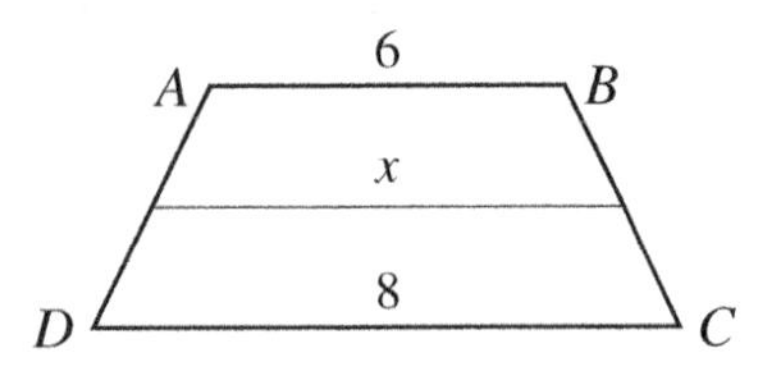


6-2)

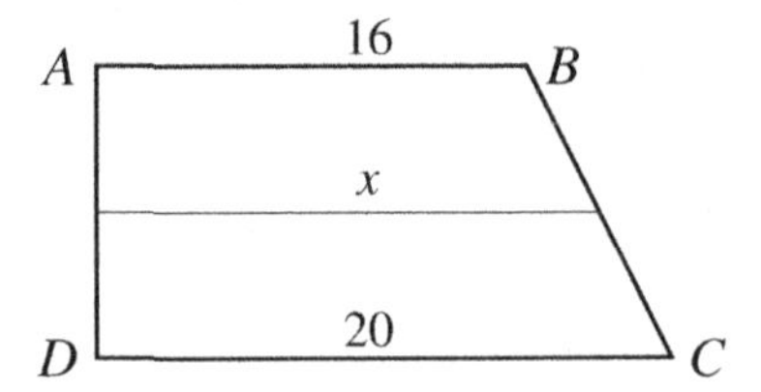


**7)** Solve for $x$. Each figure is a trapezoid.

7-1)

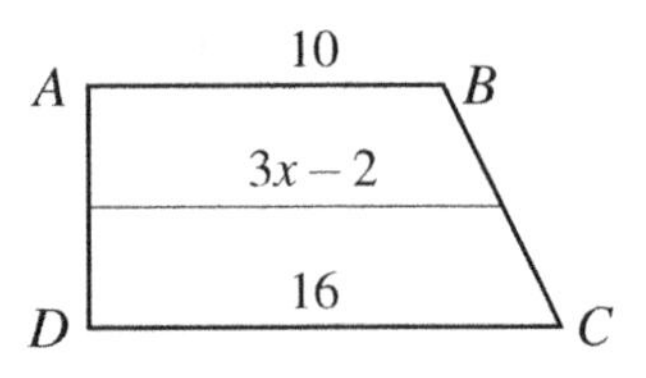


7-2)

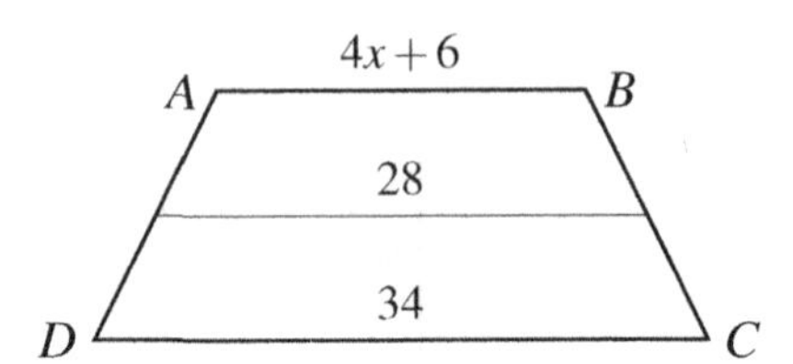


**8)** Solve for $x$ in each parallelogram.

8-1)

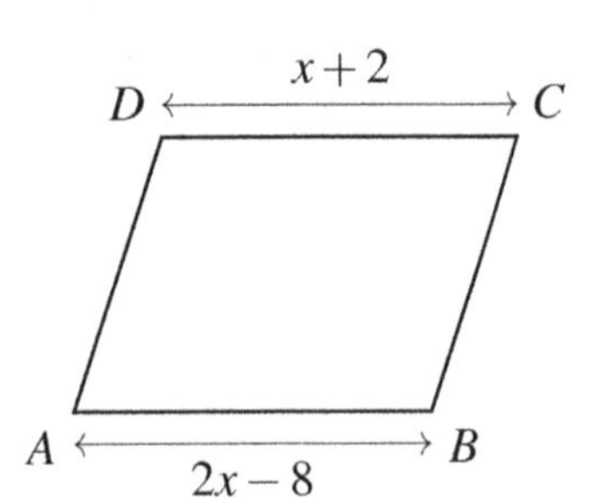


8-2)

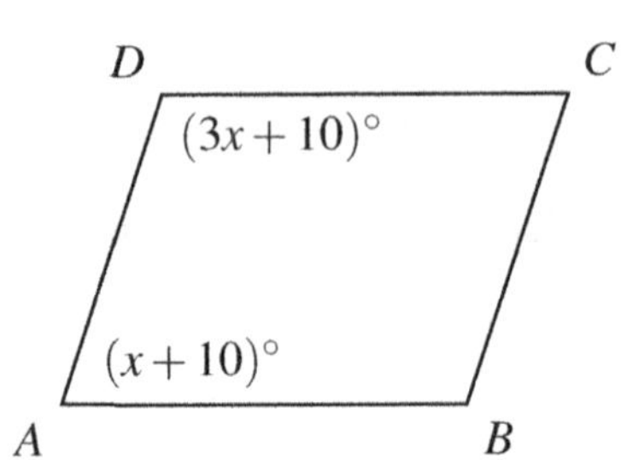

8-3)

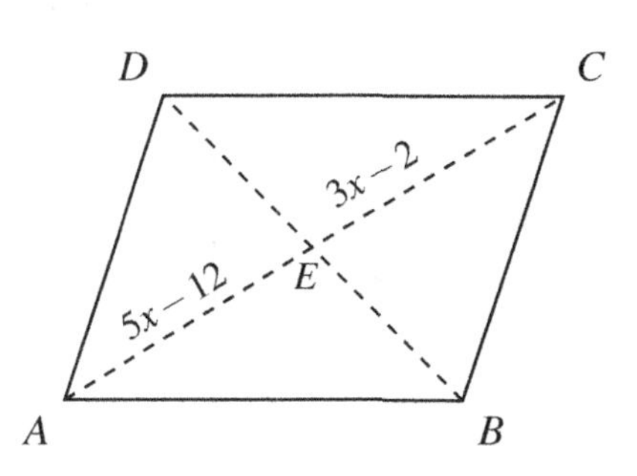


8-4)

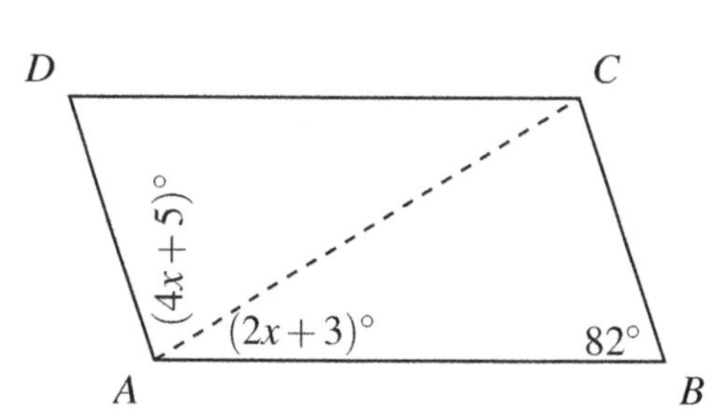


**9)** Solve for $x$. Each figure is a parallelogram.

9-1)

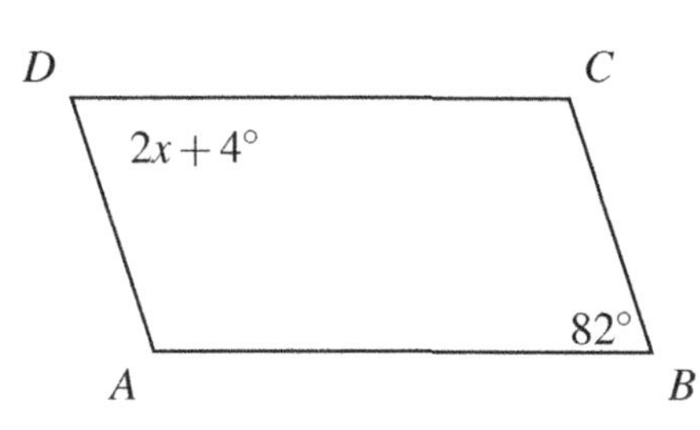


9-2)

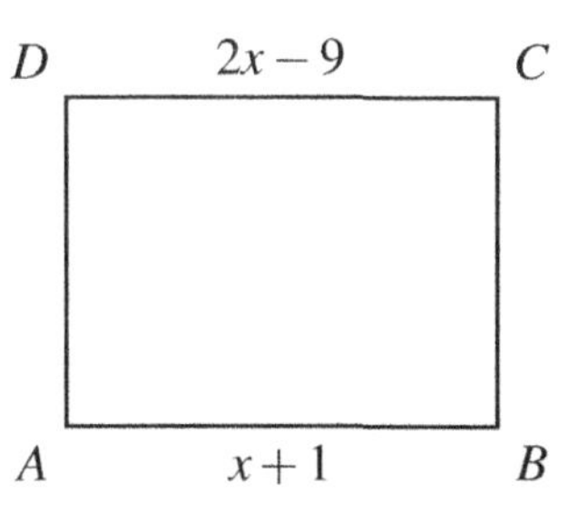


9-3)

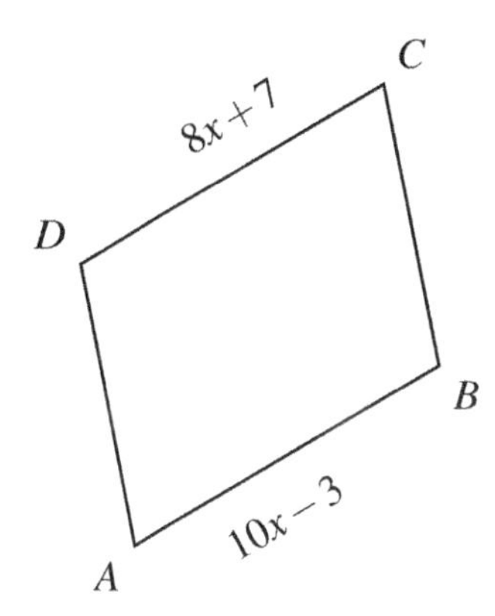


9-4)

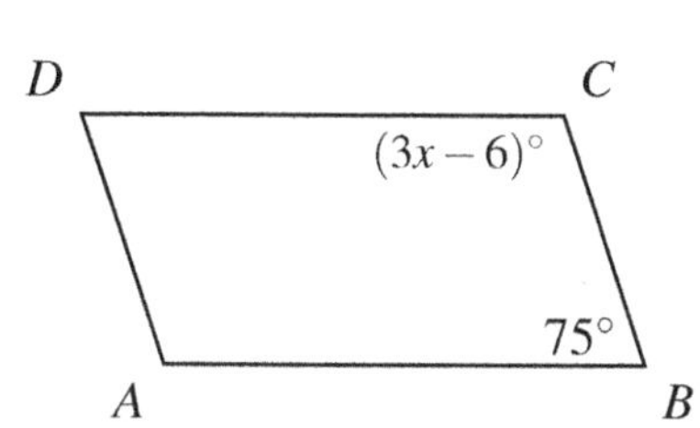


**10)** Find the measurement indicated in each Rhombus. Round your answer to the nearest tenth if necessary.

10-1) $\angle BDC = ?$

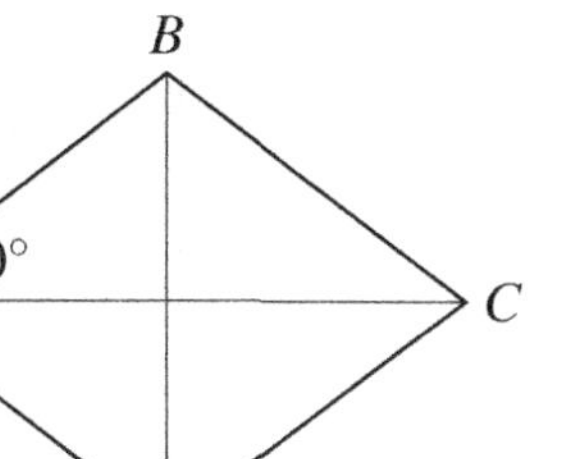


10-2) If $BD = 12$, $AD = ?$

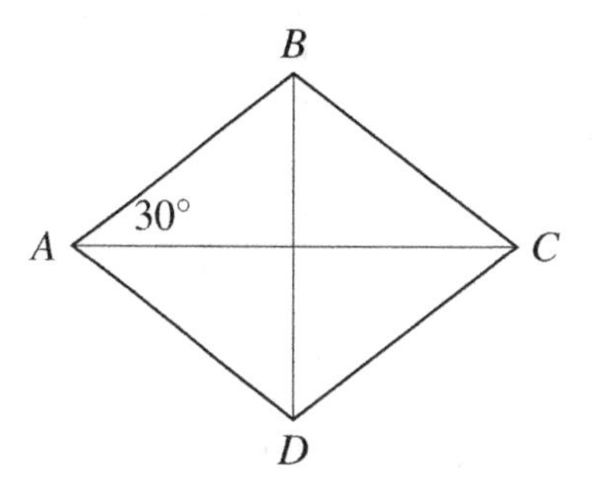

10-3) $\angle BAD = ?$

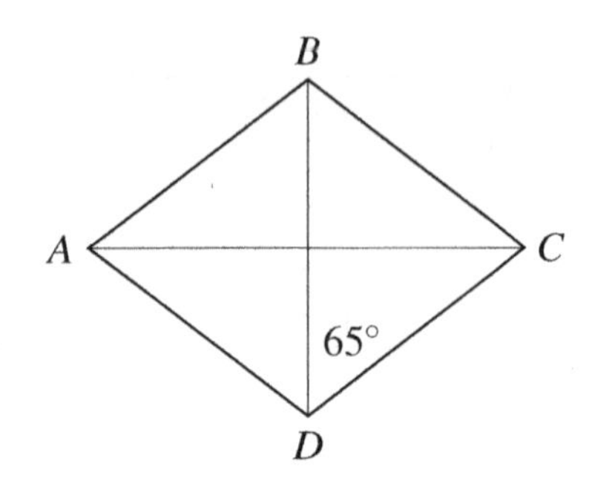

10-4) If $AC = 18$ and $BD = 8$, then $\angle BAC = ?$

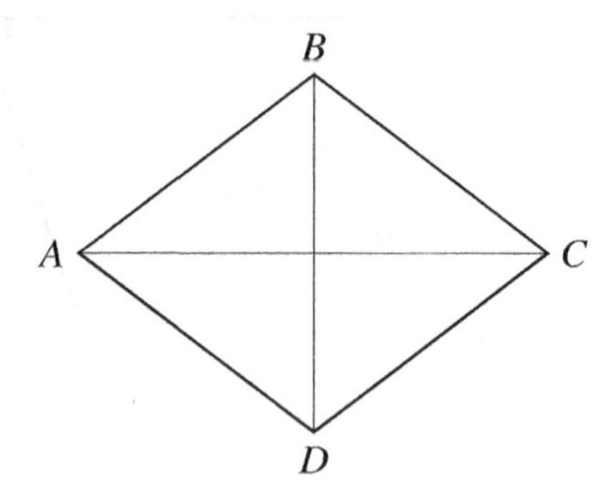


**11)** Solve for $x$. Each figure is a Rhombus.

11-1)

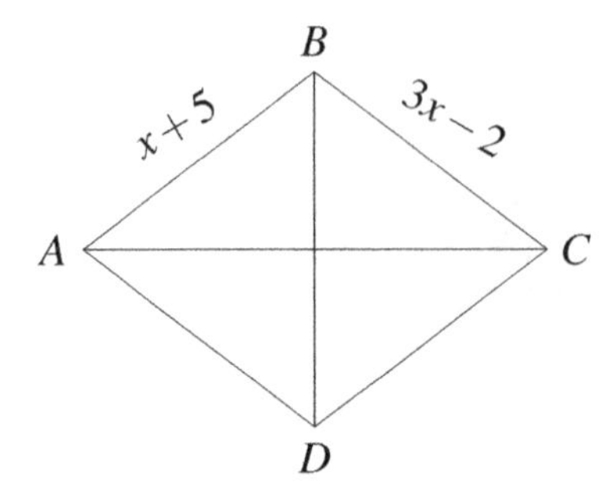


11-2)

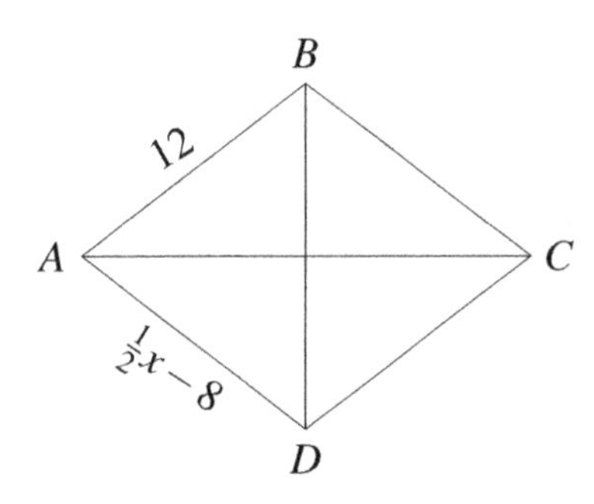


**12)** Find the measurement indicated in each square.

12-1) $d = ?$

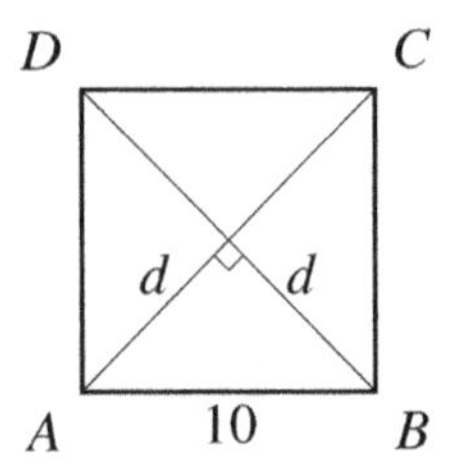


12-2) $AC = ?$

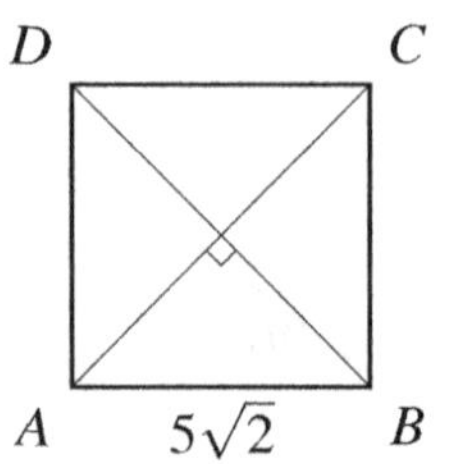


**13)** Solve for $x$. Each figure is a Square.

13-1) $BD = 2x$

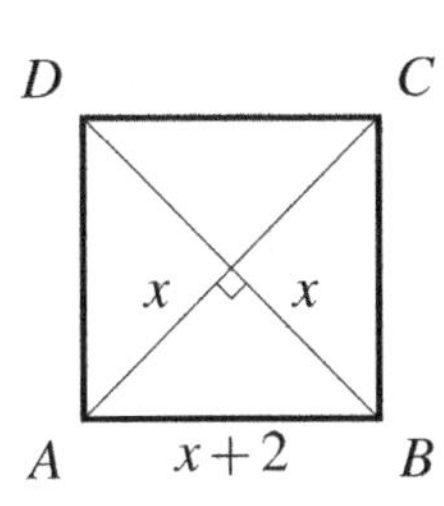


13-2) $BD = \frac{x}{2}$

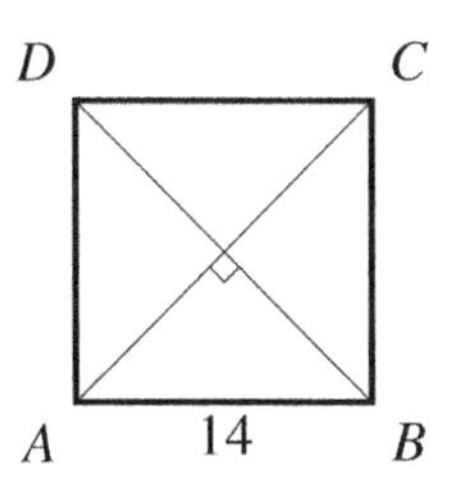


**14)** Find the area of each.

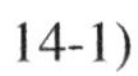

14-1)

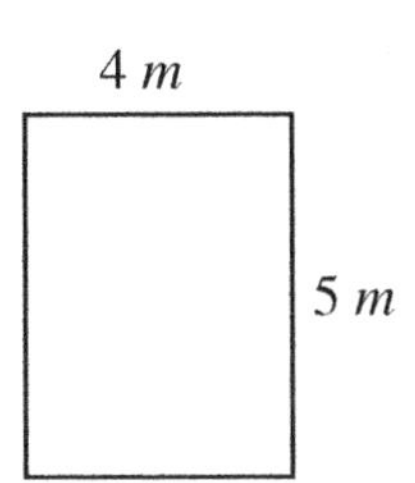


14-2)

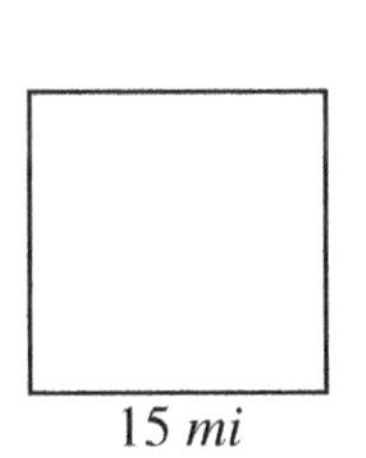


14-3)

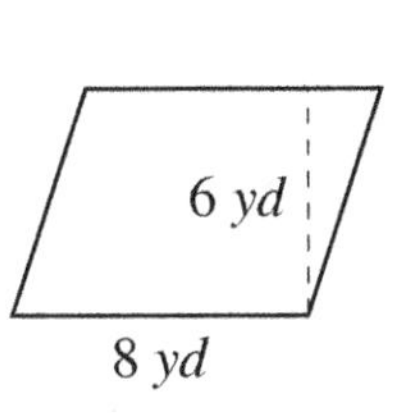


14-4)

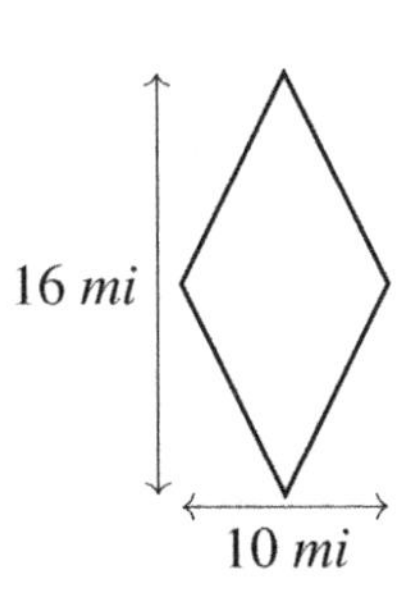


14-5)

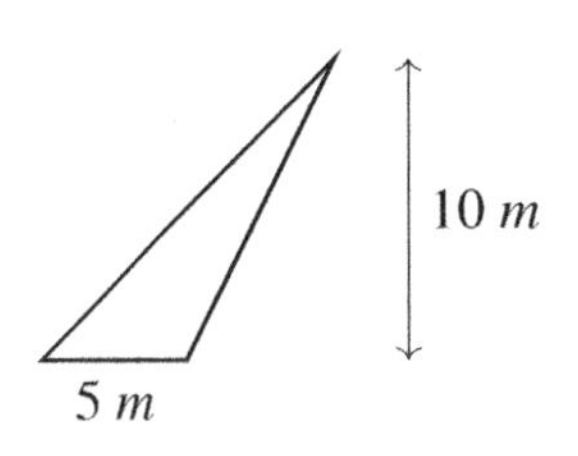


14-6)

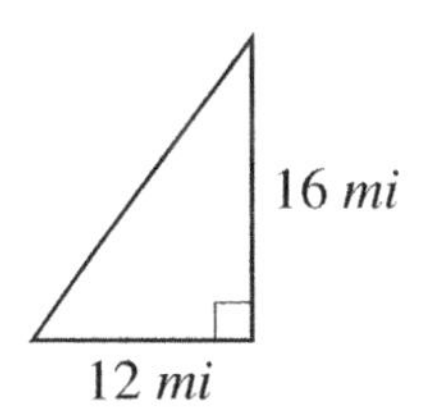


14-7)

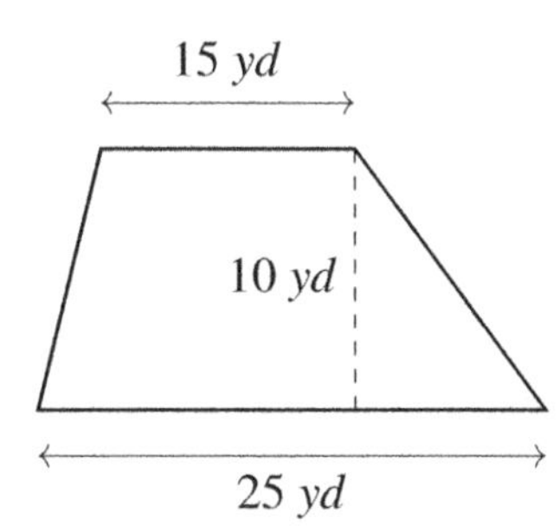


14-8)

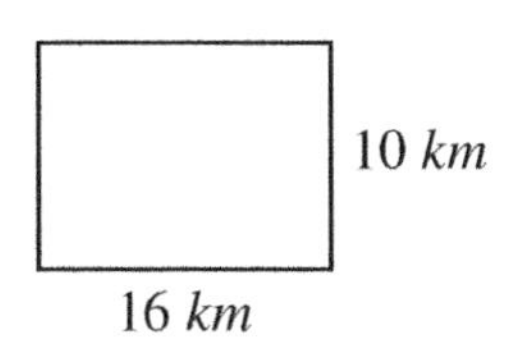


**15)** Find the perimeter or circumference of each shape.

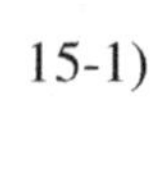

15-1)

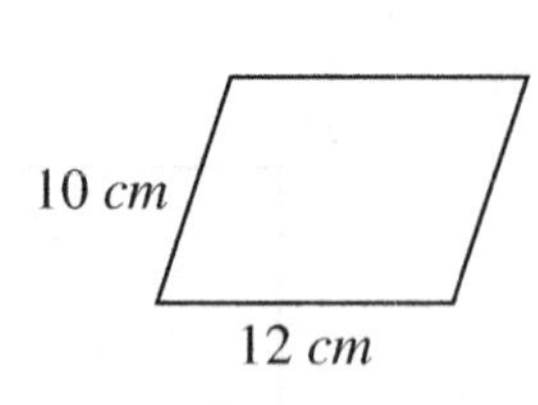


15-2)

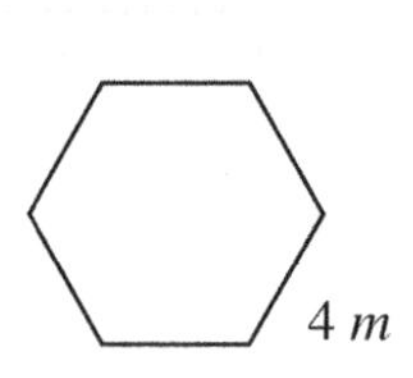


15-3)

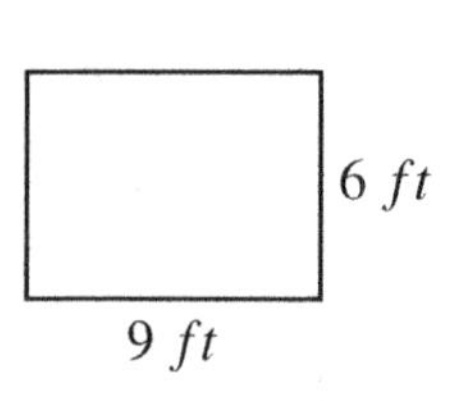


15-4)

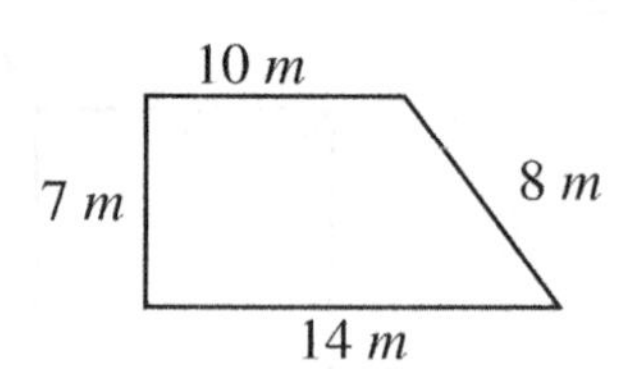


15-5)

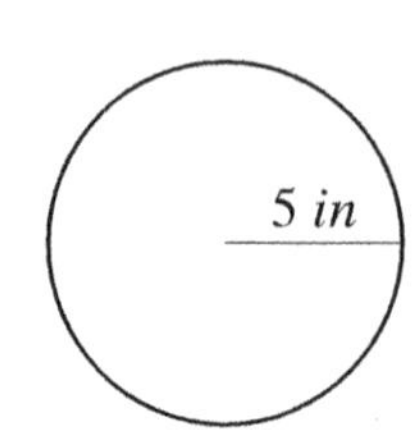


15-6)

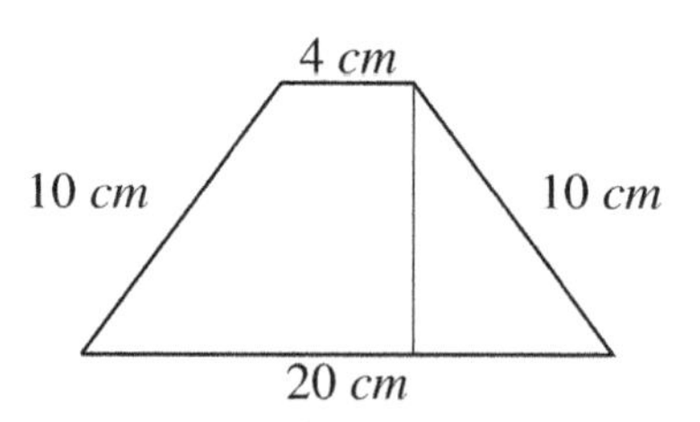


**16)** Find the measure of one interior angle in each polygon. Round your answer to the nearest tenth if necessary.

16-1)

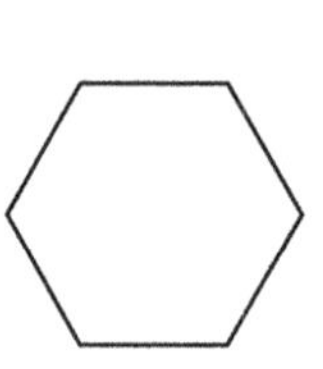

16-2)

16-3)

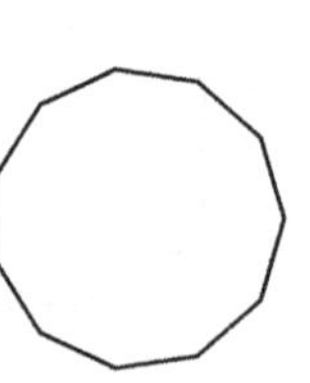

16-4)

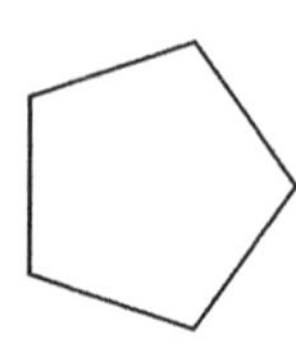

16-5)

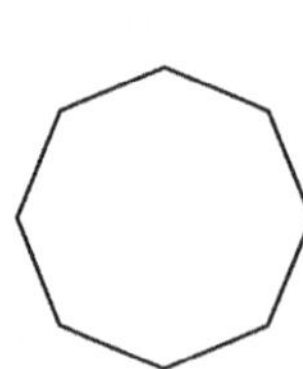

16-6)

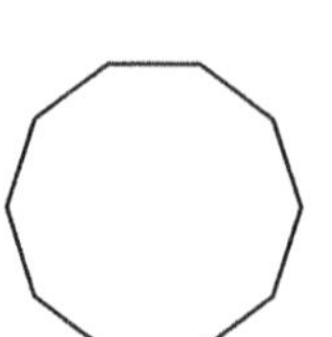

16-7) Regular 14-gon

16-8) Regular quadrilateral

16-9) Regular 17-gon

16-10) Regular 32-gon

## 5.12 Answers

**1)**

1-1) Square

1-2) Parallelogram

1-3) Rectangle

1-4) Rhombus

1-5) Trapezoid

1-6) Kite

1-7) Quadrilateral

**2)**

2-1) Quadrilateral, parallelogram, rhombus, rectangle, square

2-2) Quadrilateral, parallelogram, kite, rhombus

2-3) Quadrilateral, trapezoid

**3)**

3-1) $70^\circ$

3-2) $70^\circ$

3-3) $100^\circ$

3-4) $135^\circ$

**4)**

4-1) 5

4-2) 32

4-3) 7

4-4) 5

**5)**

5-1) $130^\circ$

5-2) $107^\circ$

**6)**

6-1) 7

6-2) 18

**7)**

7-1) 5

7-2) 4

**8)**

8-1) 10

8-2) $40^\circ$

8-3) 5

8-4) $15^\circ$

**9)**

9-1) 39

9-2) 10

9-3) 5

9-4) 37

**10)**

10-1) $60^\circ$

10-2) 12

10-3) $50^\circ$

10-4) $24^\circ$

**11)**

11-1) $\frac{7}{2} = 3.5$

11-2) 40

**12)**

12-1) $5\sqrt{2}$

12-2) 10

**13)**

13-1) $2(\sqrt{2}+1)$

13-2) $28\sqrt{2}$

**14)**

14-1) $20\ m^2$

14-2) $225\ mi^2$

14-3) $48\ yd^2$

14-4) $80\ mi^2$

14-5) $25\ m^2$

14-6) $96\ mi^2$

14-7) $200\ yd^2$

14-8) $160\ km^2$

**15)**

15-1) $44\ cm$

15-2) $24\ m$

15-3) $30\ ft$

15-4) $39\ cm$

15-5) $10\pi$ *in*

15-6) 44 *cm*

**16)**

16-1) 120°

16-2) 128.6°

16-3) 147.3°

16-4) 108°

16-5) 135°

16-6) 144°

16-7) 154.3°

16-8) 90°

16-9) 158.8°

16-10) 168.75°

# 6. Triangles

## 6.1 Triangles

Triangles play a crucial role as the simplest polygon with just three edges and three vertices. They can take various forms, but there are fundamental principles that apply to all triangles.

**Key Point**

In any triangle, the sum of the internal angles is always 180 degrees.

This principle is immensely useful in solving problems related to triangles. When two angles are known, the third can always be found by using the fact that their sum must equal 180 degrees.

**Key Point**

The area of a triangle can be calculated by the formula: $\frac{1}{2}(\text{base} \times \text{height})$.

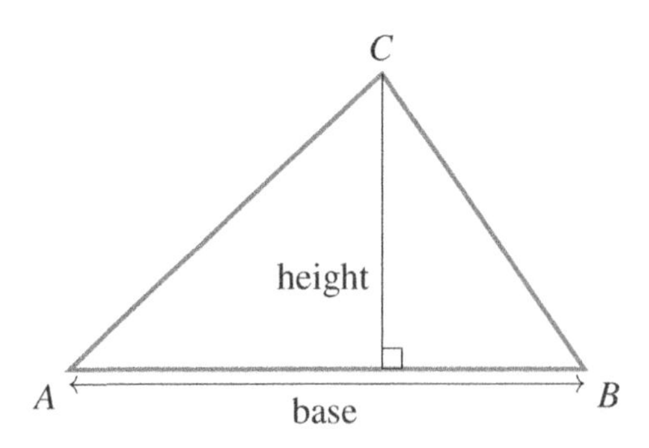


The base of a triangle can be any one of its sides, and the height is the perpendicular distance from the chosen base to the opposite vertex.

What is the area of a triangle with a base of 14 units and a height of 10 units?

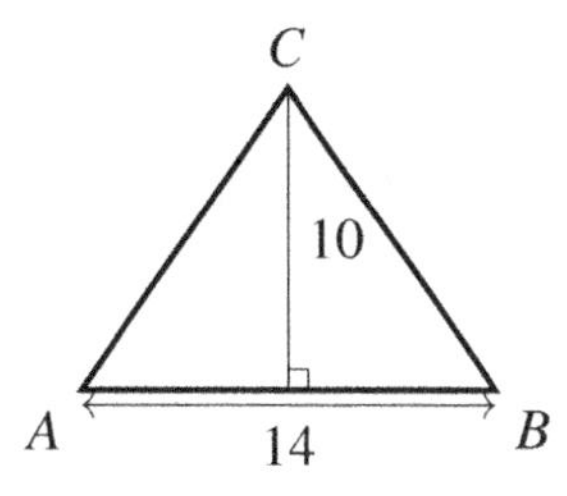


**Solution:** Using the area formula:

$$\text{Area} = \frac{1}{2}(\text{base} \times \text{height}) = \frac{1}{2}(14 \times 10) = 70.$$

The area of the triangle is 70 square units.

What is the missing angle in a triangle if two of its angles measure 55 degrees and 80 degrees?

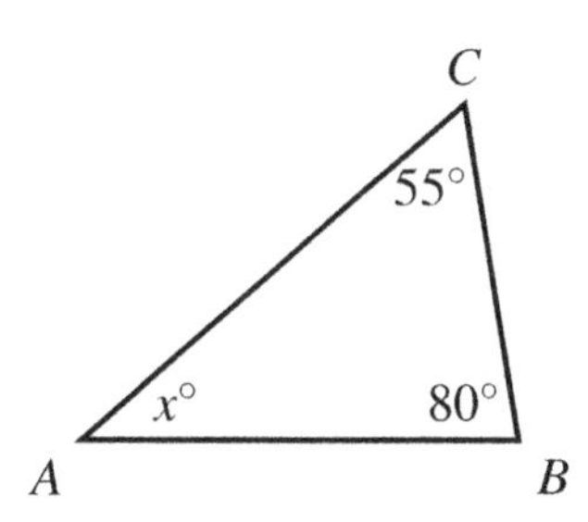


**Solution:** Let the missing angle be $x$. We can use the principle that the sum of angles in a triangle is 180 degrees:

$$55^\circ + 80^\circ + x = 180^\circ \Rightarrow 135^\circ + x = 180^\circ \Rightarrow x = 180^\circ - 135^\circ = 45^\circ.$$

The missing angle is $45^\circ$.

## 6.2 Classifying Triangles

Triangles are three-sided polygons with three internal angles that always sum to $180^\circ$. They can be classified based on side lengths and internal angles. This classification helps us understand the properties of triangles and solve related geometric problems more effectively.

**Based on Side Lengths** triangles can be categorized into three types: equilateral, where all sides are equal; isosceles, with two sides of equal length; and scalene, where each side has a different length.

**Key Point**

An equilateral triangle has three sides of equal length and three equal angles, each measuring 60°.

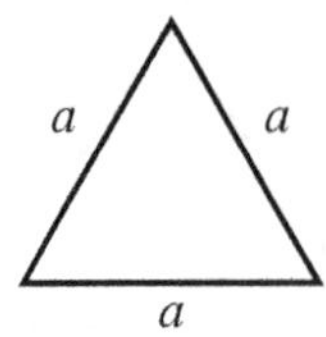


Figure 6.1: Equilateral triangle with all sides equal to $a$.

**Key Point**

An isosceles triangle has two sides of equal length. The angles opposite the equal sides are also equal.

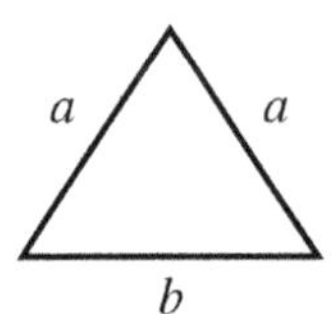


Figure 6.2: Isosceles triangle with two sides equal to $a$ and base $b$.

**Key Point**

A scalene triangle has no sides of equal length, and all angles are different.

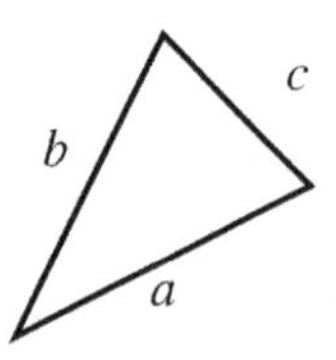


Figure 6.3: Scalene triangle with different side lengths $a$, $b$, and $c$.

**Based on Internal Angles** When classifying triangles based on their internal angles, there are three primary categories: acute, right, and obtuse.

**Key Point**

An acute triangle has all three internal angles less than 90°.

**Key Point**

A right triangle has one internal angle equal to 90°, known as the right angle.

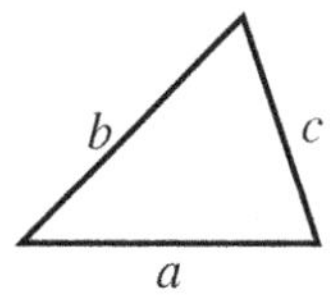


Figure 6.4: Acute triangle with sides $a$, $b$, and $c$.

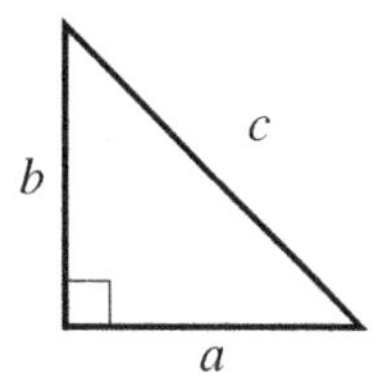


Figure 6.5: Right triangle with sides $a$, $b$, and hypotenuse $c$.

An obtuse triangle has one internal angle greater than $90^\circ$.

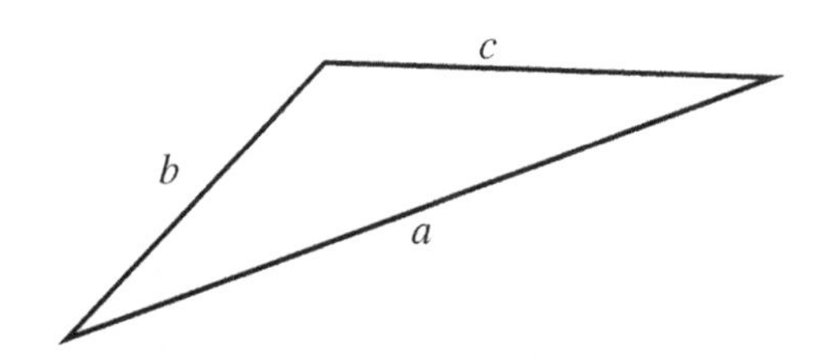


Figure 6.6: Obtuse triangle with sides $a$, $b$, and $c$.

**Example** If one side of a triangle is 7 units and all its sides are of equal length, what type of triangle is it?

**Solution:** Since all three sides are of equal length, it is an equilateral triangle.

**Example** If the largest angle in a triangle measures $105^\circ$, what type of triangle is it based on its angles?

**Solution:** Since one angle is greater than $90^\circ$, the triangle is an obtuse triangle.

**Example** If the largest angle in a triangle measures $70^\circ$, what type of triangle might it be based on its angles?

**Solution:** Since the largest angle is less than $90^\circ$ and an acute triangle has all internal angles less than $90^\circ$, this triangle is acute.

## 6.3 Triangle Angle Sum

One of the fundamental properties of a triangle is that regardless of its shape, the sum of its interior angles will always equal $180^\circ$. This concept is not only foundational in geometry but also serves as a critical tool for solving numerous problems that involve triangles.

Consider a triangle $ABC$. To better comprehend why the sum of the interior angles equals $180^\circ$, we can use a parallel line. A line $DE$ drawn through vertex $B$ and parallel to side $AC$ will help us see why the sum of the angles in triangle $ABC$ is $180^\circ$.

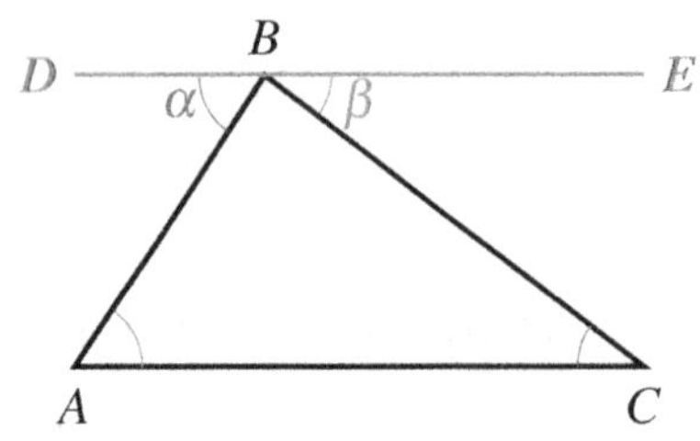


When line $DE$ is drawn parallel to side $AC$, angle $\alpha$ will equal the angle at $A$ because of the alternate interior angles theorem. Similarly, angle $\beta$ is equal to the angle at $C$. Thus, the sum of the angles in triangle $ABC$ can be written as:

$$\angle A + \angle B + \angle C = \alpha + \angle B + \beta.$$

Since $\alpha + \angle B + \beta$ forms a straight line, their sum is $180^\circ$. Therefore, the sum of the angles in triangle $ABC$ also equals $180^\circ$.

If two angles of a triangle measure $60^\circ$ and $50^\circ$, what is the measure of the third angle?

**Solution:** To find the third angle, use the triangle angle sum property:

$$\text{Third angle} = 180^\circ - (60^\circ + 50^\circ) = 180^\circ - 110^\circ = 70^\circ.$$

## 6.4 Triangle Midsegment

A midsegment in a triangle is a line segment connecting the midpoints of two sides of the triangle. Every triangle will have three such possible midsegments, creating a smaller triangle within the original one. Equally important is the understanding of the unique properties of the midsegments, which are quite useful in geometric proofs and problem-solving.

Consider triangle $ABC$ with points $D$ and $E$ as the midpoints of sides $AB$ and $AC$, respectively. The midsegment is parallel to the third side of the triangle. Draw $DE$, the midsegment. Since $D$ and $E$ are midpoints, $AD = DB$ and $AE = EC$. By the Alternate Interior Angles Theorem, $DE$ is parallel to $BC$.

The length of the midsegment is half the length of the third side of the triangle.

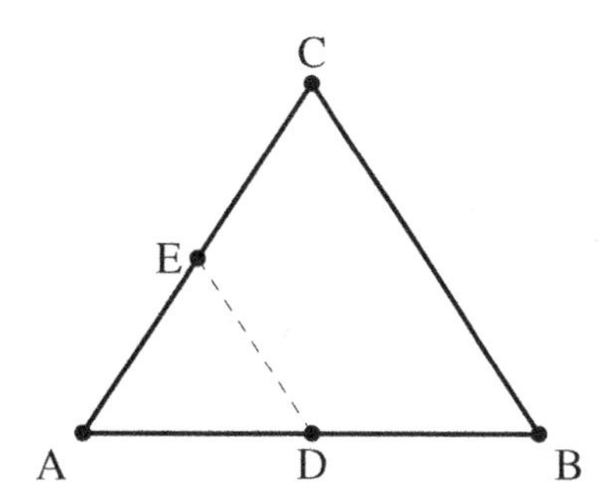


Figure 6.7: Midsegment DE is parallel with side BC.

Example In triangle $XYZ$, point $M$ is the midpoint of side $XY$, and point $N$ is the midpoint of side $XZ$. If the length of side $YZ$ is 14 units, what is the length of midsegment $MN$?

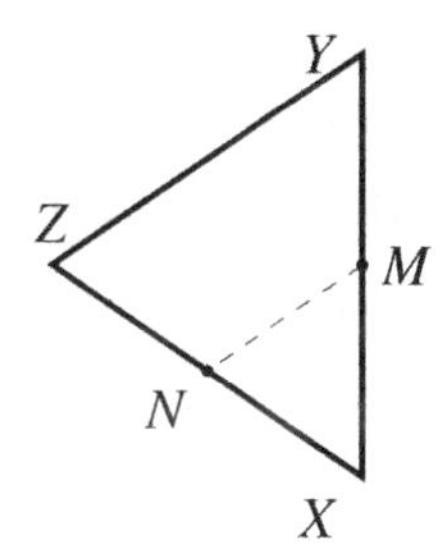


**Solution:** Using the properties of the midsegment: $MN = \frac{1}{2}YZ = \frac{1}{2}(14) = 7$ units.

Example In triangle $PQR$, if the midsegment connecting the midpoints of sides $PQ$ and $PR$ measures 9 units, what is the length of side $QR$?

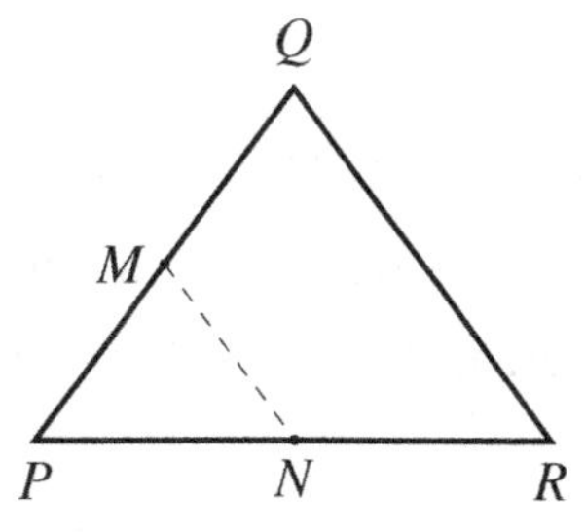


**Solution:** Using the properties of the midsegment: $QR = 2 \times 9 = 18$ units.

## 6.5 Angle Bisectors of Triangles

An angle bisector in a triangle is a ray that divides an angle into two congruent angles, each possessing equal measure. This geometric feature not only enhances our understanding of angular relationships but also leads to useful proportional relationships within the triangle.

**Key Point**

An angle bisector in a triangle splits the opposite side into segments proportional to the other two sides of the triangle, forming the basis for the Angle Bisector Theorem.

To comprehend this in practical terms, consider a triangle $ABC$ with $AD$ as the bisector of angle $A$ intersecting side $BC$ at $D$. The Angle Bisector Theorem simplifies this to a straightforward proportion: $\frac{AB}{AC} = \frac{BD}{CD}$.

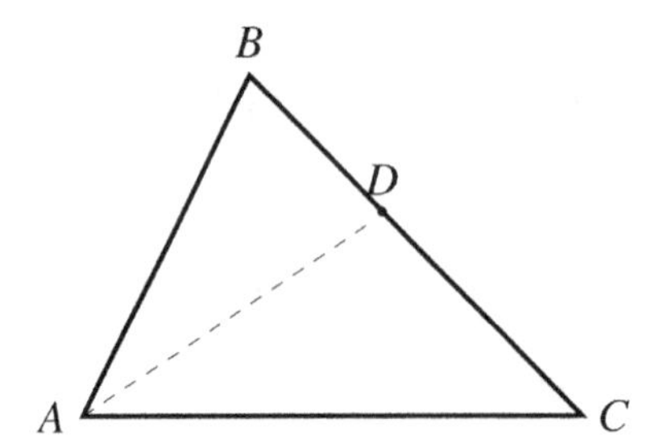


Further intrigue is found at the incenter of a triangle, where all three angle bisectors intersect. The incenter is remarkable for its equidistance from all three sides of the triangle, forming the center of the incircle which grazes each side at a neat right angle. The radius of the incircle, known as the inradius, is equally drawn to all three sides, and as we will later see, has clear applications in problem solving.

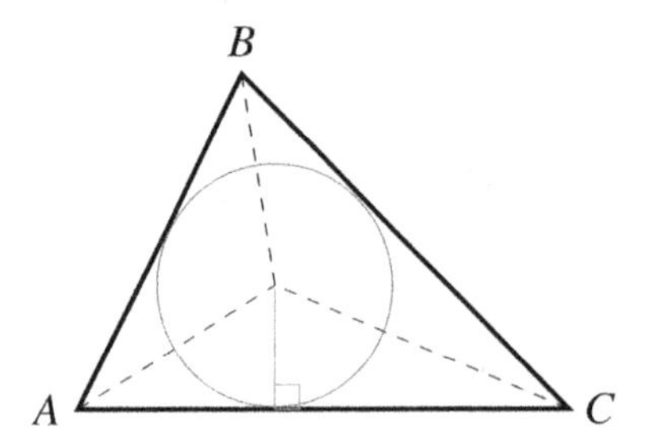


**Example** Given that in triangle $DEF$, side $DE$ measures 6 units, side $DF$ measures 8 units, and $DM$ is the angle bisector of angle $D$, which meets $EF$ at $M$ with $MF = 4$ units. What is the length of $ME$?

**Solution:** Apply the Angle Bisector Theorem: $\frac{DE}{DF} = \frac{ME}{MF}$. Insert the given values to find $ME$:

$$\frac{6}{8} = \frac{ME}{4} \implies ME = \frac{6 \times 4}{8} \implies ME = 3.$$

Hence, $ME = 3$ units.

**Example** In triangle $GHI$, if $GH = 5$ units, $GI = 10$ units, and angle bisector $GN$ meets side $HI$ at $N$ with $HN = x$ units, what is the length of $IN$ in terms of $x$?

**Solution:** Utilize the Angle Bisector Theorem: $\frac{GH}{GI} = \frac{HN}{IN}$. Insert the given values:

$$\frac{5}{10} = \frac{x}{IN} \implies IN = \frac{10x}{5} \implies IN = 2x.$$

Thus, $IN = 2x$ units.

## 6.6 Isosceles and Equilateral Triangles

An *Isosceles Triangle* has at least two sides of equal length. These equal sides are known as the legs, and the third side is commonly referred to as the base of the triangle. The angles opposite these legs are equal, known as the base angles.

**Key Point**

In an isosceles triangle, the angles opposite the equal sides, called the base angles, are congruent.

These unique features of isosceles triangles allow simplifications in geometric computations. For instance, when an *altitude* (a perpendicular line from the vertex opposite the base) is drawn to the base, it bisects both the base and the vertex angle. This results in two congruent right triangles.

**Key Point**

The altitude in an isosceles triangle bisects the base and the vertex angle, forming two congruent right triangles.

Moving on to *Equilateral Triangles*, these are triangles where all sides are of equal length, which also means all interior angles are equal. According to the properties of triangles, each angle in an equilateral triangle measures $60°$.

**Key Point**

All angles of an equilateral triangle equal $60°$, and any altitude bisects the side it is drawn to and forms two $30° - 60° - 90°$ triangles.

In an isosceles triangle $ABC$, where $AB = AC$, and angle $B$ is $50°$, what is angle $C$?

**Solution:** Since triangle $ABC$ is isosceles with $AB = AC$, we have $\angle B = \angle C$. Therefore, angle $C$ is also $50°$.

Triangle $DEF$ is isosceles with $DE = DF$. If angle $E$ is $40°$, what is angle $D$?

**Solution:** Since $DE = DF$, we know that angles $E$ and $F$ are congruent; hence, angle $F$ is also $40°$. With the triangle angle sum property, $\angle D = 180° - (40° + 40°) = 100°$.

If one side of an equilateral triangle is 6 units, what is the length of an altitude drawn to that side?

**Solution:** An altitude in an equilateral triangle creates two $30^\circ - 60^\circ - 90^\circ$ triangles. With $c$ being the hypotenuse and $b$ the shorter leg, the longer leg $a$ (the altitude) can be found using the Pythagorean theorem:

$$a^2 + b^2 = c^2 \Rightarrow a^2 + 3^2 = 6^2 \Rightarrow a^2 = 36 - 9 \Rightarrow a = \sqrt{27} \Rightarrow a = 3\sqrt{3}.$$

Therefore, the length of the altitude is $3\sqrt{3}$ units.

## 6.7 Right Triangles; Pythagorean Theorem

A right triangle is a geometric figure featuring one angle of exactly $90^\circ$. This characteristic angle is named the right angle, and the side opposite this angle is called the hypotenuse. The other two sides are referred to as the legs of the triangle.

**Key Point**

A right triangle's hypotenuse is the triangle's longest side and is opposite the right angle.

The Pythagorean theorem represents a fundamental relationship within right triangles. It states that the sum of the squares of the legs' lengths equals the square of the hypotenuse's length. In the formula $a^2 + b^2 = c^2$, $a$ and $b$ represent the legs, while $c$ denotes the hypotenuse.

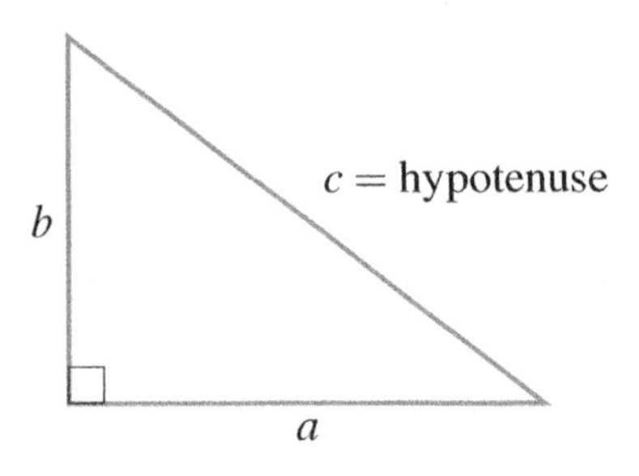


**Key Point**

The Pythagorean theorem: $a^2 + b^2 = c^2$, where $a$ and $b$ are the legs, and $c$ is the hypotenuse of a right triangle.

Let us consider some examples to illustrate these concepts.

**Example** In a right triangle $ABC$ with angle $C$ being $90^\circ$, if leg $AC = 5$ units and leg $BC = 12$ units, determine the length of the hypotenuse $AB$.

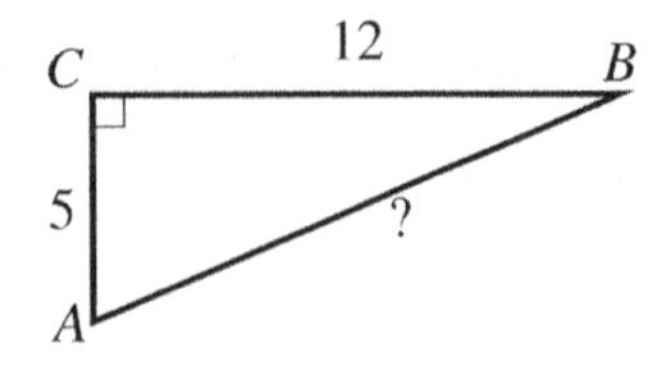

**Solution:** Applying the Pythagorean theorem:

$$AC^2 + BC^2 = AB^2 \Rightarrow 5^2 + 12^2 = AB^2 \Rightarrow 25 + 144 = AB^2 \Rightarrow AB^2 = 169,$$

therefore: $AB = \sqrt{169} = 13$ units.

**Example** A triangle has a hypotenuse of 15 $cm$ and one leg measuring 12 $cm$. Calculate the length of the other leg.

**Solution:** Apply the Pythagorean Theorem $a^2 + b^2 = c^2$:

$$12^2 + b^2 = 15^2 \Rightarrow 144 + b^2 = 225 \Rightarrow b^2 = 225 - 144 \Rightarrow b^2 = 81 \Rightarrow b = \sqrt{81} = 9.$$

The missing leg length is 9 $cm$.

## 6.8 Special Right Triangles

We now focus on two important types of right triangles: the $45^\circ - 45^\circ - 90^\circ$ and $30^\circ - 60^\circ - 90^\circ$ triangles. These triangles have side lengths that follow specific ratios, making it easier to solve problems involving these geometric figures.

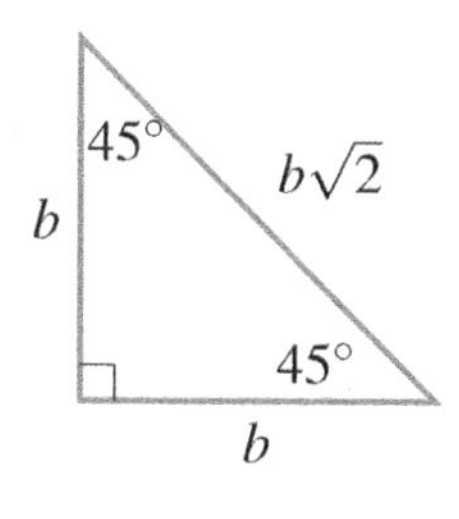


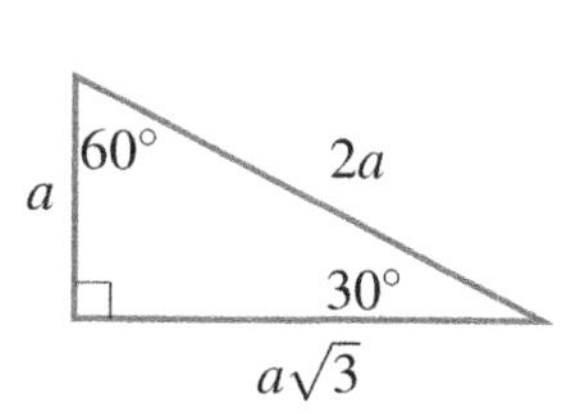


**Key Point**

The sides of a $45^\circ - 45^\circ - 90^\circ$ triangle are in the ratio $1 : 1 : \sqrt{2}$. This means that the legs of the triangle are congruent, and the hypotenuse is $\sqrt{2}$ times as long as one leg.

**Key Point**

The sides of a $30^\circ - 60^\circ - 90^\circ$ triangle are in the ratio $1 : \sqrt{3} : 2$. The side opposite the $30^\circ$ angle is the shortest, the side opposite the $60^\circ$ angle is $\sqrt{3}$ times longer, and the hypotenuse is twice as long as the shortest side.

**Example** Find the length of the hypotenuse of a right triangle if the length of the other two sides are both 4 inches.

**Solution:** Since the legs are equal, this is a $45^\circ - 45^\circ - 90^\circ$ triangle. Using the side ratio of $1 : 1 : \sqrt{2}$, and setting each leg to 4 inches, we find the hypotenuse length is $4\sqrt{2}$ inches.

**Example** Given a right triangle whose hypotenuse is 6 inches and one angle is 30°, find the lengths of the other two sides.

**Solution:** With the hypotenuse being 6 inches, this is a $30^\circ - 60^\circ - 90^\circ$ triangle. By the side ratio, the shortest side (opposite the 30° angle) is half the hypotenuse, so 3 inches. The side opposite the 60° angle is $3\sqrt{3}$ inches.

## 6.9 Pythagorean Theorem Converse: Is This a Right Triangle?

The converse of the Pythagorean theorem provides us with a useful test to identify whether a triangle is a right triangle when we know the lengths of its sides.

The Pythagorean Theorem states that for a right triangle, the square of the length of the hypotenuse ($c$) is equal to the sum of the squares of the lengths of the other two sides ($a$ and $b$). The converse asserts that any triangle for which this relationship holds is a right triangle. In symbolic form, we have:

$$c^2 = a^2 + b^2 \Longrightarrow \text{Triangle with sides } a, b, c \text{ is a right triangle.}$$

Since both the Pythagorean theorem and its converse are true, we can state them as an 'if and only if' statement.

**Key Point**

A triangle is a right triangle if and only if the Pythagorean Theorem ($c^2 = a^2 + b^2$) holds true for it.

**Example** A triangle has sides measuring 7 units, 24 units, and 25 units. Is it a right triangle?

**Solution:** First, identify the longest side, which in this case is 25 units. Now apply the converse of the Pythagorean theorem: $7^2 + 24^2 = 49 + 576 = 625$. Now, square the longest side: $25^2 = 625$. The sum of the squares of the two shorter sides is equal to the square of the longest side ($625 = 625$) which confirms that the triangle is a right triangle.

**Example** A triangle has sides with lengths of 9 units, 40 units, and 42 units. Can we classify it as a right triangle based on the converse of the Pythagorean theorem?

**Solution:** Identify the longest side, which is 42 units. Apply the converse of the Pythagorean theorem: $9^2 + 40^2 = 81 + 1600 = 1681$. Square of the longest side: $42^2 = 1764$. The sum of the squares of the two shorter sides 1681 does not equal the square of the longest side 1764, indicating that the triangle is not a right triangle.

## 6.10 Geometric Mean in Triangles

A key property of right triangles involving the geometric mean states that if an altitude is drawn to the hypotenuse, then the length of the altitude is the geometric mean of the lengths of the two segments it creates.

**Key Point**

If a right triangle $ABC$ has a right angle at $C$ and an altitude $CD$ is drawn to the hypotenuse $AB$, then $CD^2 = AD \cdot DB$, hence $CD = \sqrt{AD \cdot DB}$.

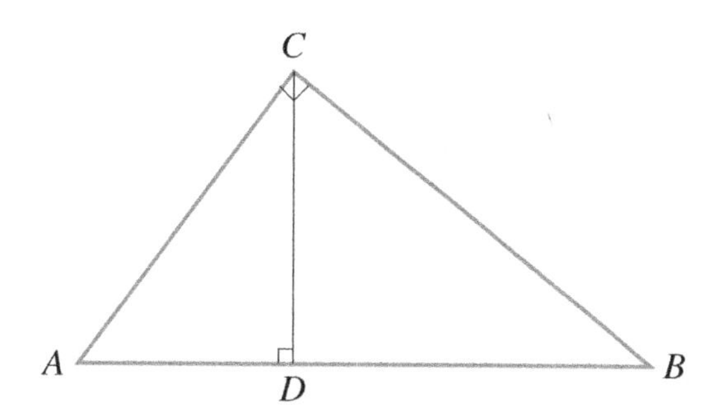


This relationship is also evident when considering the small triangles $ADC$ and $BDC$ formed by the altitude.

**Key Point**

Triangles $ADC$ and $BDC$ are similar to the original triangle $ABC$ (AA similarity criterion), leading to proportional relationships which give $AD \cdot AB = AC^2$ and $DB \cdot AB = BC^2$.

Lastly, combining the above proportionalities further validates the Pythagoras theorem:

$$AC^2 + BC^2 = AD \cdot AB + DB \cdot AB = (AD + DB) \cdot AB = AB \cdot AB = AB^2.$$

**Example** In a right triangle, the hypotenuse measures 16 units, and one of its segments after drawing an altitude from the right angle is 9 units. What is the length of the other segment, and what is the length of the altitude?

**Solution:** Let the length of the other segment be $x$, hence $x = 16 - 9 = 7$ units. By the geometric mean property: Altitude$^2 = 9 \times 7$. Therefore, the altitude is $\sqrt{63}$ or $3\sqrt{7}$ units in length.

## 6.11 Exterior Angle Theorem

The Exterior Angle Theorem is one of the cornerstone concepts in triangle geometry. It explains that the measure of any exterior angle of a triangle is equal to the sum of the measures of the two interior opposite angles. The exterior angle is created by extending one side of the triangle.

**Key Point**

The Exterior Angle Theorem states that the measure of any exterior angle of a triangle is equal to the sum of the measures of the two interior opposite angles.

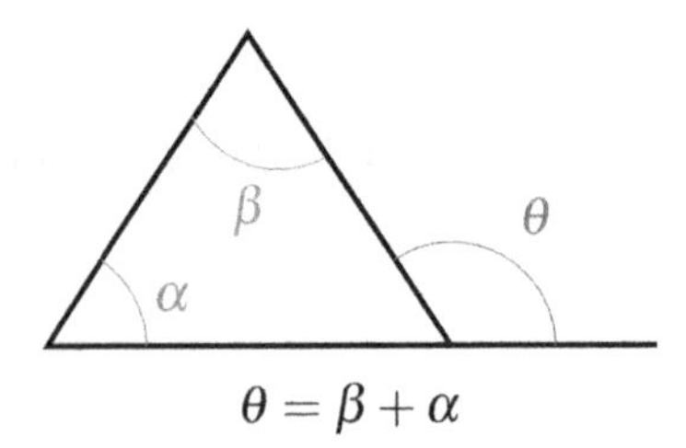


$$\theta = \beta + \alpha$$

This theorem is exceptionally useful when trying to find unknown angles in triangles or verifying the validity of a triangle's angle measurements.

**Example** Given a triangle with interior angles measuring $35^\circ$ and $70^\circ$, determine the measure of the exterior angle adjacent to the third interior angle.

**Solution:** By the Exterior Angle Theorem, the measure of an exterior angle of a triangle is equal to the sum of the measures of the two non-adjacent interior angles. Given two interior angles are $35^\circ$ and $70^\circ$, the exterior angle adjacent to the third angle is $35^\circ + 70^\circ = 105^\circ$.

Now let us apply this theorem to find a missing interior angle when given an exterior angle and one interior angle.

**Example** Find the value of $x$ according to the figure where the exterior angle measures $123^\circ$ and one of the remote interior angles measures $56^\circ$.

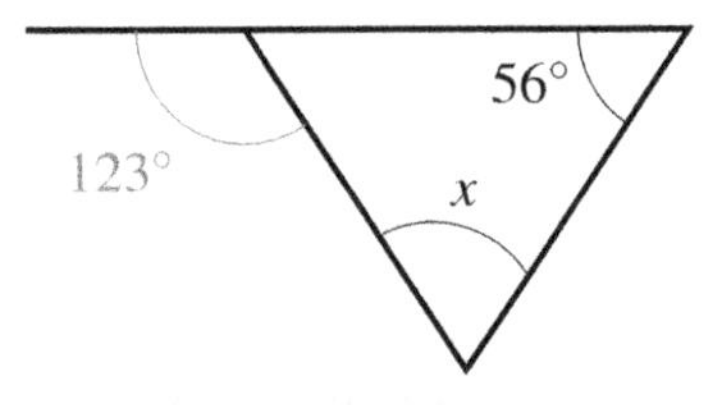


**Solution:** Applying the Exterior Angle Theorem: $123^\circ = x + 56^\circ$. Simplifying, we get $x = 123^\circ - 56^\circ = 67^\circ$.

## 6.12 Medians

A *median* of a triangle is a line segment that joins a vertex of the triangle to the midpoint of the opposing side. By definition, every triangle will have exactly three medians, one from each vertex.

The three medians in a triangle are concurrent; they intersect at a singular point known as the *centroid*.

**Key Point**

The centroid divides each median in a 2:1 ratio, where the longer segment is the part connected to the vertex.

This means that medians not only bisect areas but also provide a balance point for the triangle—the centroid.

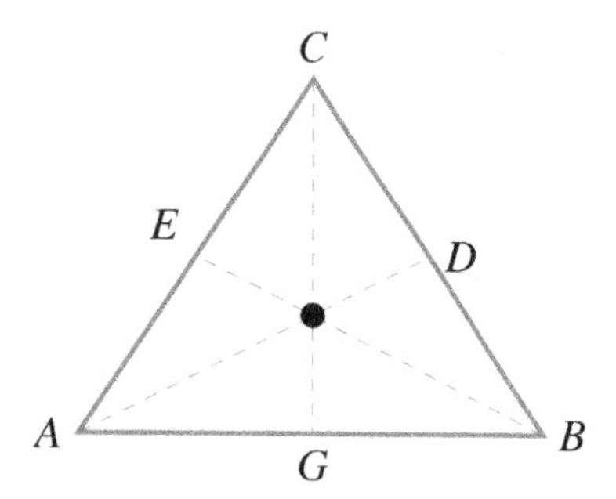


Additionally, the distances from the vertices to the centroid and from the centroid to the midpoints have specific proportions:

**Key Point**

The distance from a vertex to the centroid is $\frac{2}{3}$ the length of the median, while the distance from the centroid to the midpoint is $\frac{1}{3}$ the total median length.

**Example** In triangle $ABC$, let $AD$ and $BE$ be medians with lengths 9 units and 12 units respectively, and $G$ be the centroid. Find the lengths of $AG$, $GD$, and the distance between $B$ and $G$. This is visually represented in the following graph:

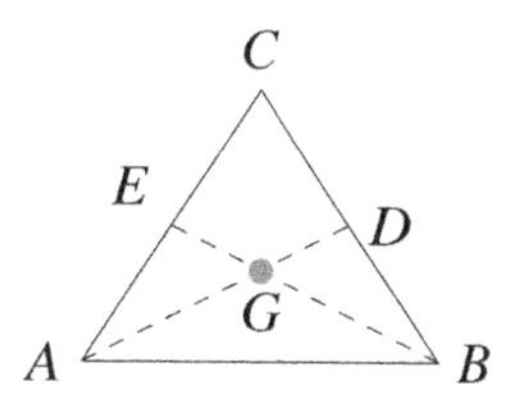


**Solution:** To address this, we leverage the properties of medians in a triangle. Initially, for median $AD$, the segment $AG$ (from vertex $A$ to the centroid $G$) is calculated as two-thirds of the length of $AD$, leading to:

$$AG = \frac{2}{3} \times AD = \frac{2}{3} \times 9 = 6 \text{ units.}$$

In a similar manner, the segment $GD$ (from the centroid $G$ to the midpoint $D$ of side $BC$) is one-third of the length of $AD$, which is:

$$GD = \frac{1}{3} \times AD = \frac{1}{3} \times 9 = 3 \text{ units.}$$

Additionally, for the median $BE$, the length of segment $BG$ (from vertex $B$ to the centroid $G$) is two-thirds of $BE$, thus:

$$BG = \frac{2}{3} \times BE = \frac{2}{3} \times 12 = 8 \text{ units.}$$

## 6.13 Centroid

The centroid of a triangle is a point of great interest in geometry due to its unique and significant properties.

**Key Point**

The centroid is the point where the triangle's three medians intersect, known as the point of concurrency.

When we deal with triangles on a coordinate plane, the position of the centroid can be calculated precisely.

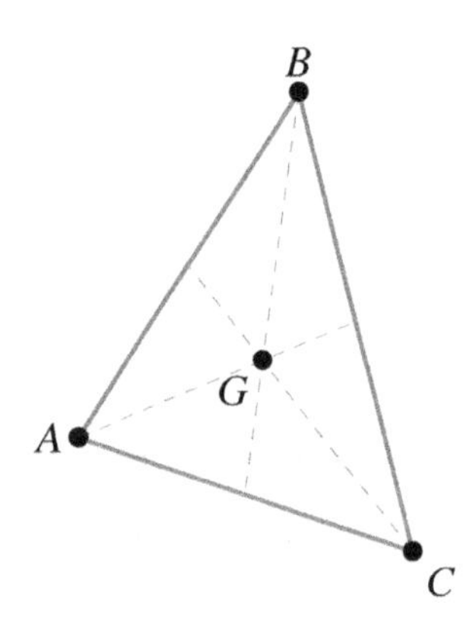


**Key Point**

The coordinates of the centroid $(G_x, G_y)$ are the average of the $x$-coordinates and the $y$-coordinates of the triangle's vertices.

Now, let us look at a couple of examples illustrating these concepts.

**Example** Given triangle $ABC$ with vertices $A(1,4)$, $B(5,10)$, and $C(7,2)$, find the coordinates of the centroid $G$.

**Solution:** Using the formula for finding the centroid's coordinates:

$$G_x = \frac{1+5+7}{3} = \frac{13}{3}, \text{ and } G_y = \frac{4+10+2}{3} = \frac{16}{3}.$$

Hence, the coordinates of $G$ are $(\frac{13}{3}, \frac{16}{3})$.

## 6.14 The Triangle Inequality Theorem

We delve into the Triangle Inequality Theorem. This theorem is crucial in determining whether three line segments can form a triangle.

**Key Point**

The Triangle Inequality Theorem states that in any triangle, the sum of the lengths of any two sides must be greater than the length of the remaining side.

This theorem prevents us from creating a "triangle" with sides that cannot possibly meet. For any triangle $ABC$, the theorem is formalized as:

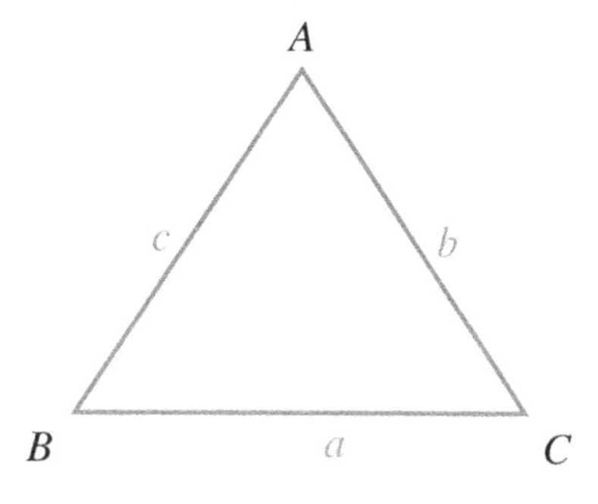


$$AB + BC > AC$$
$$BC + AC > AB$$
$$AC + AB > BC$$

**Key Point**

In any triangle with sides $a$, $b$, and $c$, the Triangle Inequality Theorem states that the length of any side $c$ must satisfy: $|a-b| < c < a+b$. This theorem ensures that the given lengths can form a valid triangle.

We will first explore a scenario where two sides are given, and we need to find the possible lengths for the third side.

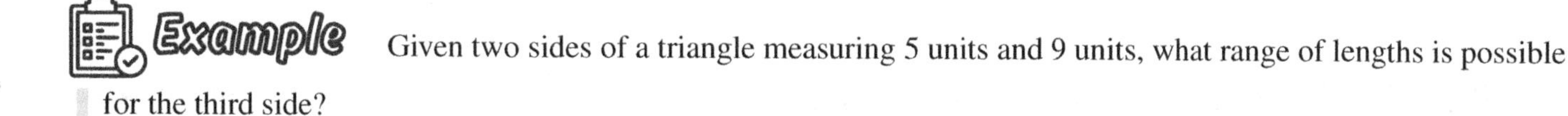

**Example** Given two sides of a triangle measuring 5 units and 9 units, what range of lengths is possible for the third side?

**Solution:** For our triangle with sides 5 units and 9 units, the third side must be greater than $|9-5| = 4$ units and less than $9+5 = 14$ units. Therefore, the possible range for the third side is from greater than 4 units to less than 14 units.

Now consider a scenario where you are given all three sides and are asked to determine if they can form a triangle.

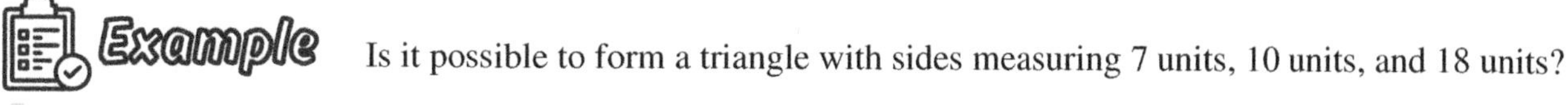

**Example** Is it possible to form a triangle with sides measuring 7 units, 10 units, and 18 units?

**Solution:** Using the Triangle Inequality Theorem: $7+10 = 17$, which is less than 18. Therefore, because the sum of two of the sides is not greater than the third side, these lengths cannot form a triangle.

## 6.15 SSS and SAS Congruence

Move on to the conditions that allow us to conclude that two triangles are completely identical, known as congruence. This means every side and every angle of one triangle matches perfectly with those of the other. Key concepts in determining triangle congruence include the Side-Side-Side (SSS) Congruence Postulate and the Side-Angle-Side (SAS) Congruence Postulate.

**Key Point**

The SSS Congruence Postulate states that if three sides of one triangle are congruent to three sides of another triangle, then the triangles are congruent.

**Key Point**

The SAS Congruence Postulate asserts that if two sides and the included angle of one triangle are congruent to the two sides and the included angle of another triangle, then the triangles are congruent.

**Example** Given triangle $ABC$ and triangle $DEF$ with $AB = DE = 6\ cm$, $BC = EF = 5\ cm$, and $AC = DF = 4\ cm$, can we conclude the triangles are congruent?

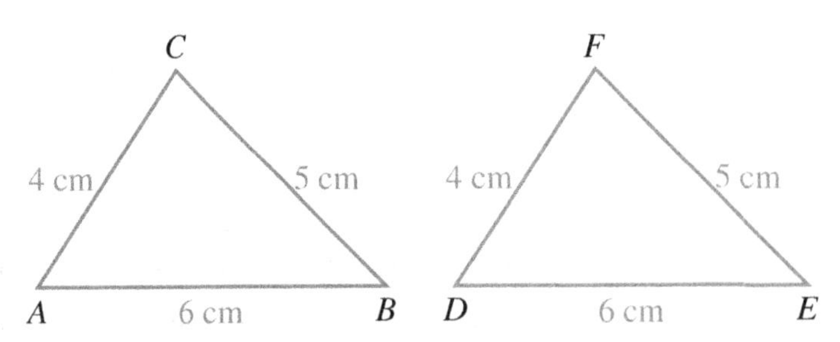


**Solution:** Here, all three sides of triangle $ABC$ are equal to the corresponding sides of triangle $DEF$. Hence, by the SSS congruence postulate, triangle $ABC$ is congruent to triangle $DEF$.

**Example** Given triangle $PQR$ and triangle STU with $PQ = ST = 7\ cm$, $QR = TU = 8\ cm$, and $\angle PQR = \angle STU = 60^\circ$, are the triangles congruent?

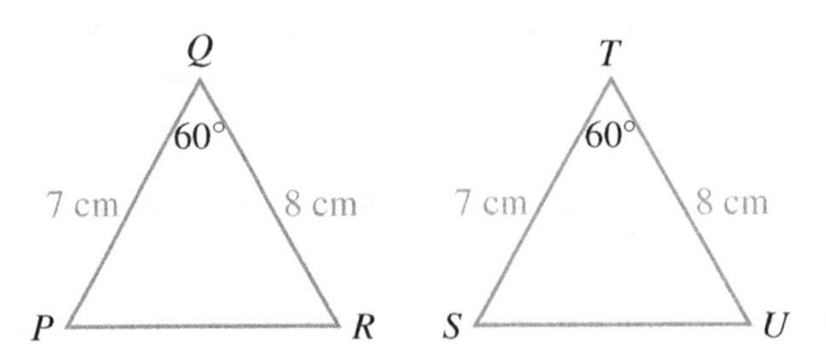


**Solution:** Here, two sides and the included angle of triangle $PQR$ are equal to the corresponding sides and angle of triangle $STU$. Hence, by the SAS congruence postulate, triangle $PQR$ is congruent to triangle $STU$.

## 6.16 ASA and AAS Congruence

The *Angle-Side-Angle (ASA) Congruence Postulate* comes into play when we have two angles and the included side of a triangle.

Key Point

If two angles and the included side of one triangle are congruent to two angles and the included side of another triangle, the two triangles are congruent.

The *Angle-Angle-Side (AAS) Congruence Postulate* is used when we know two angles and a non-included side of a triangle.

Key Point

If two angles and a non-included side of one triangle are congruent to two angles and the non-included side of another triangle, the two triangles are congruent.

Example Given triangle $ABC$ and triangle $DEF$ with $\angle A = \angle D$, $\angle B = \angle E$, and $\overline{AB} = \overline{DE}$, can we say that the two triangles are congruent?

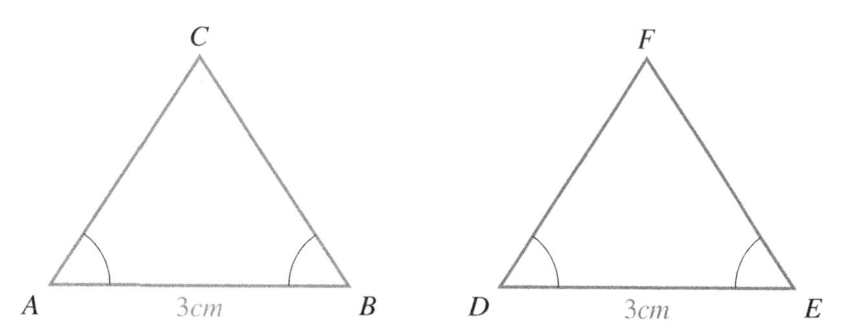


**Solution:** As $\angle A$ and $\angle D$, $\angle B$ and $\angle E$, and $\overline{AB}$ and $\overline{DE}$ are congruent, by the ASA congruence postulate, triangle $ABC$ is congruent to triangle $DEF$.

Consider triangles $GHI$ and $JKL$ where $\angle G = \angle J$, $\angle I = \angle L$, and $\overline{GI} = \overline{JL}$. Are the triangles congruent?

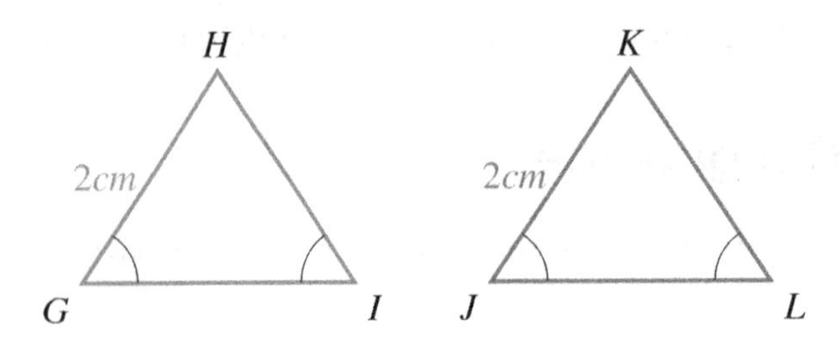


**Solution:** Since $\angle G$ and $\angle J$, $\angle I$ and $\angle L$, and $\overline{GI}$ and $\overline{JL}$ are congruent, by the AAS congruence postulate, triangle $GHI$ is congruent to triangle $JKL$.

## 6.17 Hypotenuse-Leg Congruences

When determining if two triangles are congruent, it means that they are identical in terms of shape and size. Their corresponding angles are equal, and their corresponding sides are of the same length. However, we do not always require all the information about the sides and angles to conclude that two triangles are congruent.

Among the several postulates and theorems that allow us to prove the congruence of triangles, there is a specific one for right triangles known as the Hypotenuse-Leg (HL) Congruence theorem.

**Key Point**

The HL Congruence theorem states that two right triangles are congruent if their hypotenuses are congruent and one corresponding leg is congruent.

**Example** Consider right triangle $ABC$ with hypotenuse $AC = 13\ cm$ and leg $AB = 5\ cm$. A second triangle, $DEF$, is also right-angled with hypotenuse $DF = 13\ cm$ and leg $FE = 12\ cm$. Can we conclude the triangles are congruent?

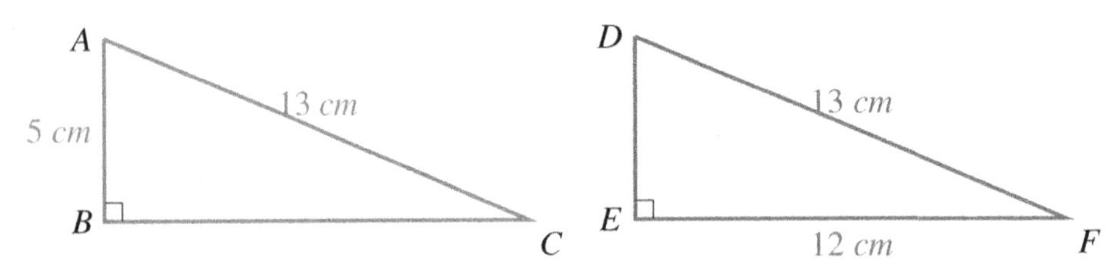


**Solution:** The information provided does not allow us to immediately conclude the triangles are congruent. However, if we can determine the length of the other leg of triangle $ABC$ or $DEF$, we would have sufficient information to apply the HL Congruence theorem. For triangle $ABC$, according to the Pythagorean Theorem we have:

$$AB^2 + BC^2 = AC^2,\ AB = 5,\ \text{and}\ AC = 13.$$

Upon substitution, we obtain:

$$5^2+BC^2=13^2 \Rightarrow 25+BC^2=169 \Rightarrow BC^2=169-25 \Rightarrow BC=\sqrt{144} \Rightarrow BC=12\ cm.$$

Since both triangle $ABC$ and triangle $DEF$ have a hypotenuse of 13 $cm$ and one leg of 12 $cm$, by the HL Congruence theorem, we can conclude that triangle $ABC$ is congruent to triangle $DEF$.

## 6.18 Practices

**1)** Find the measure of the unknown angle in each triangle.

1-1)

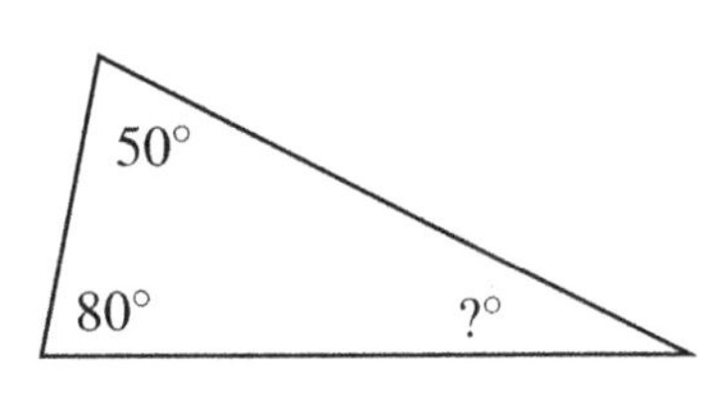


1-3)

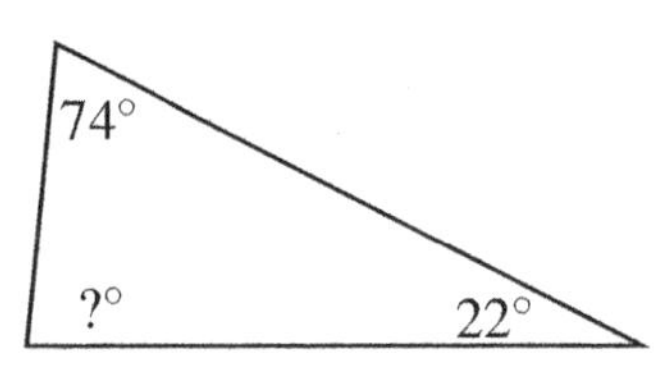


1-2)

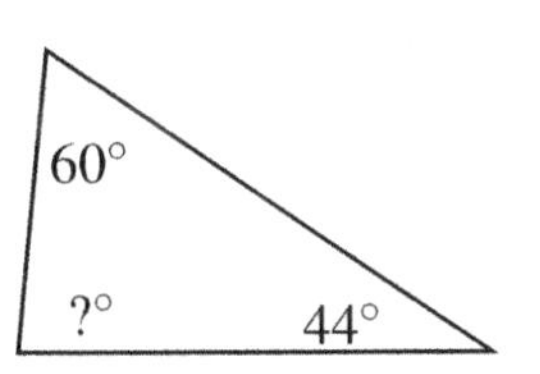


1-4)

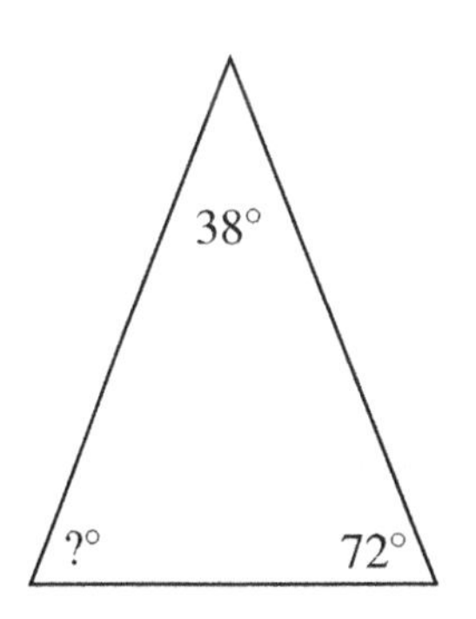


**2)** Find the area of each triangle.

2-1)

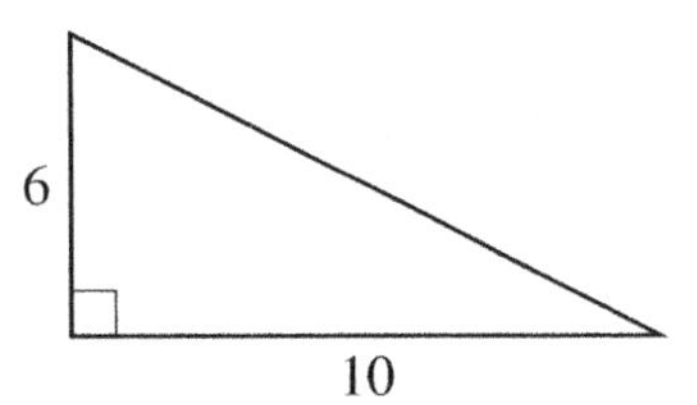


2-2)

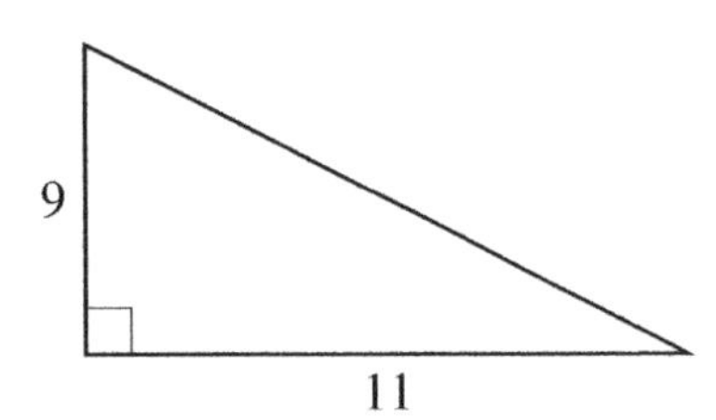

2-3)

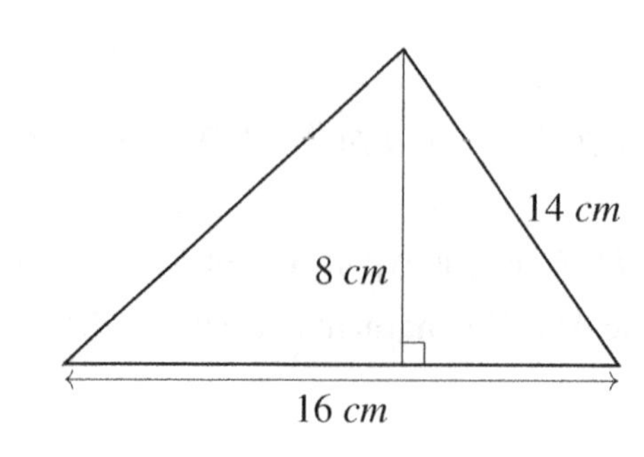


2-4)

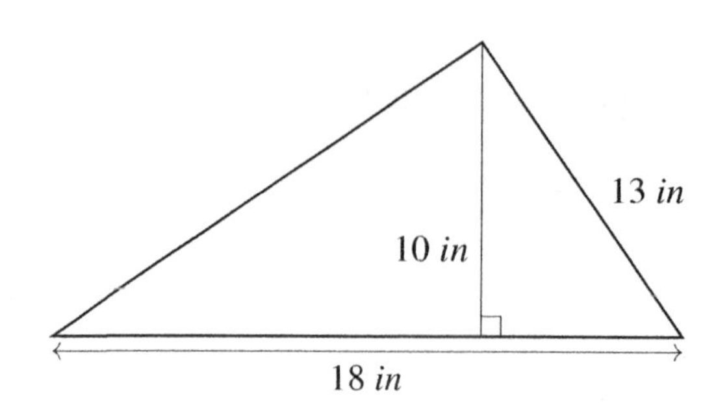


**3)** Classify each triangle by each angle and side. Base your decision on the actual lengths of the sides and the measures of the angles.

3-1)

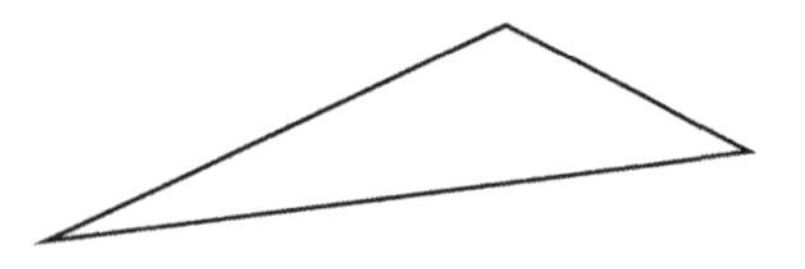

3-2)

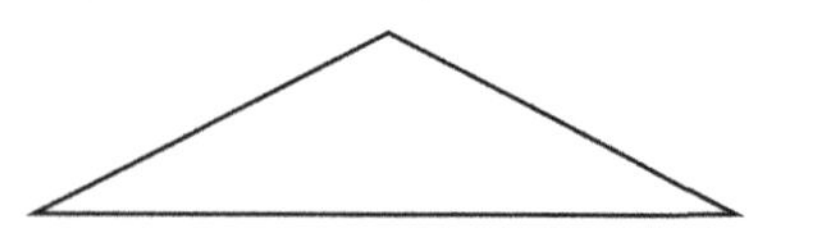

3-3)

3-4)

**4)** Classify each triangle by angles and sides.

4-1)

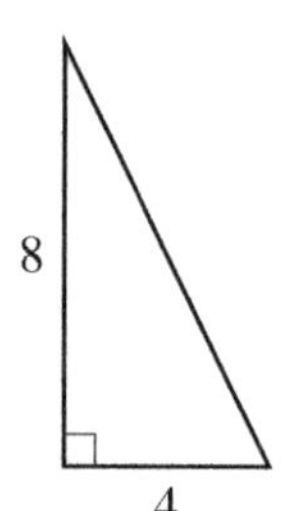


4-2)

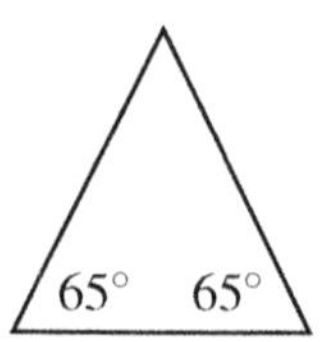

4-3)

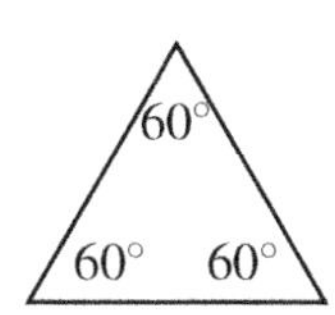


4-4)

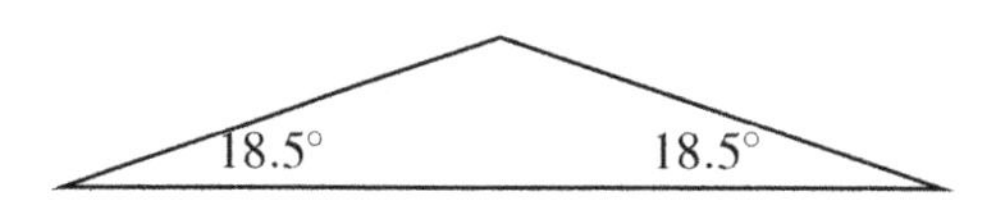


**5)** Find the measure of each angle indicated.

5-1)

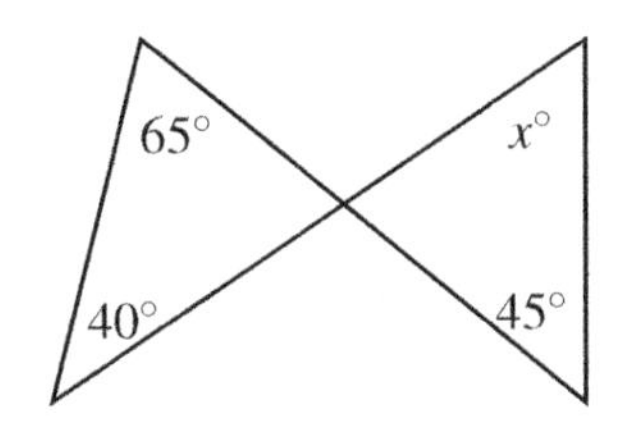


5-2)

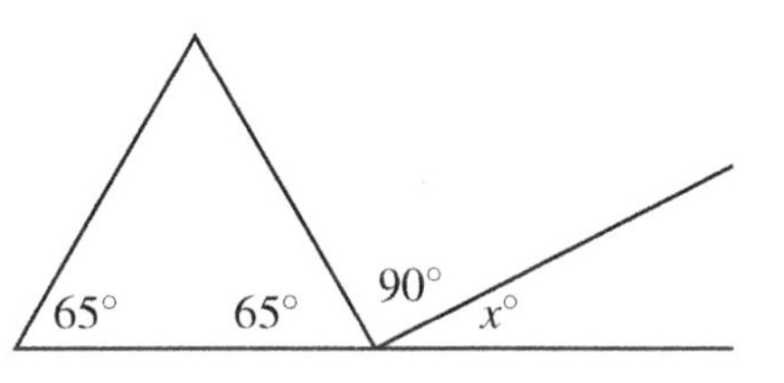


5-3)

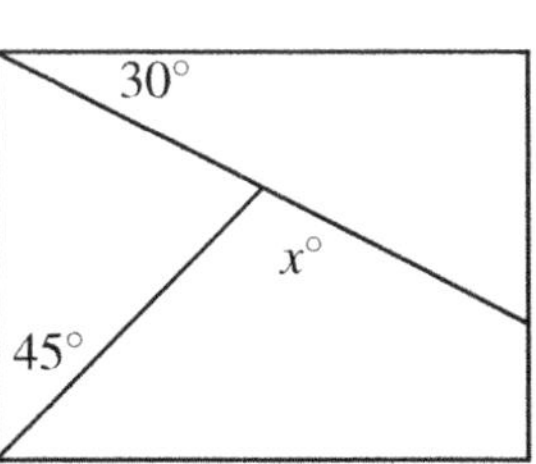

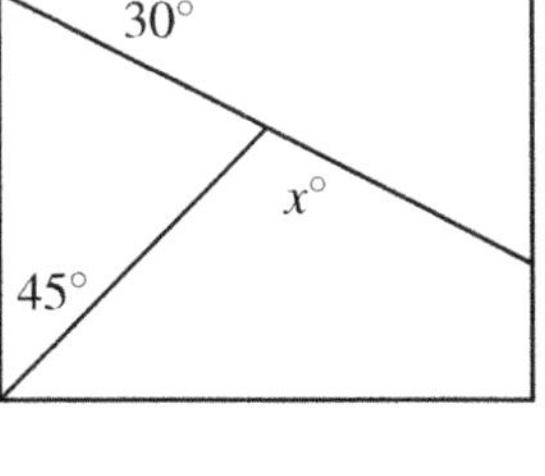


5-4)

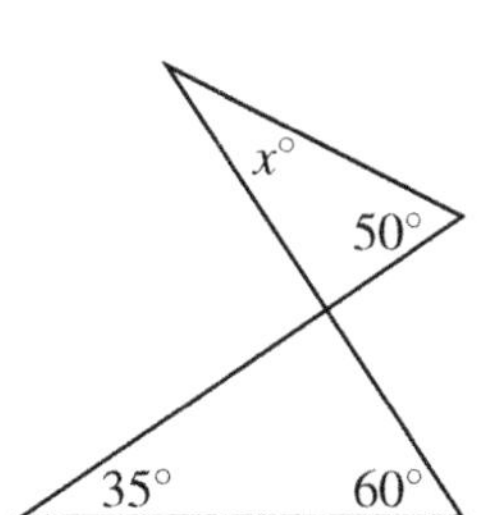


5-5)

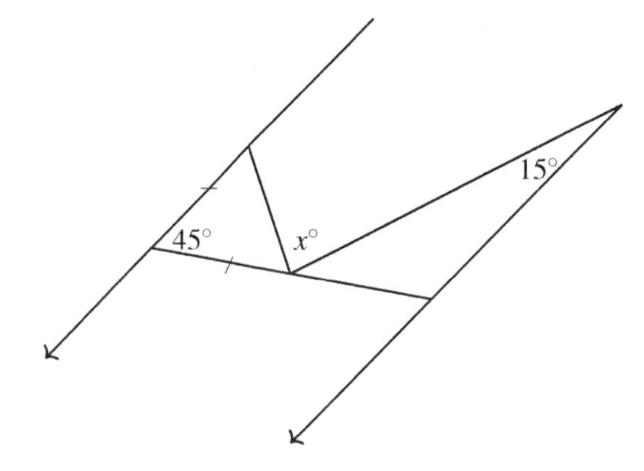


5-6)

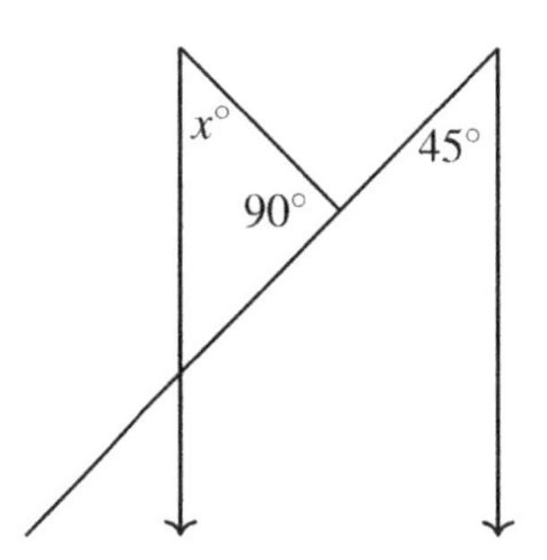


**6)** Solve for $x$.

6-1)

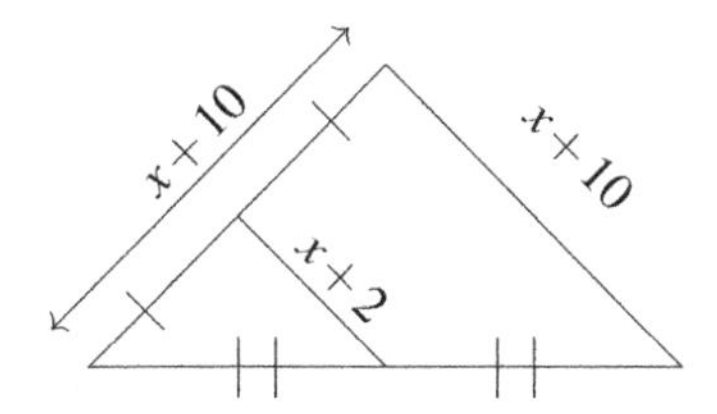

6-2)

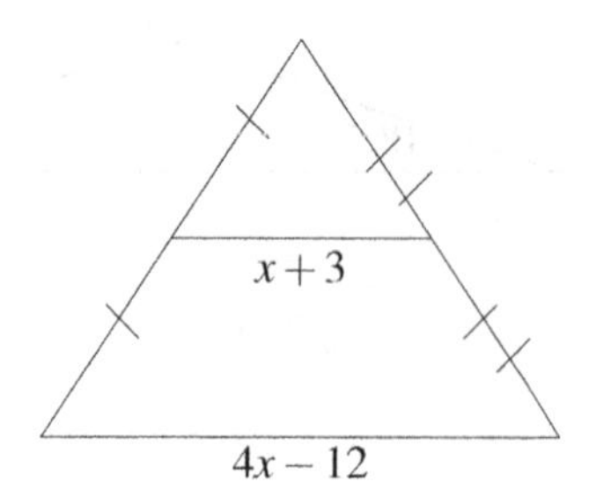


**7)** Find the missing length indicated.

7-1)

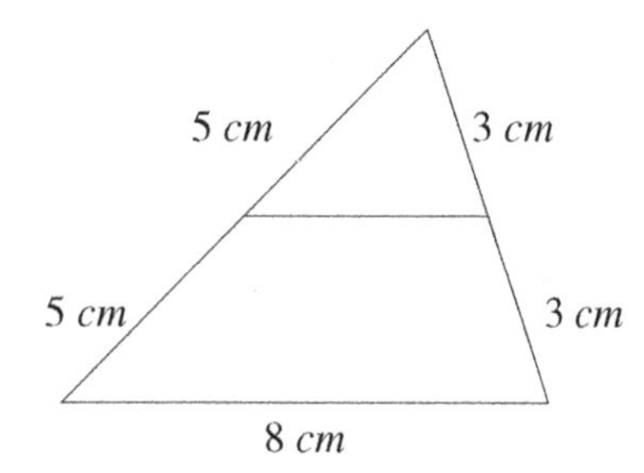


7-2)

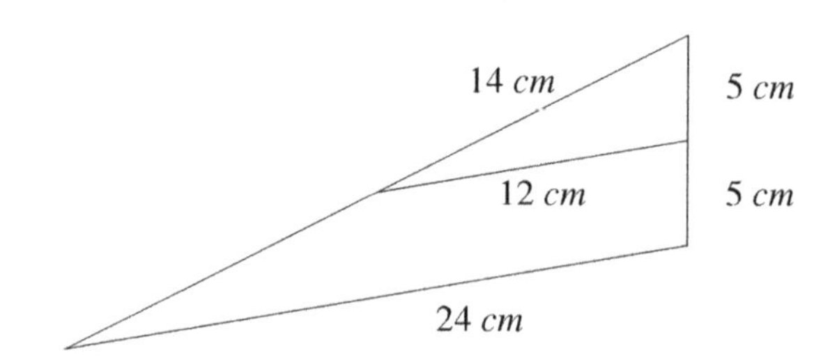


**8)** In the following triangle, $F$, $E$, and $D$ are the midpoints of the sides. Name a segment parallel to $FE$.

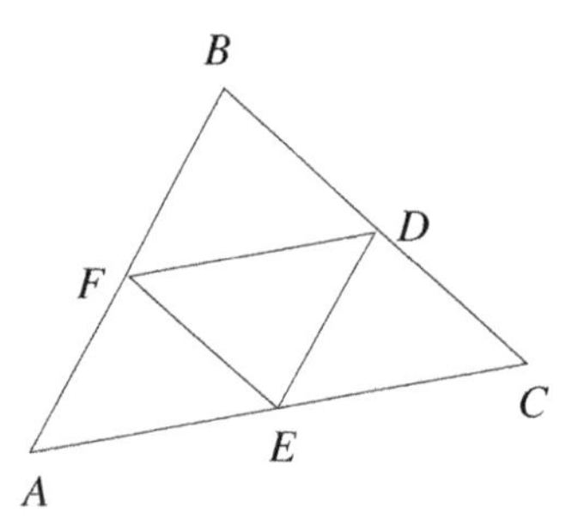


**9)** For each triangle, construct the angle bisector of angle $A$.

9-1)

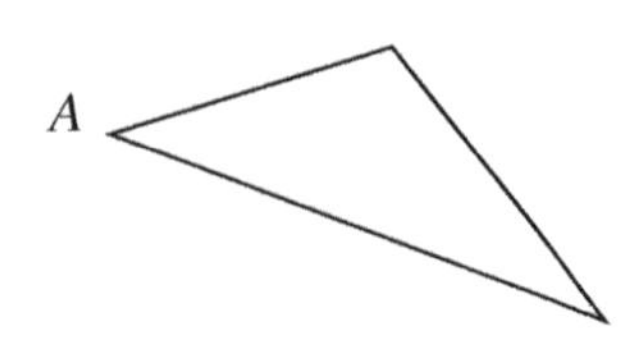


9-2)

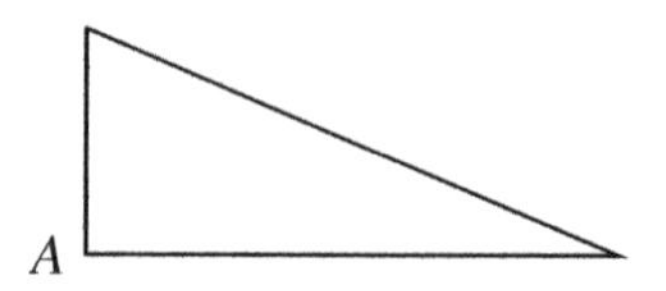


**10)** Locate the incenter of each triangle.

10-1)

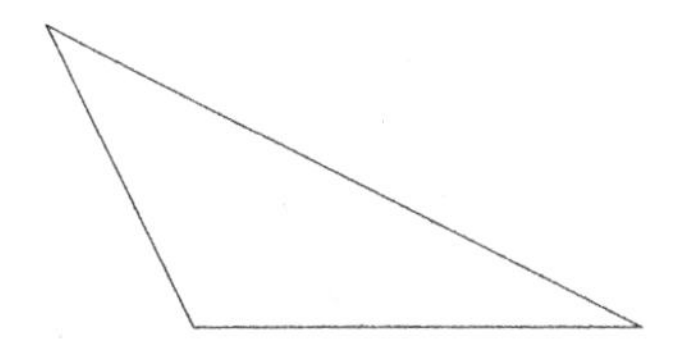

10-1)

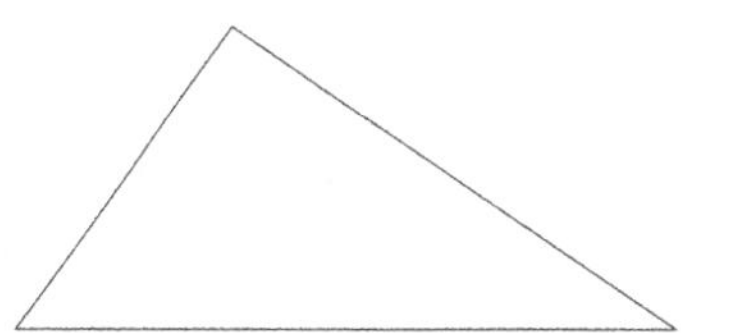

**11)** For each triangle, construct all three angle bisectors to show they are concurrent.

11-1)

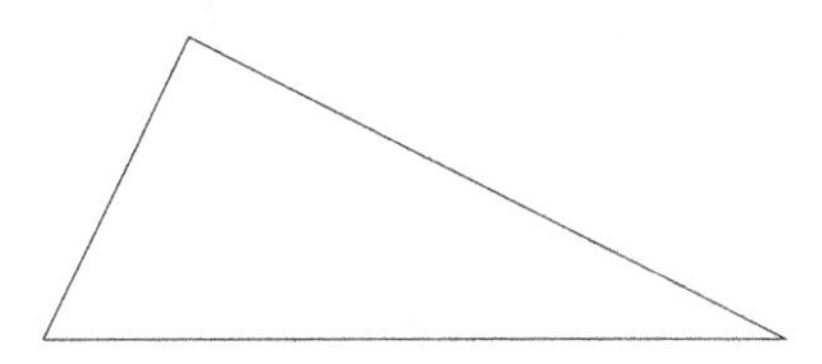

10-1)

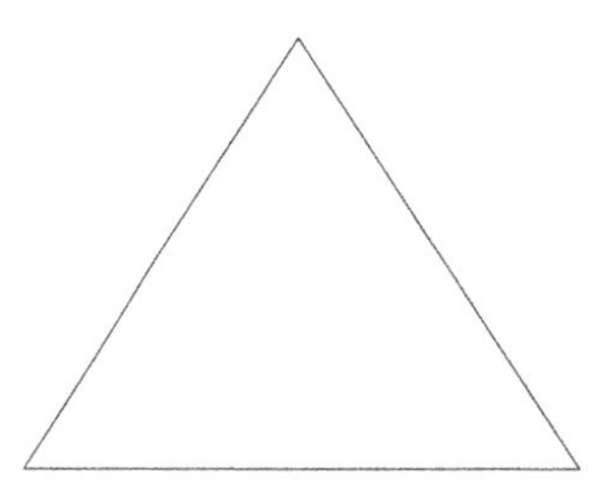

**12)** Find the value of $x$.

12-1)

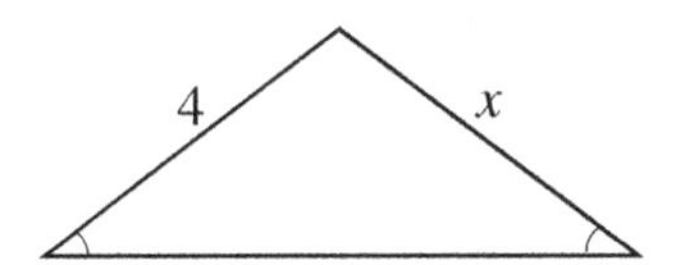


12-2)

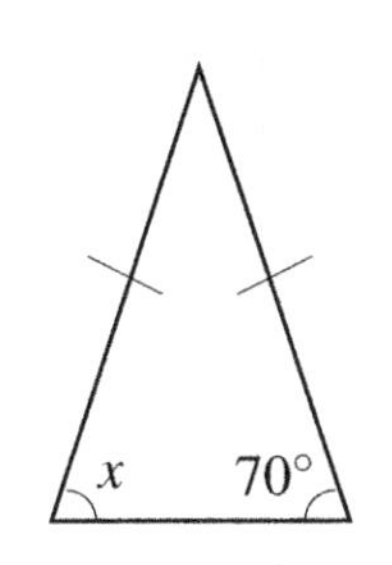


12-3)

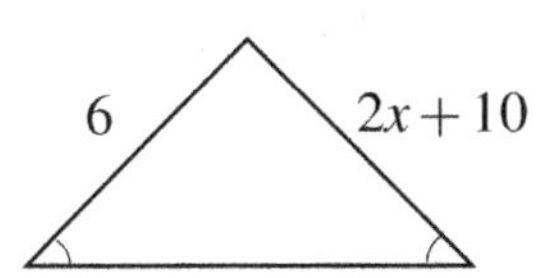


12-4)

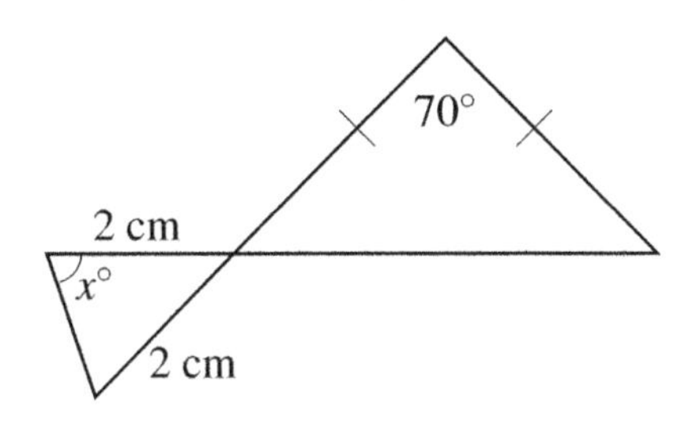


12-5)

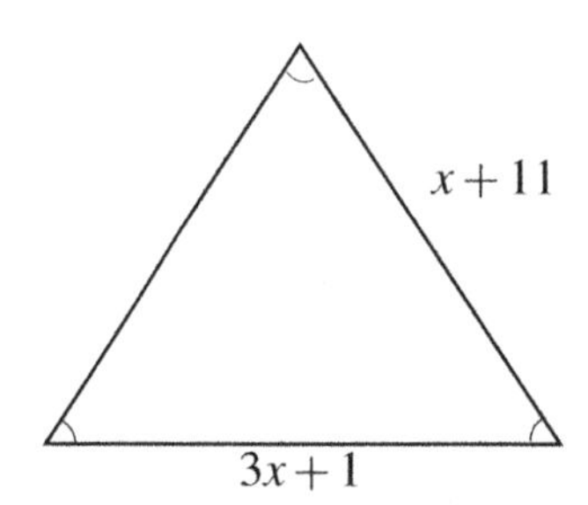


12-6)

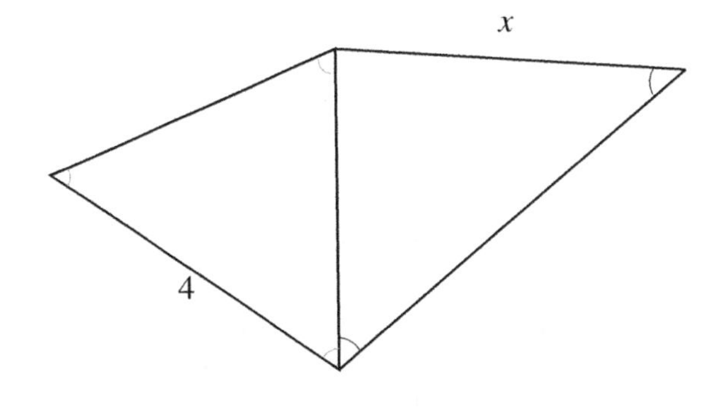

12-7)

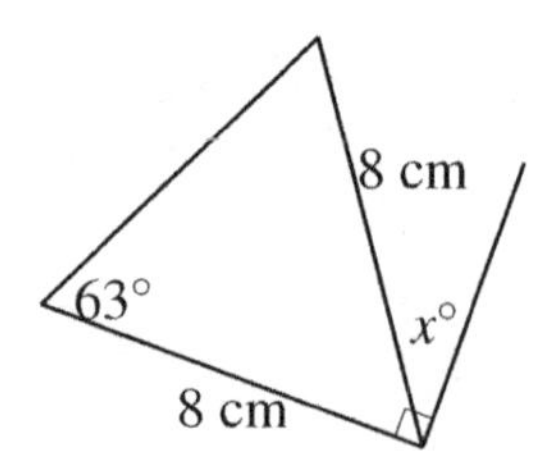


12-8)

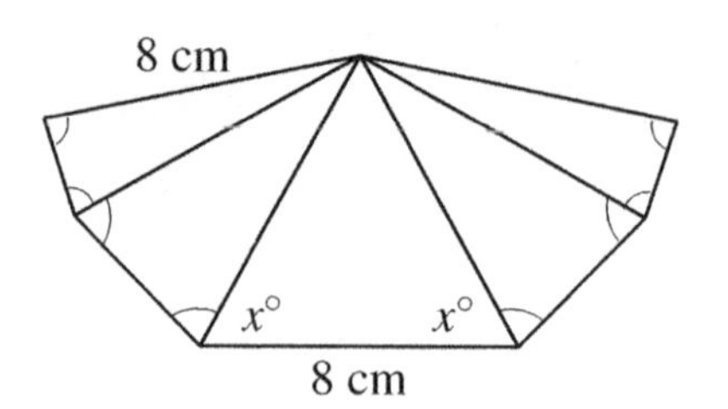


**13)** Find the missing side. Round to the nearest tenth.

13-1)

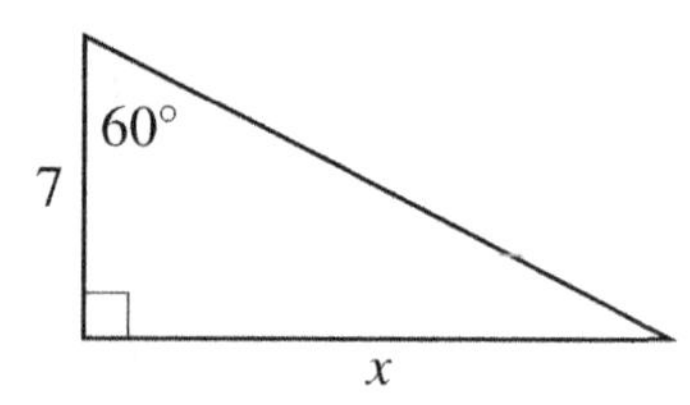


13-2)

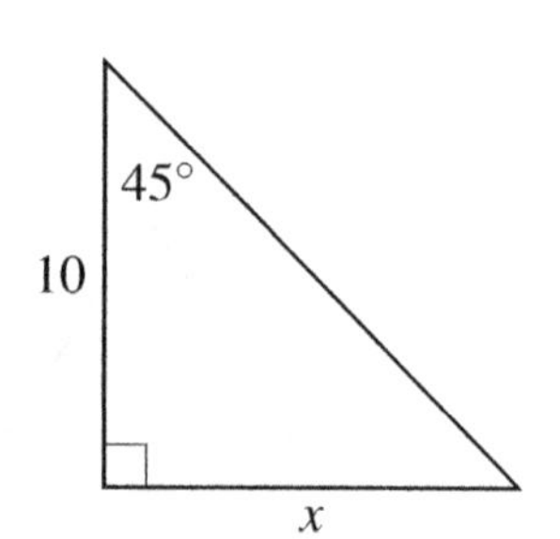


13-3)

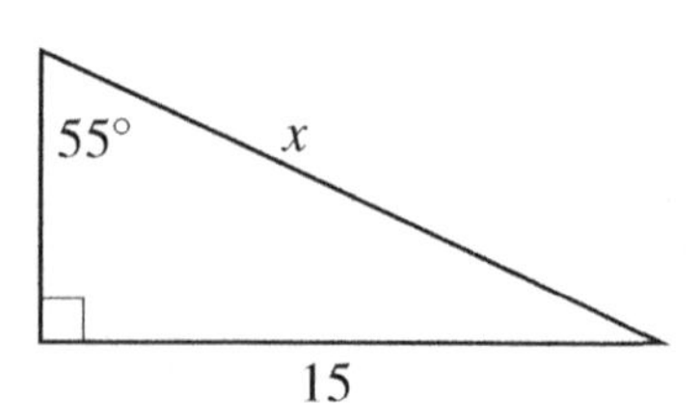


13-4)

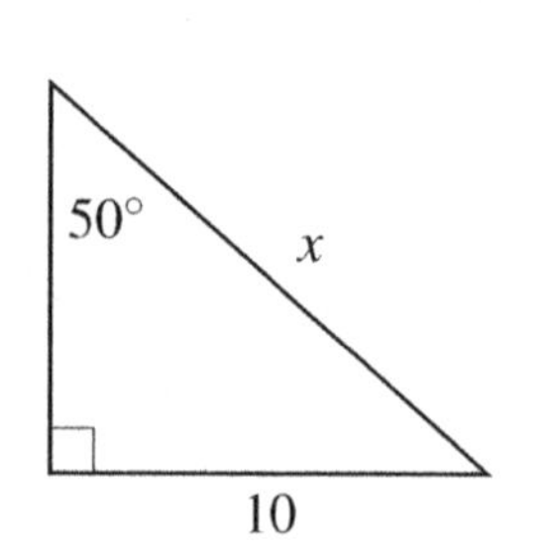


13-5)

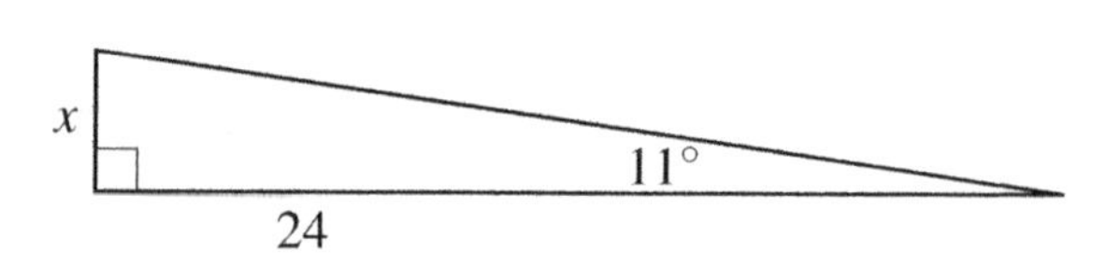


13-6)

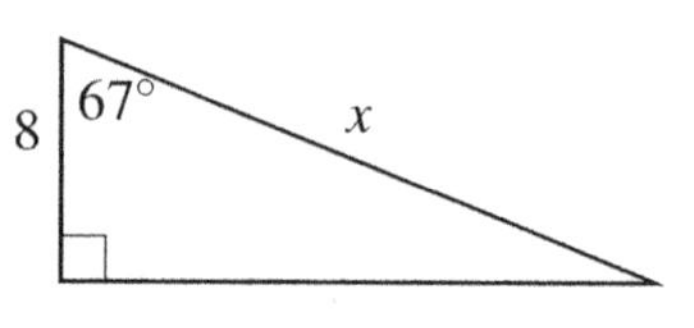


13-7)

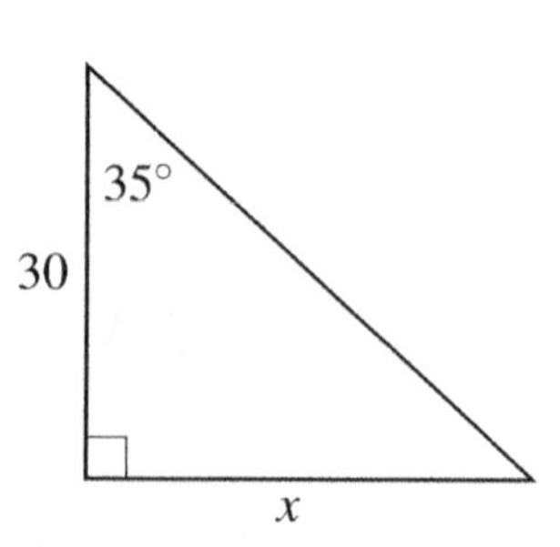


13-8)

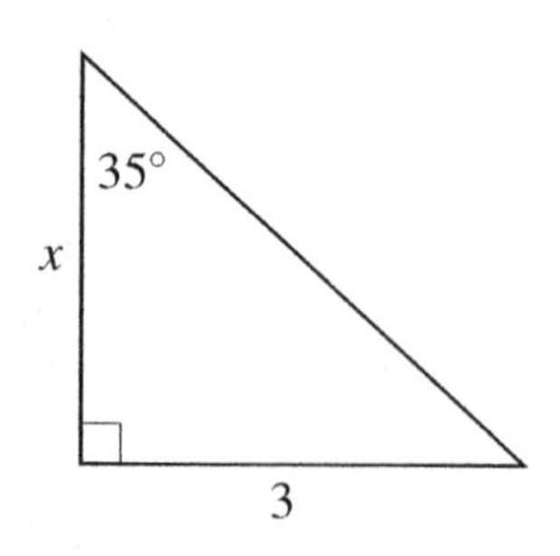

13-9)

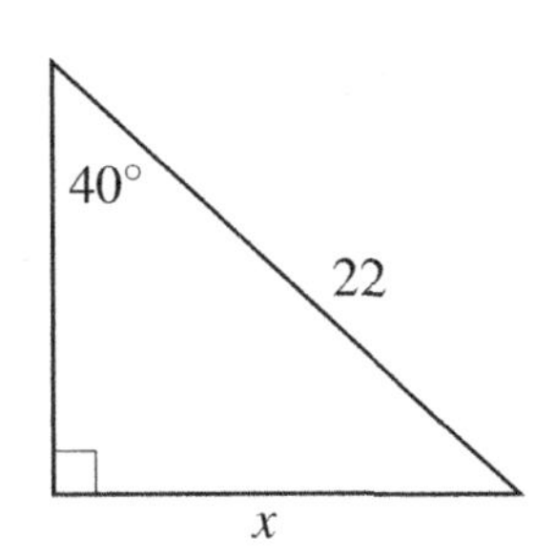


13-10)

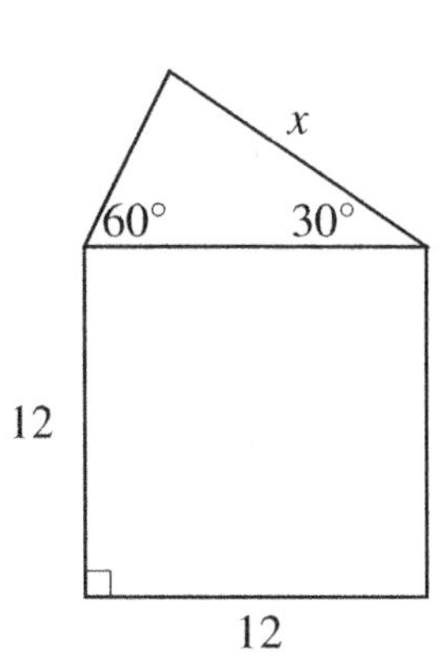


**14)** Solve

14-1) If one of the acute angles in a right triangle is 45°, and the length of the hypotenuse is $10\sqrt{2}$ units, what are the lengths of the other two sides?

14-2) A right triangle has one leg that measures 5 units, and the two acute angles measure 30° and 60°, respectively. What are the lengths of the hypotenuse and the other leg?

14-3) In a right triangle with angles of 45, 45, and 90 degrees, the length of one leg is "a" units. Express the lengths of the hypotenuse and the other leg in terms of "a".

14-4) The length of the hypotenuse of a right triangle is 12 inches. What are the lengths of the other two sides if one angle of the triangle is 60°?

**15)** Find the missing side lengths. Leave your answers as radicals in the simplest form.

15-1)

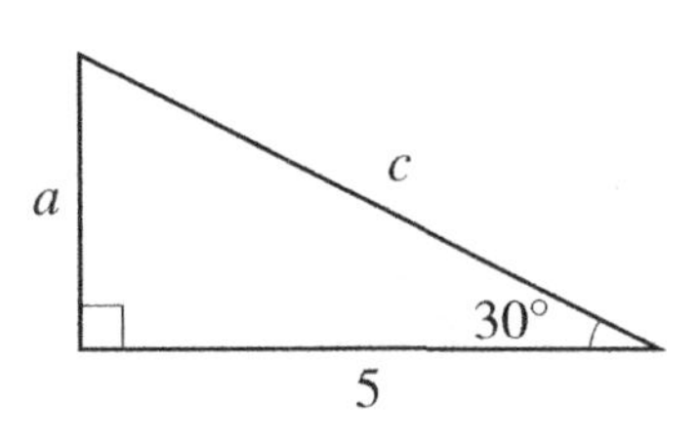


15-3)

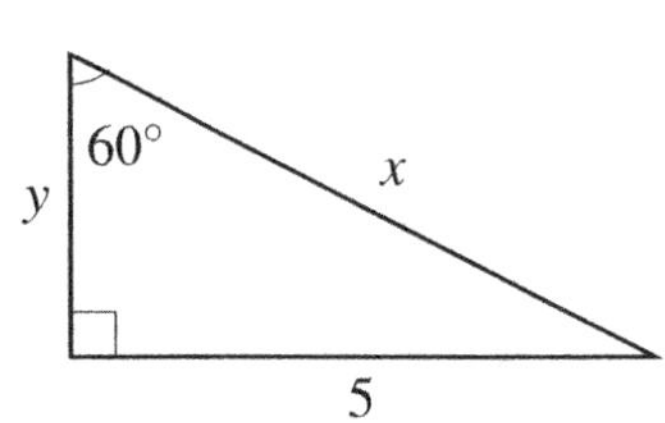


15-2)

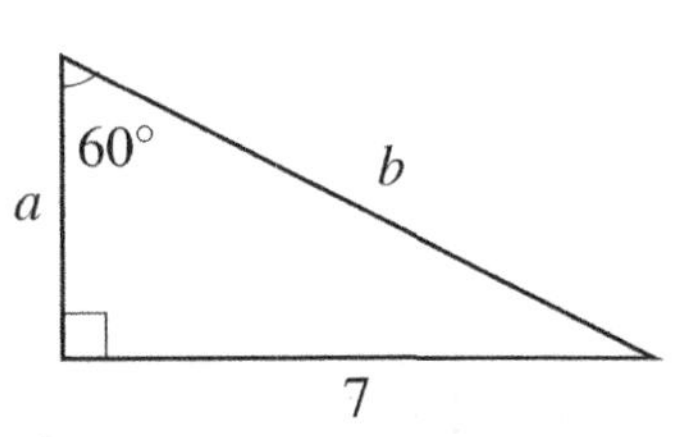


15-4)

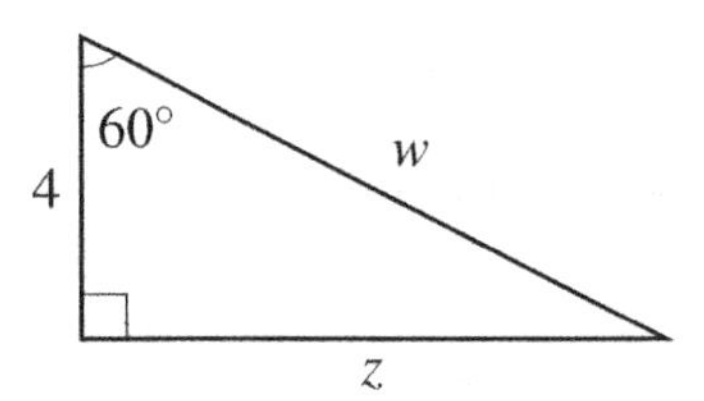

15-5)

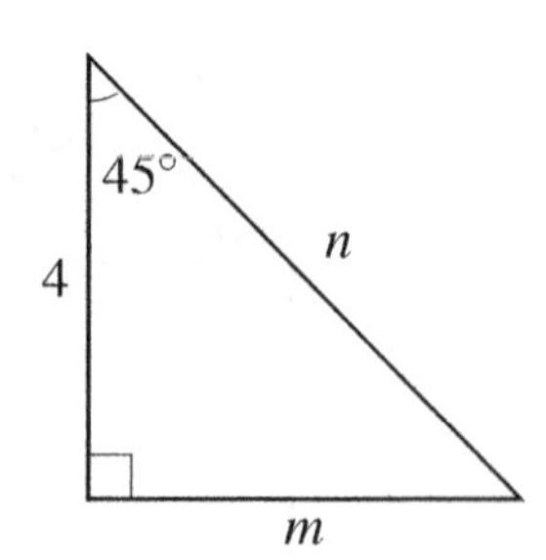


15-6)

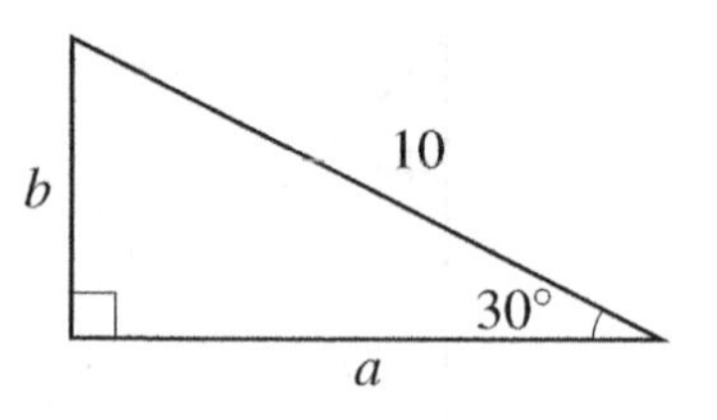


**16)** Find the missing side lengths. Leave your answers as radicals in the simplest form.

16-1)

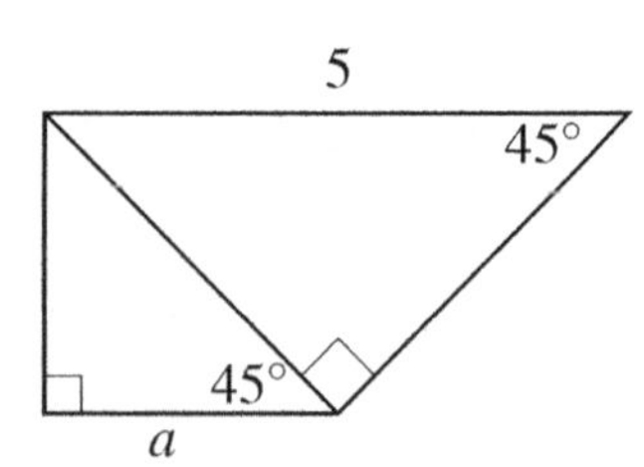


16-4)

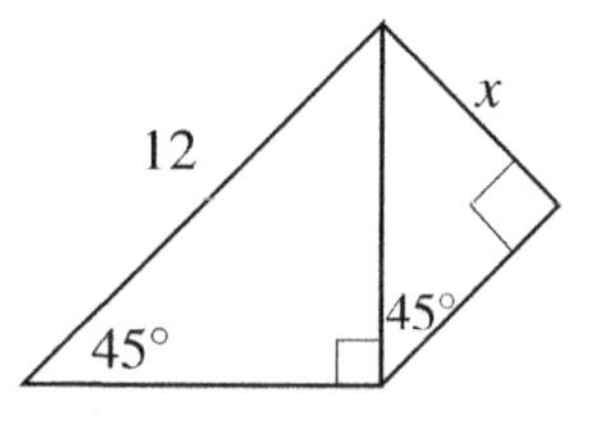


16-2)

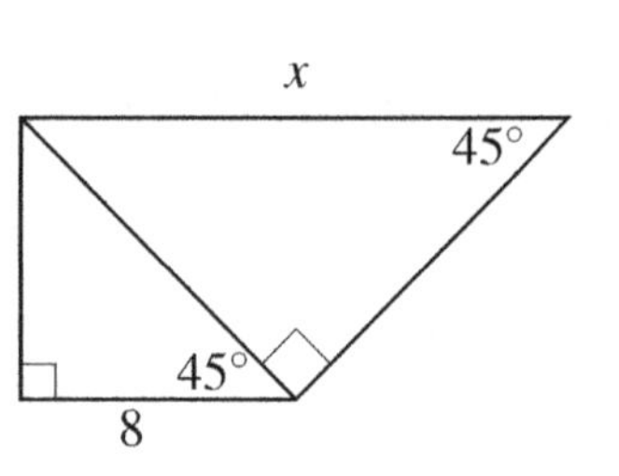


16-5)

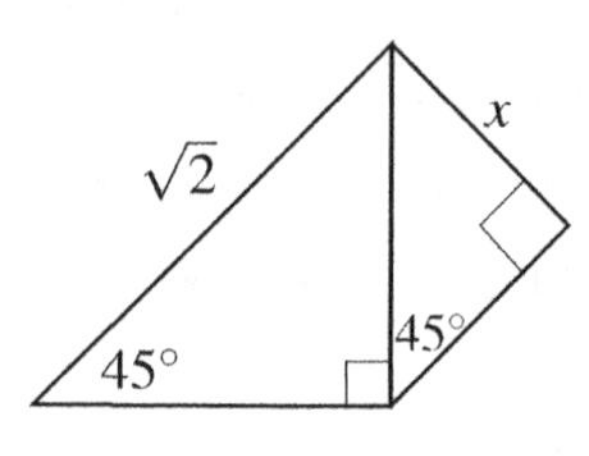


16-3)

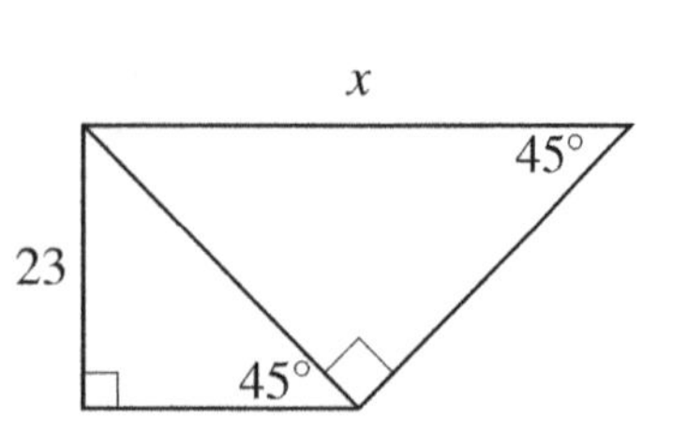


16-6)

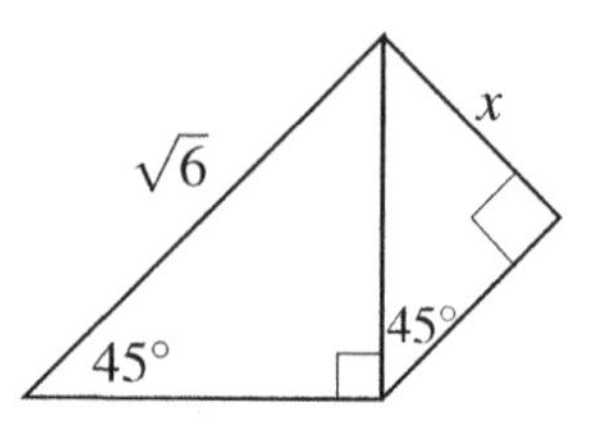


**17)** Find the missing side.

17-1)

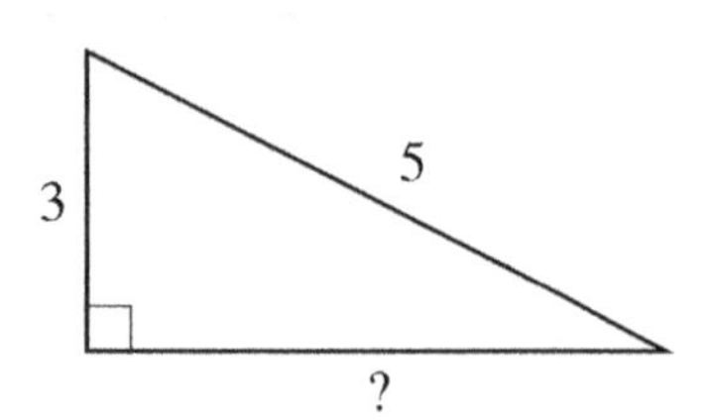


17-2)

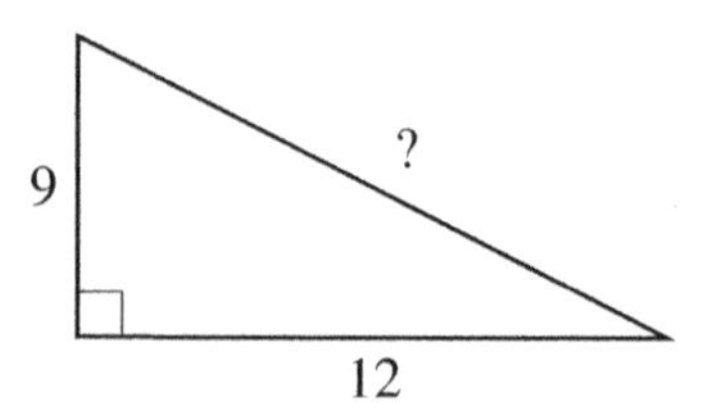

17-3)

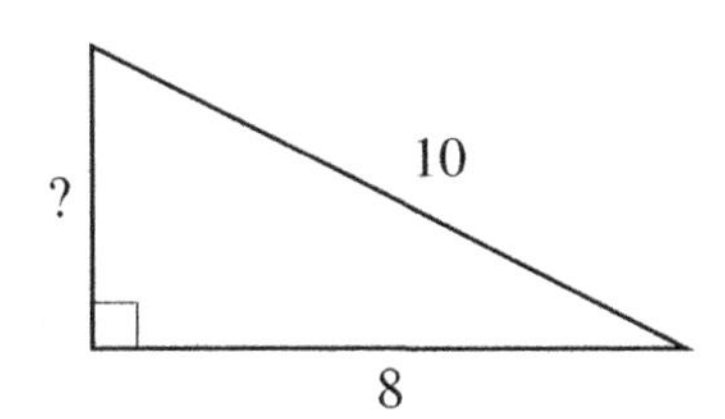


17-4)

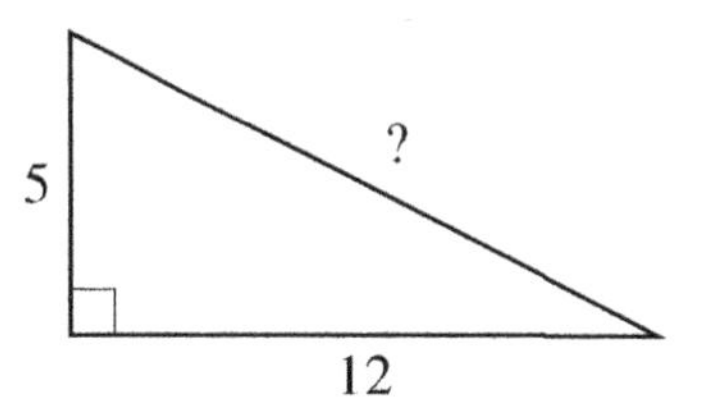


**18)** Find.

18-1) Determine the type of triangle (whether the triangle is right-angled or not) formed by the sides with lengths 9 cm, 14 cm, and 15 cm, using the Pythagorean converse theorem.

18-2) Given a triangle with sides of length 7 cm, 24 cm, and 27 cm, determine whether the triangle is right-angled or not using the Pythagorean converse theorem.

18-3) A triangle has sides of length 8 cm, 15 cm, and 17 cm. Determine whether this triangle is right-angled or not using the Pythagorean converse theorem.

18-4) A triangle has sides of length 6 cm, 8 cm, and 10 cm. Determine whether this triangle is right-angled or not using the Pythagorean converse theorem.

**19)** Find the value of $x$. Round to the nearest tenth.

19-1)

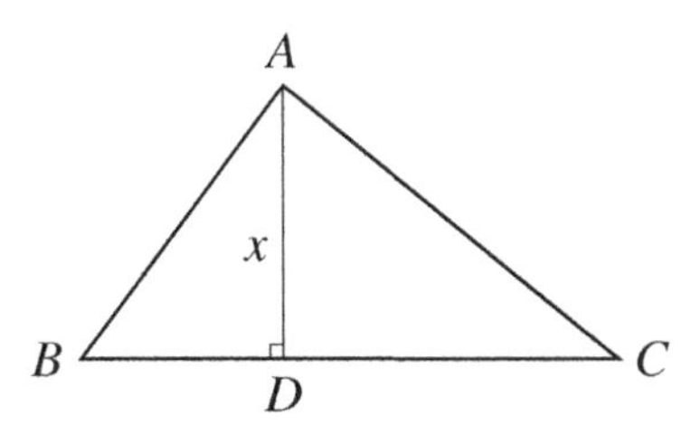


19-2)

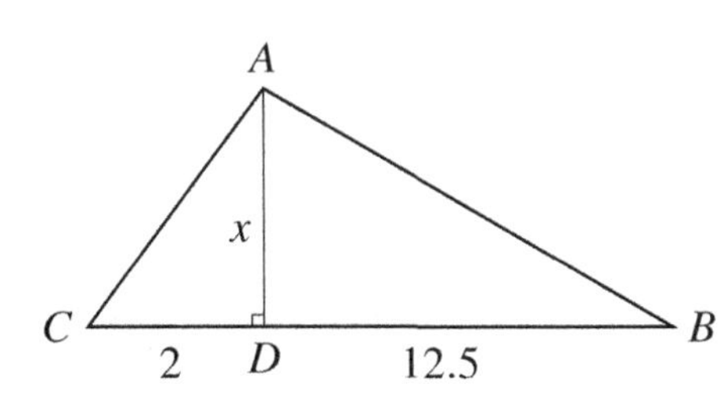


19-3)

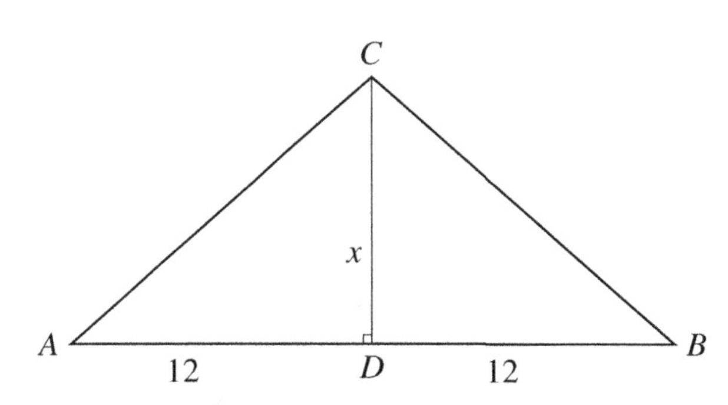


19-4)

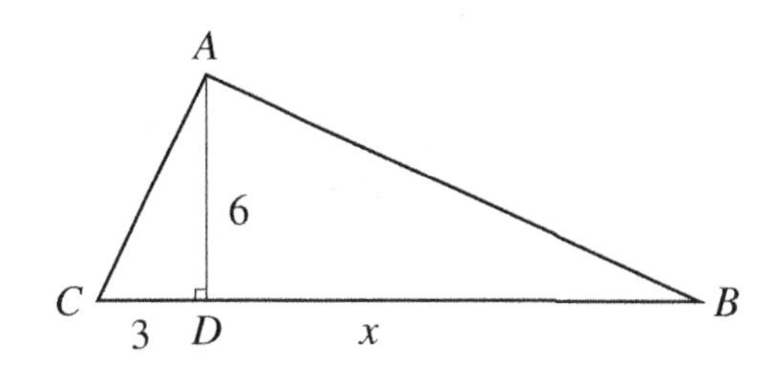

19-5)

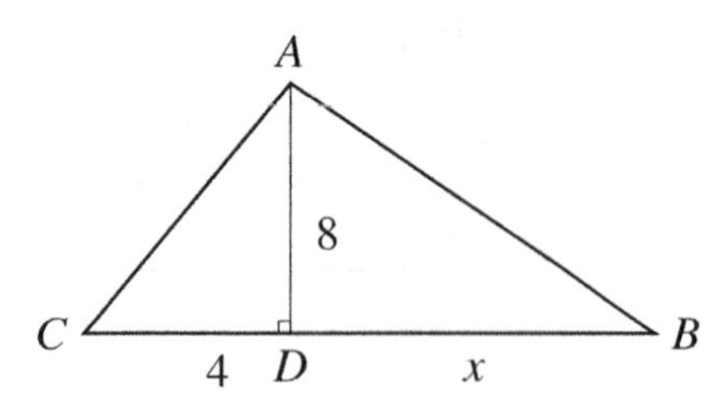


19-6)

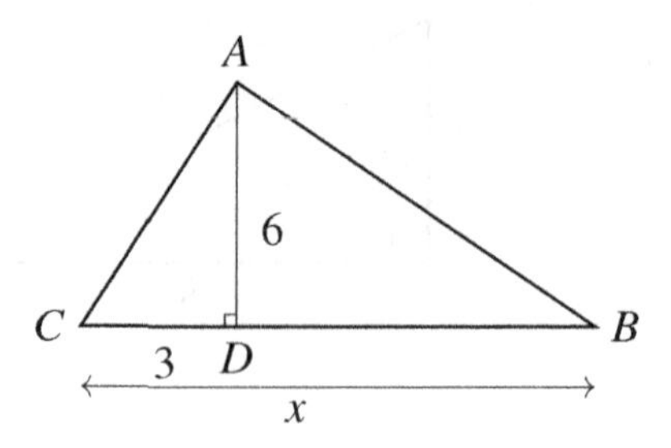


**20)** Solve for $x$.

20-1)

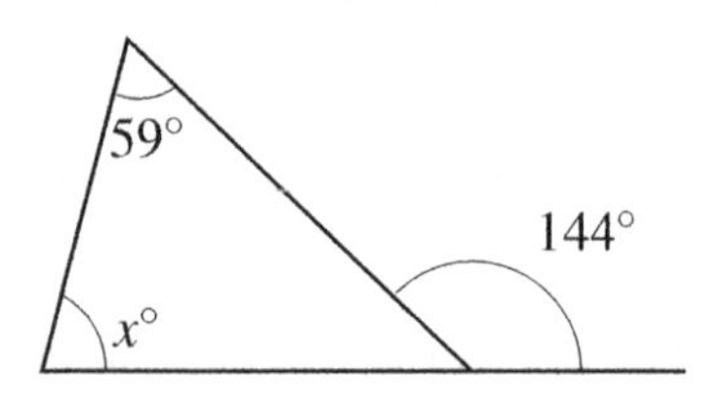


20-2)

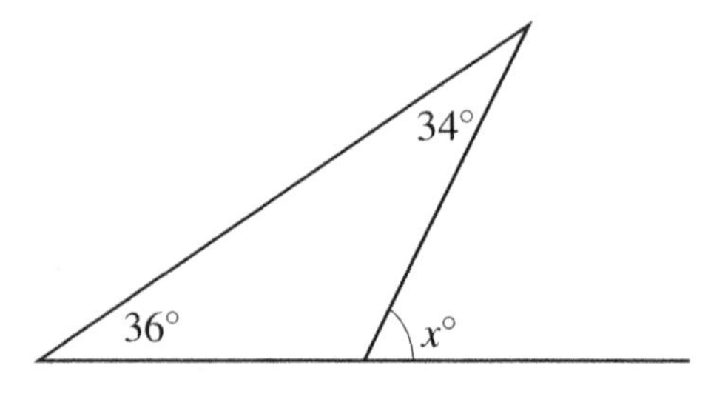


20-3)

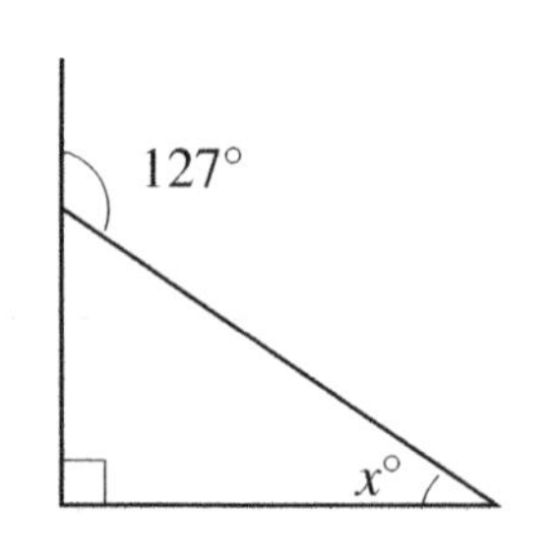


20-4)

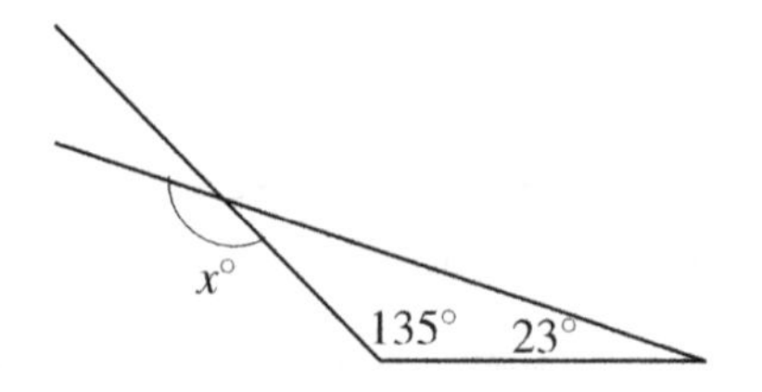


20-5)

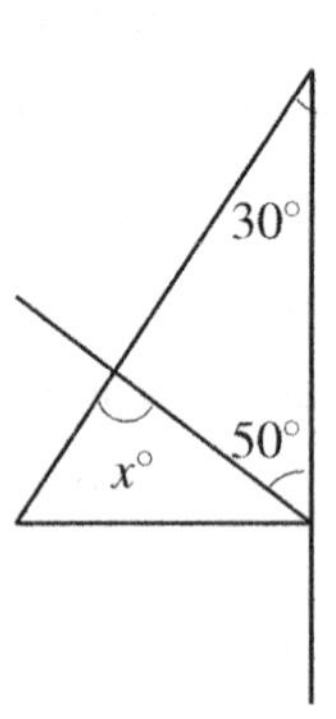


20-6)

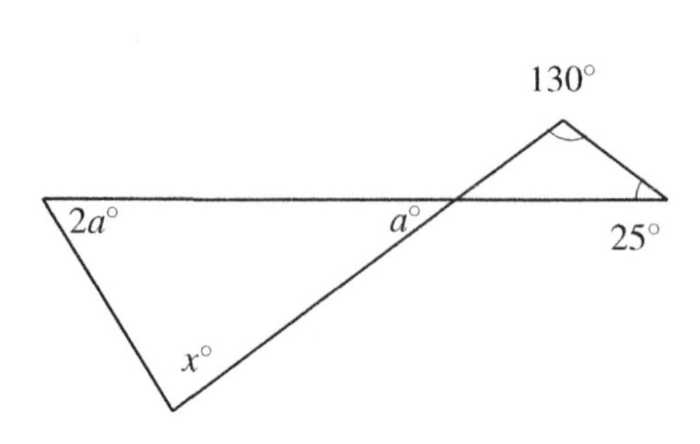


20-7)

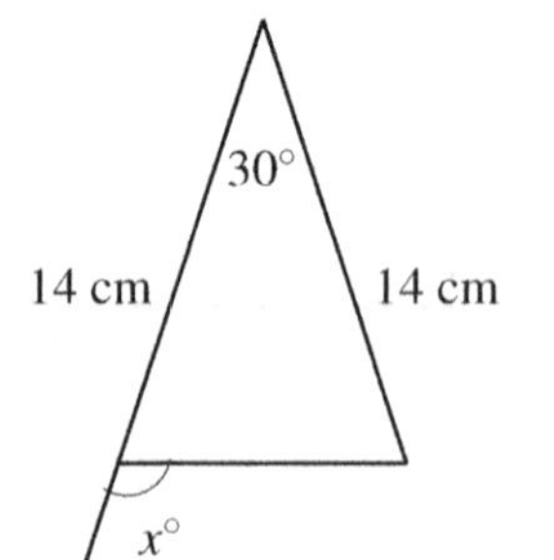

20-8)

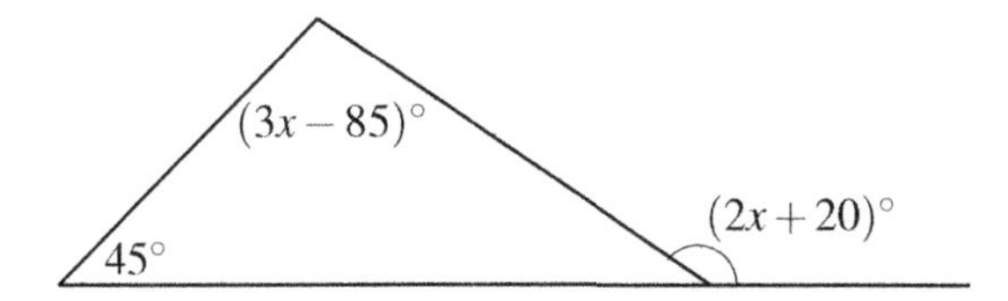


**21)** Each figure shows a triangle with one or more of its medians.

21-1) Find $DM$ if $CM = 6.6$

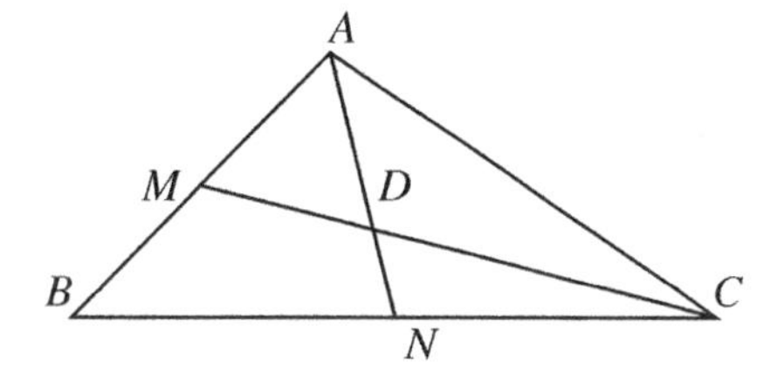


21-2) Find $AD$ if $AM = 4.5$

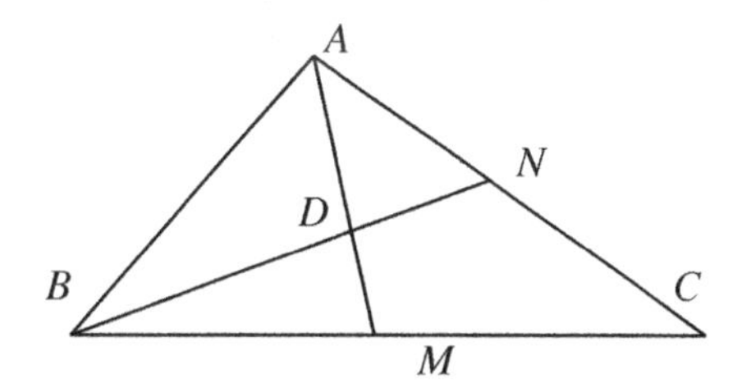


21-3) Find $AD$ if $DM = 5$

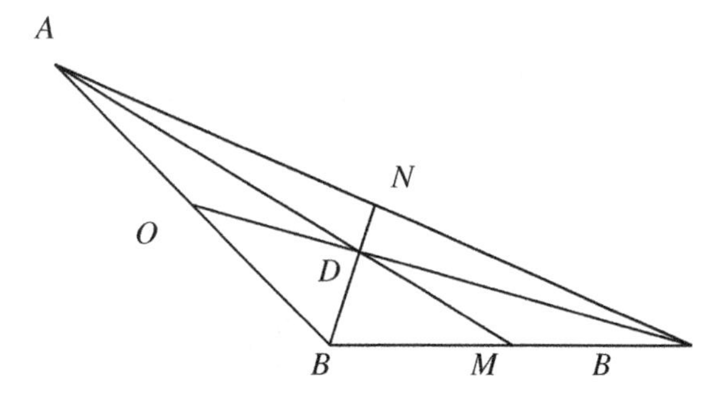


21-4) Find $x$ if $BN = 6x - 9$ and $DN = x$

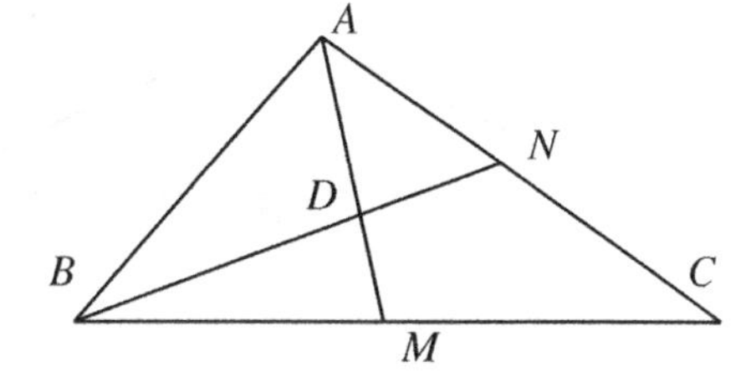

21-5) Find $CD$ if $DN = 6x$ and $CN = 15x + 6$

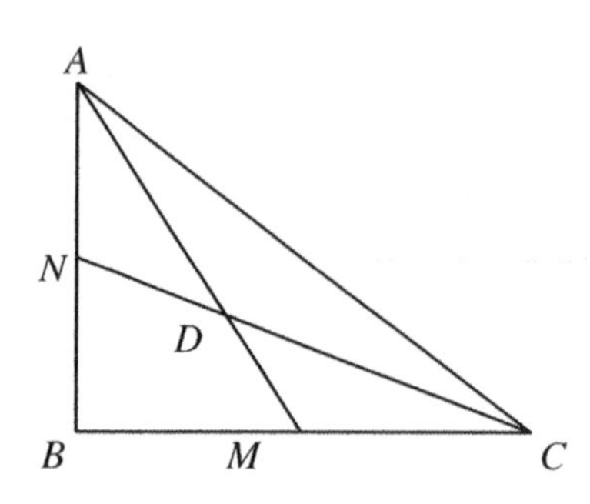


21-6) Find $AD + ND$ if $BD = 4$ and $DM = 2$

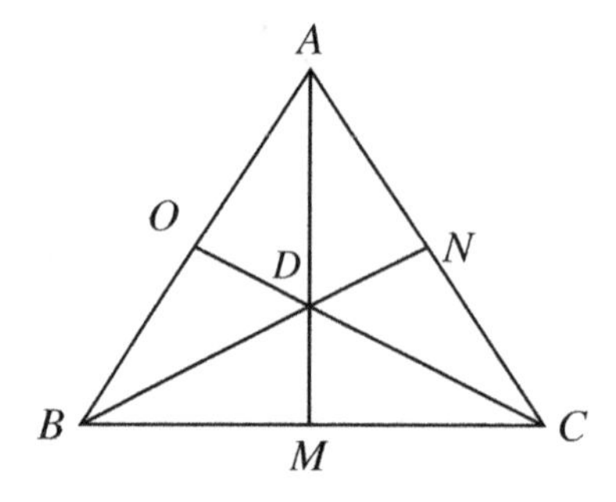


21-7) Find $x$ if $BN = 6x - 3$ and $DN = x + 7$

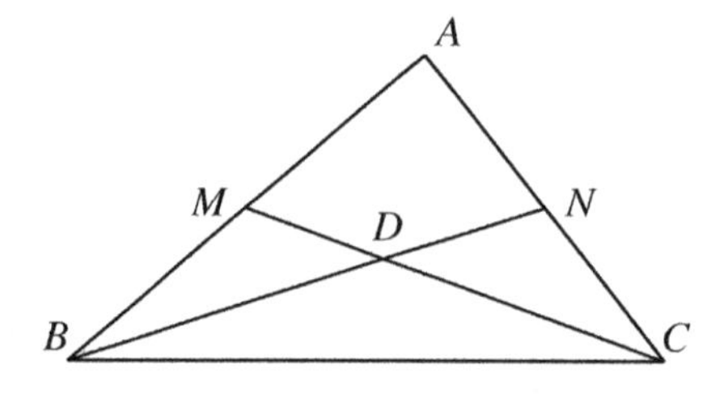


21-8) Find $DM$ if $DN = 6$ and $AN = 8$

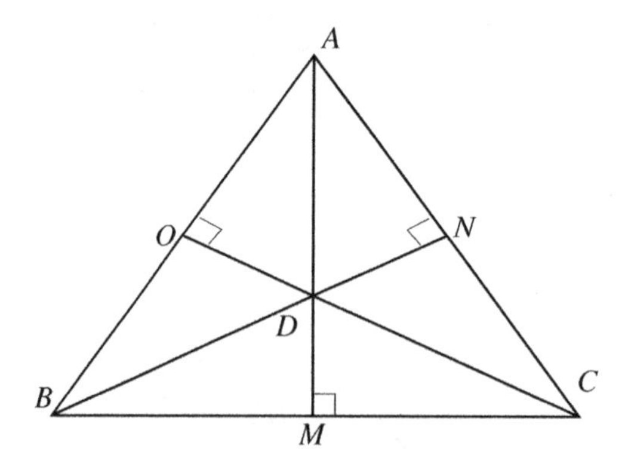

**22)** Find the coordinates of the centroid of each triangle given the three vertices.

22-1) $Y(1,-4)$, $X(2,10)$, $W(2,-1)$

22-2) $T(2,-1)$, $U(3,3)$, $V(1,4)$

22-3) $S(1,2)$, $R(-1,6)$, $Q(2,3)$

22-4) $E(2,-9)$, $D(5,0)$, $C(1,-1)$

**23)** Find the coordinates of the centroid of each triangle.

23-1)

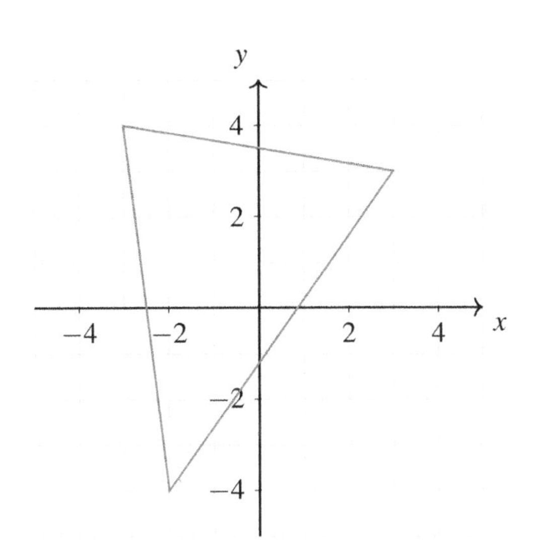


23-2)

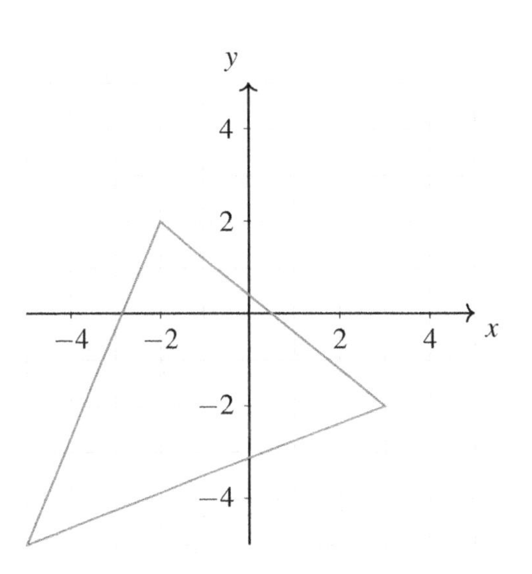


**24)** Two sides of a triangle have the following measures. Find the range of possible measures for the third side.

24-1) 4, 10

24-2) 16,4

24-3) 16, 24

24-4) 1, 3

24-5) 13, 20

24-6) 8, 11

**25)** State if the three numbers can be the measures of the sides of a triangle.

25-1) 3, 5, 13

25-2) 8, 11, 9

25-3) 3, 9, 8

25-4) 4, 12, 8

25-5) 3, 3, 3

25-6) 3, 9, 8

25-7) 1, 5, 5

25-8) 4, 7, 2

25-9) 15, 21, 7

**26)** State what additional information is required in order to know that the triangles are congruent for the reason given.

26-1) SSS

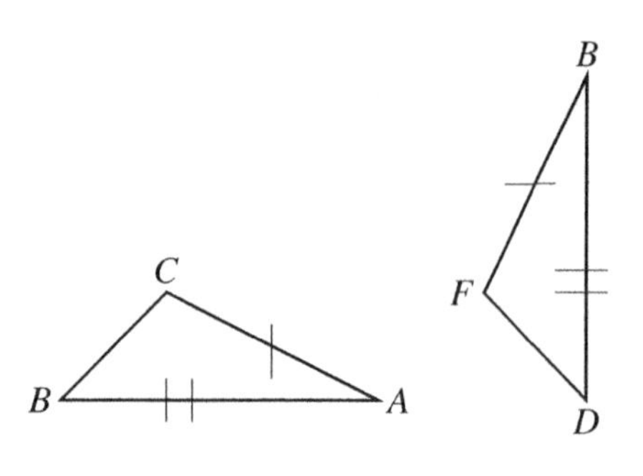


26-2) AAS

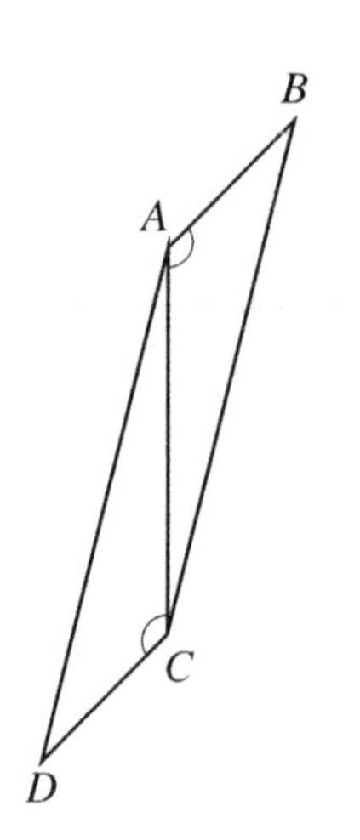


26-3) AAS

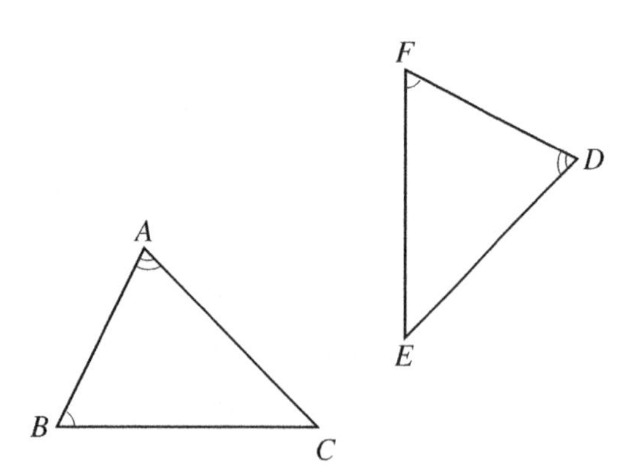


26-4) SAS

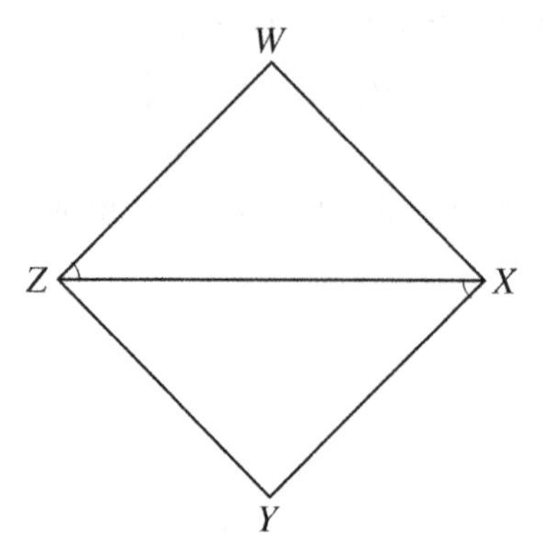


26-5) ASA

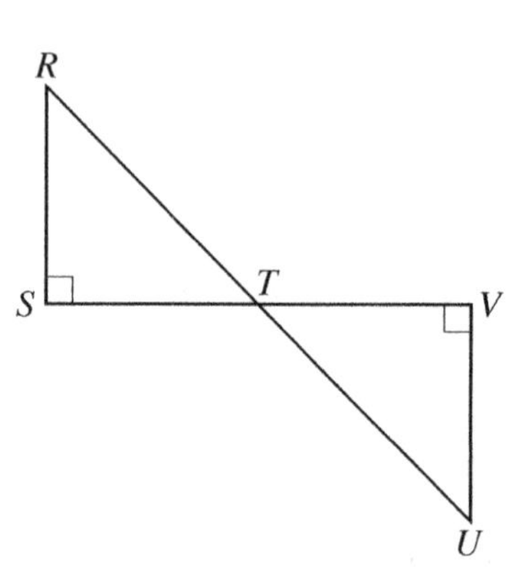


26-6) SAS

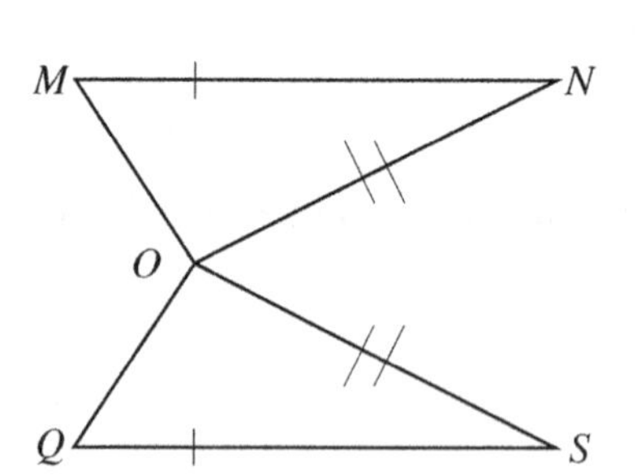

26-7) HL

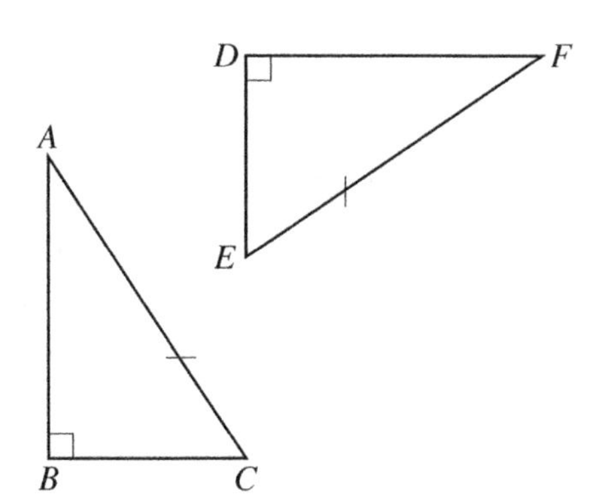


26-8) HL

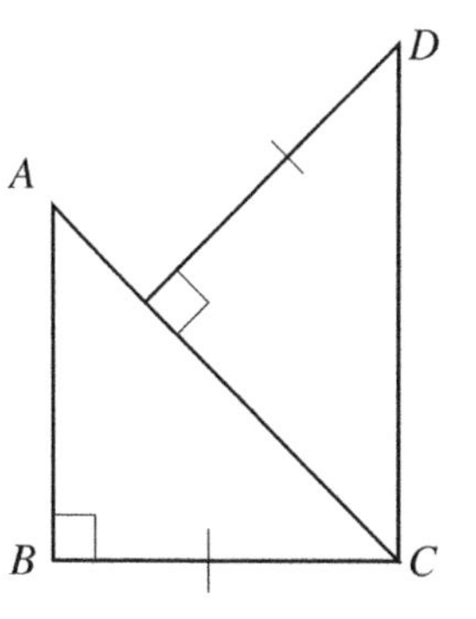

## 6.19 Answers

**1)**

1-1) 50

1-2) 76

1-3) 84

1-4) 70

**2)**

2-1) 30

2-2) 49.5

2-3) 64

2-4) 90

**3)**

3-1) Obtuse scalene

3-2) Obtuse isosceles

3-3) Right scalene

3-4) Acute isosceles

**4)**

4-1) Right scalene

4-2) Acute isosceles

4-3) Equilateral

4-4) Obtuse isosceles

**5)**

5-1) $60^\circ$

5-2) $25^\circ$

5-3) $105^\circ$

5-4) $55^\circ$

5-5) $82.5^\circ$

5-6) $45^\circ$

**6)**

6-1) 6

6-2) 9

**7)**

7-1) 4 *cm*

7-2) 14 *cm*

**8)** $BC$

**9)**

9-1)

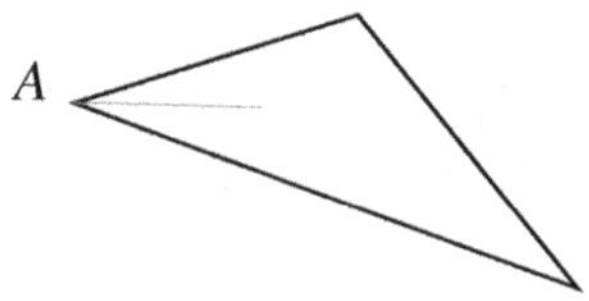

9-2)

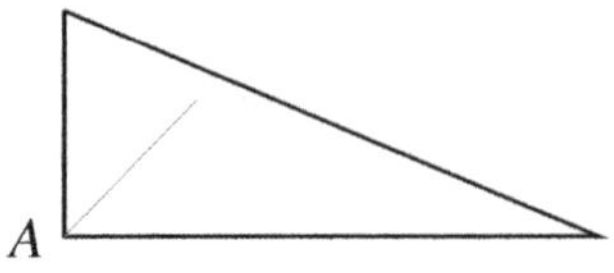

**10)**

10-1)

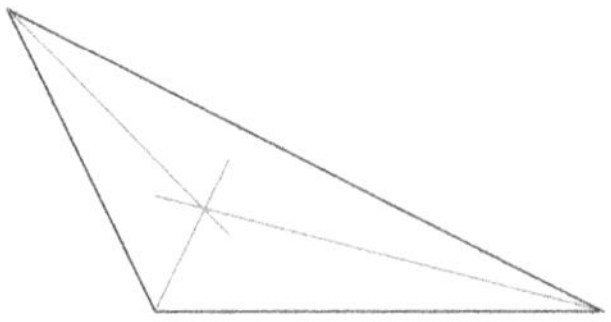

10-2)

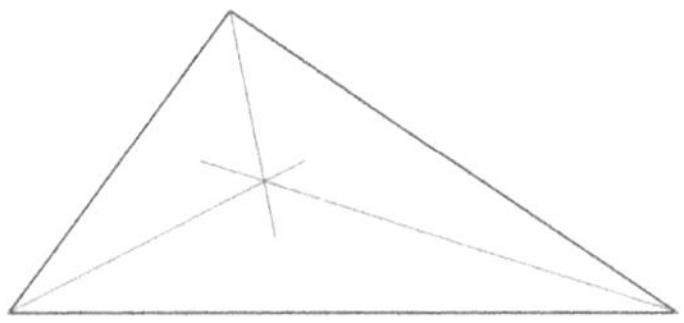

**11)**

11-1)

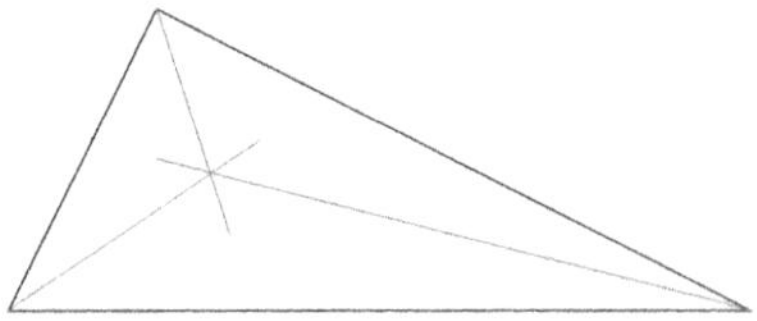

11-2)

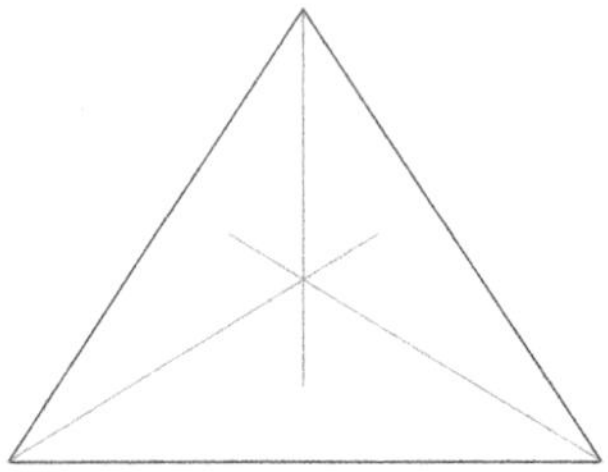

**12)**

12-1) 4

12-2) $70^\circ$

12-3) $-2$

12-4) $62.5^\circ$

12-5) 5

12-6) 4

12-7) $36^\circ$

12-8) $60^\circ$

**13)**

13-1) 12.1

13-2) 10

13-3) 18.3

13-4) 13

13-5) 4.7

13-6) 20.5

13-7) 21

13-8) 4.3

13-9) 14.1

13-10) 10.4

**14)**

14-1) 10

14-2) 10 and $5\sqrt{3}$

14-3) a$\sqrt{2}$

14-4) 6 and $6\sqrt{3}$

**15)**

15-1) $c = 10\frac{\sqrt{3}}{3}$ and $a = 5\frac{\sqrt{3}}{3}$

15-2) $a = \frac{7}{\sqrt{3}}$ and $b = \frac{14}{\sqrt{3}}$

15-3) $x = \frac{10}{\sqrt{3}}$ and $y = \frac{5}{\sqrt{3}}$

15-4) $z = 4\sqrt{3}$ and $w = 8$

15-5) $m = 4$ and $n = 4\sqrt{2}$

15-6) $a = 5\sqrt{3}$ and $b = 5$

**16)**

16-1) 2.5

16-2) 16

16-3) 46

16-4) 6

16-5) $\frac{\sqrt{2}}{2}$

16-6) $\frac{\sqrt{6}}{2}$

**17)**

17-1) 4

17-2) 15

17-3) 6

17-4) 13

**18)**

18-1) The triangle is not a right triangle

18-2) The triangle is not a right triangle

18-3) The triangle is a right triangle

18-4) The triangle is a right triangle

**19)**

19-1) $x = \sqrt{BD \times CD}$

19-2) 5

19-3) 12

19-4) 12

19-5) 16

19-6) 15

**20)**

20-1) 85

20-2) 70

20-3) 37

20-4) 158

20-5) 80

20-6) 105

20-7) 105

20-8) 60

**21)**

21-1) 2.2

21-2) 3

21-3) 10

21-4) 3

21-5) 24

21-6) 6

21-7) 8

21-8) 5

**22)**

22-1) $(\frac{5}{3}, \frac{5}{3})$

22-2) $(2,2)$

22-3) $(\frac{2}{3},\frac{11}{3})$

22-4) $(\frac{8}{3},-\frac{10}{3})$

**23)**

23-1) $(-\frac{2}{3},1)$

23-2) $(-\frac{4}{3},-\frac{5}{3})$

**24)**

24-1) $6 < x < 14$

24-2) $12 < x < 20$

24-3) $8 < x < 40$

24-4) $2 < x < 4$

24-5) $7 < x < 33$

24-6) $3 < x < 19$

**25)**

25-1) No

25-2) Yes

25-3) Yes

25-4) No

25-5) Yes

25-6) Yes

25-7) Yes

25-8) No

25-9) Yes

**26)**

26-1) $BC = FD$

26-2) $\angle B = \angle D$

26-3) $DE = AC$ or $FE = BC$

26-4) $WZ = XY$

26-5) $ST = TV$

26-6) $\angle N = \angle S$

26-7) $AB = FD$ or $BC = DF$

26-8) $DC = AC$

# 7. Dilation and Similarity

## 7.1 Dilations

Dilation is a transformation in geometry that changes the size of a figure without altering its shape. A figure can be enlarged or reduced based on two important factors: the center of dilation and the scale factor $(k)$.

**Key Point**

The center of dilation is the fixed point on the plane from which all points of the figure are expanded or contracted.

**Key Point**

The scale factor $(k)$ determines how much the figure grows or shrinks.

- If $k > 1$, the dilation is an enlargement.
- If $0 < k < 1$, the dilation is a reduction.
- If $k = 1$, the figure remains unchanged.

Properties of dilations include:

- Preserving the shape but not necessarily the size.
- Maintaining the angle measures.
- Multiplying the lengths of sides by the scale factor.
- Increasing the area of 2-dimensional figures by $k^2$.
- Amplifying the volume of 3-dimensional figures by $k^3$.

**Example** A rectangle with dimensions 4 cm by 6 cm undergoes a dilation with a scale factor of 2.5. What are the new dimensions?

**Solution:** Multiply each side length by the scale factor 2.5:

- New length $= 4 \times 2.5 = 10$ cm.

- New width $= 6 \times 2.5 = 15$ cm.

Thus, the new dimensions of the rectangle are 10 cm by 15 cm.

**Example** Given a triangle with vertices $A(2,3)$, $B(5,3)$, and $C(4,6)$, find the coordinates of the vertices after a dilation with a scale factor $k = \frac{1}{2}$ centered at the origin (0, 0).

**Solution:** Dilate each vertex by multiplying its coordinates by the scale factor of $\frac{1}{2}$, with the dilation centered at the origin (0,0):

- $A'\left(2 \times \frac{1}{2}, 3 \times \frac{1}{2}\right) = A'(1,1.5)$
- $B'\left(5 \times \frac{1}{2}, 3 \times \frac{1}{2}\right) = B'(2.5,1.5)$
- $C'\left(4 \times \frac{1}{2}, 6 \times \frac{1}{2}\right) = C'(2,3)$

Thus, the vertices of the dilated triangle are at $A'(1,1.5)$, $B'(2.5,1.5)$, and $C'(2,3)$.

## 7.2 Dilations and Angles

Dilation is a transformation that alters the size of a figure while retaining its shape. We now focus on understanding *Dilations and Angles.*

**Key Point**

Angles Preserve Their Measure: A fundamental aspect of dilation is the preservation of angles. No matter the scale factor, the angle measures in the pre-image and the dilated image remain constant.

The preservation of angles leads to an important consequence regarding parallel lines in polygons.

**Key Point**

Parallel Lines Remain Parallel: When dilating polygons, if sides were parallel before the transformation, they will continue to be parallel afterward. This property also ensures that angles between these lines, created by transversals, remain congruent.

Why are angles preserved under dilation? Imagine two rays emanating from the center of dilation, forming an angle. As these rays stretch outwards or shrink inwards proportionally, they retain their relative positions, keeping the angle intact. When lines do not pass through the center, their orientation relative to each other remains unchanged, making their angles consistent.

Now, let us examine some examples to reinforce our understanding.

**Example** A rectangle with a length of 4 units and a width of 2 units has been dilated with a scale factor of 1.5. Determine the angle measures of the dilated rectangle.

**Solution:** Regardless of the scale factor, the angles in a rectangle are all 90°. After the dilation, all angles in the rectangle will still measure 90°, being unaffected by the change in scale.

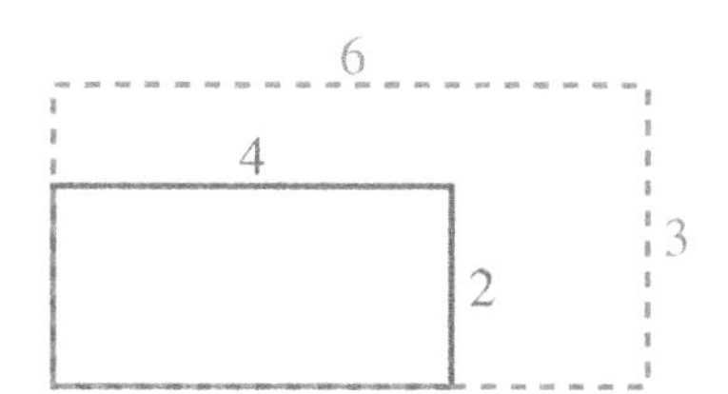


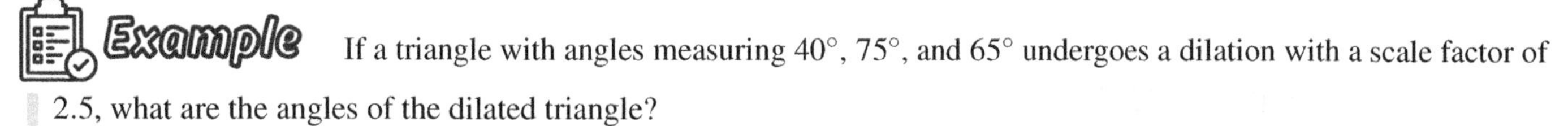

**Example** If a triangle with angles measuring 40°, 75°, and 65° undergoes a dilation with a scale factor of 2.5, what are the angles of the dilated triangle?

**Solution:** The angles remain unchanged under dilation. Therefore, the angles of the dilated triangle are still 40°, 75°, and 65°.

## 7.3 Similarity

Similarity occurs when two figures have the same shape but may differ in size. This notion is fundamental in geometry as it extends beyond mere visual resemblance to encompass precise mathematical relationships.

**Key Point**

Dilation is a pivotal transformation in establishing similarity between figures, preserving the shape while changing the size.

**Key Point**

Similar figures have corresponding angles that are congruent, which means they have equal measures.

**Key Point**

The lengths of corresponding sides in similar figures are proportional to each other.

Let us explore these concepts through some examples.

**Example** Triangles ABC and DEF are similar. Side AB measures 8 *cm* and corresponds to side DE that measures 12 *cm*, and side AC measures 6 *cm*. Find the length of side DF.

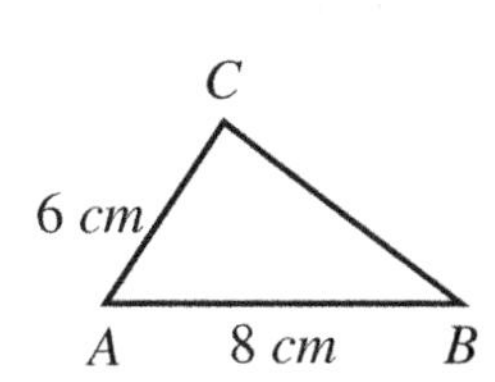


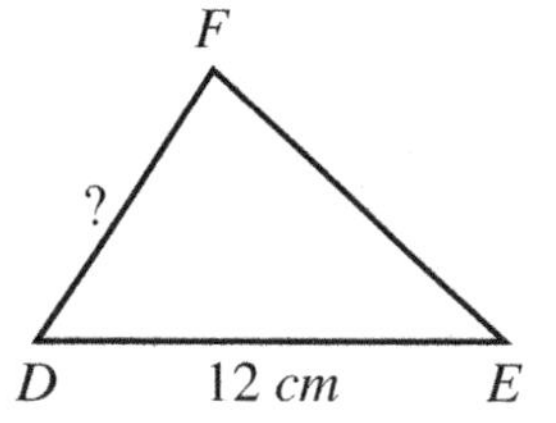


**Solution:** Since the triangles are similar, the side length ratios are equal. Thus, $\frac{AB}{DE} = \frac{AC}{DF}$. Plugging in the values, we get $\frac{8}{12} = \frac{6}{DF}$. By cross-multiplying and solving for $DF$, we find $DF = 9\ cm$.

If the perimeter of a smaller similar figure is 15 $cm$ and the perimeter of the larger similar figure is 25 $cm$, find the scale factor.

**Solution:** The scale factor is the ratio of any two corresponding lengths in similar figures. Therefore, the scale factor from the smaller to the larger is $\frac{25}{15} = \frac{5}{3}$.

## 7.4 Similarity Criteria

Similarity differs from congruence in that it pertains solely to the shape of triangles, not their size. Triangles are considered similar when they share the same shape, even though their sizes may not be identical.

Two triangles are similar if they have congruent corresponding angles and proportional corresponding sides.

The criteria for triangle similarity are widely used to prove that two triangles are similar without having to measure all angles and sides. Here we discuss these criteria.

**AA (Angle-Angle) Similarity:** If two angles of one triangle are congruent to two angles of another triangle, the third angle is automatically congruent, as all triangles have angle sums of 180°. Hence, both triangles are similar.

**SAS (Side-Angle-Side) Similarity:** Two triangles are similar if the ratios of the lengths of two corresponding sides are equal and the included angle is congruent.

**SSS (Side-Side-Side) Similarity:** If the ratios of the lengths of all three pairs of corresponding sides are equal between two triangles, then they are similar regardless of the angle measures.

Let us look at some examples that illustrate these concepts.

Example Triangle $ABC$ has angles measuring 40°, 50°, and 90°. Triangle $DEF$ has angles measuring 50°, 40°, and 90°. Are the two triangles similar?

**Solution:** Using the AA criterion, since two angles (50° and 40°) of triangle $ABC$ are congruent to two angles of triangle $DEF$, the two triangles are similar.

Example You have triangle $GHI$ with sides 6, 8, and 10 units long, and triangle $JKL$ with sides 3, 4, and 5 units long. Determine if they are similar.

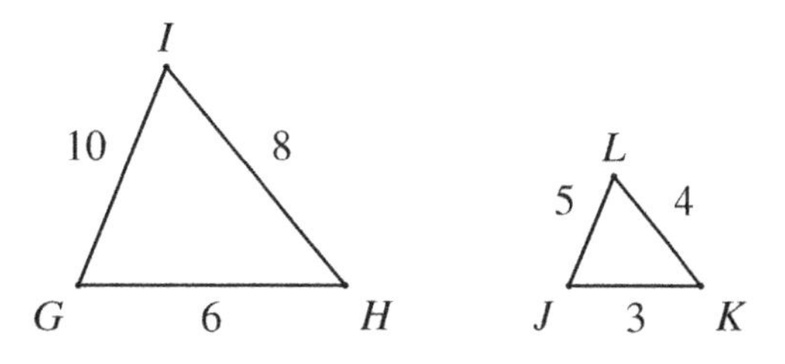


**Solution:** Comparing the ratio of the sides:

$$\frac{GH}{JK} = \frac{6}{3} = 2, \frac{HI}{KL} = \frac{8}{4} = 2, \text{ and } \frac{IG}{LJ} = \frac{10}{5} = 2.$$

All the ratios are equal to 2, so by the SSS similarity criterion, triangles $GHI$ and $JKL$ are similar.

## 7.5 Congruent and Similar Figures

Congruency and similarity are foundational concepts in geometry that help us understand how shapes relate to one another. While they share similarities, they have distinct differences that are important for a variety of applications.

**Key Point**

Congruent figures are identical in shape and size, whereas similar figures have the same shape but may differ in size, maintaining proportionality.

Similar figures have angles that are congruent and sides that are proportional. Congruent figures, however, extend this criteria: not only must the angles be congruent, and the sides be proportional, but the sides must also be equal in length.

Are rectangles $ABCD$ and $EFGH$ shown below congruent?

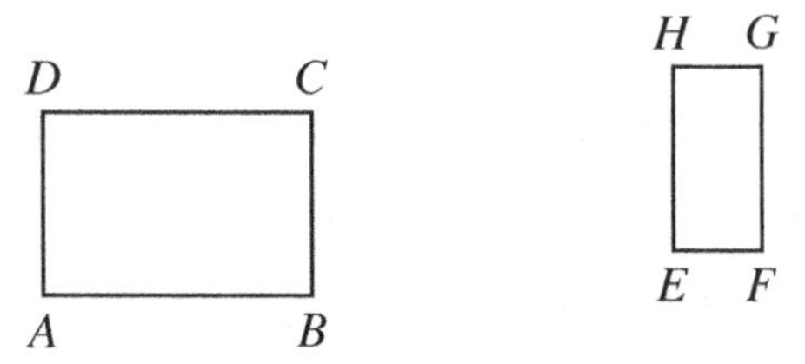


**Solution:** To determine congruency, we need to compare the dimensions of the rectangles. If all corresponding sides and angles are equivalent, then the rectangles are congruent. As seen in the figure, the dimensions of $ABCD$ and $EFGH$ are different, which implies that they are not congruent. They may be similar if the sides are proportional, but additional information would be needed to confirm that.

**Example** Is triangle $PQR$ congruent to triangle $STU$, given that the sides of $PQR$ measure 5 cm, 7 cm, and 10 cm, and the sides of $STU$ measure 10 cm, 14 cm, and 20 cm?

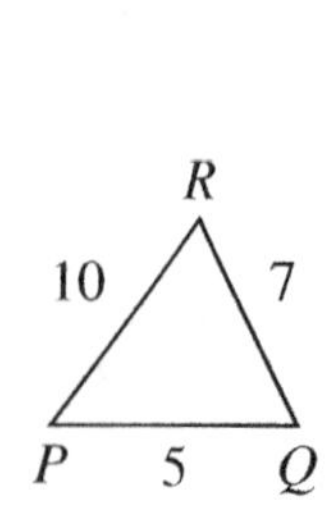


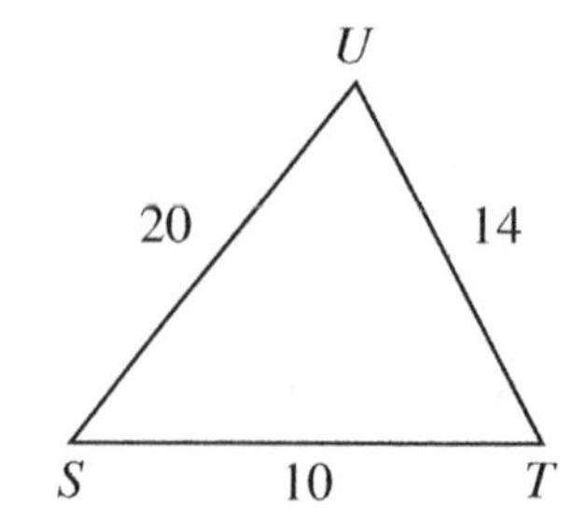


**Solution:** The triangles are not congruent because their corresponding sides are not equal in length. However, the sides of $STU$ are exactly twice the length of the corresponding sides of $PQR$, so they have the same shape but different sizes. Therefore, the triangles are similar since the ratios of corresponding sides are equal.

## 7.6 Area and Perimeter: Scale Changes

In geometric studies, the impact of dilation and similarity on a figure's dimensions, particularly its area and perimeter, is a crucial consideration. The key to altering a figure's size while preserving its shape lies in the use of a scale factor. This factor is a multiplier applied to the original dimensions to obtain the new size.

**Key Point**

The perimeter of a scaled figure changes by the same scale factor, whereas the area changes by the square of the scale factor.

The formula to calculate the scale factor is:

$$\text{Scale factor} = \frac{\text{New shape's dimensions}}{\text{Original shape's dimensions}}.$$

Let us explore the effect of changing the scale factor on the area and perimeter through an example.

**Example** A rectangle's length and width are 10 and 5 units, respectively. We scale the rectangle up by a factor of 2. What are the new length, width, area, and perimeter?

**Solution:** To find the new dimensions, we multiply the original by our scale factor of 2:

$$\text{New length}: \ 10 \times 2 = 20\,\text{units}, \text{ and New width}: \ 5 \times 2 = 10\,\text{units}.$$

The perimeter of the original rectangle is $(10+5) \times 2 = 30$ units. The new perimeter will be $(20+10) \times 2 = 60$ units, which shows the perimeter has doubled, just like the scale factor.

The area of the original rectangle is $10 \times 5 = 50$ square units. The new area will be $20 \times 10 = 200$ square units. Notice how the area has increased by a factor of $2^2 = 4$, showing that the area changes by the square of the scale factor. The following figure illustrates the scale change of a rectangle.

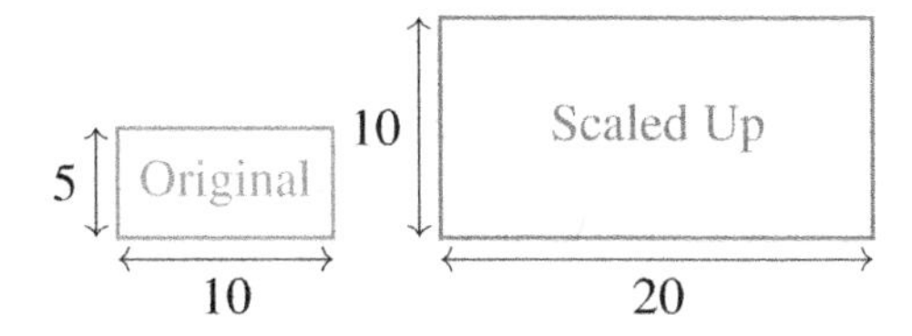


Let us consider one final example to conclude.

A square with a side length of 3 units is scaled down by a factor of 0.5. The new dimensions are:

- **Side Length:** $3 \times 0.5 = 1.5$ units.
- **Perimeter:** Reduced from 12 units to $1.5 \times 4 = 6$ units.
- **Area:** Decreases from 9 to $1.5^2 = 2.25$ square units.

These changes illustrate how scaling affects the square's area and perimeter proportionally.

## 7.7 The Side Splitter Theorem

Understanding the Side Splitter Theorem is a key step in mastering geometric concepts related to similarity and proportions within triangles.

Key Point

If a line segment $DE$ is drawn parallel to side $AC$ of $\triangle ABC$, intersecting sides $AB$ and $BC$ at $D$ and $E$ respectively, then $\frac{BD}{DA} = \frac{BE}{EC}$.

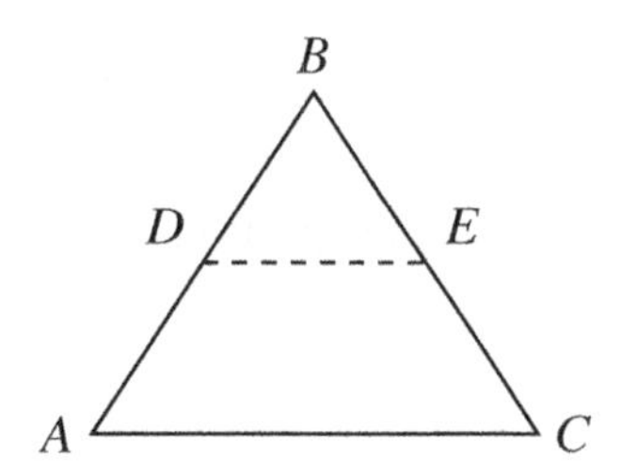


The theorem maintains that any set of parallel lines cut by two transversals will create divided segments that are proportional on the transversals. This concept extends the Side Splitter Theorem beyond triangles to other polygonal shapes, as long as the parallel and transversal conditions are met.

Example In $\triangle XYZ$, line segment $MN$ is drawn parallel to side $XZ$, intersecting side $XY$ at $M$ and side $YZ$ at $N$. Given that $XM = 6$ units, $MY = 18$ units, and $YN = 24$ units, find the length of $NZ$.

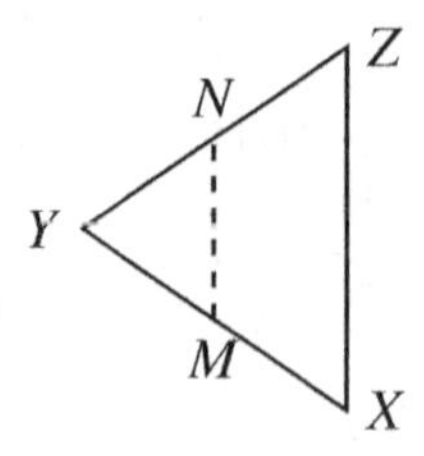


**Solution:** According to the Side Splitter Theorem,

$$\frac{YM}{MX} = \frac{YN}{NZ}.$$

Plugging in the given values,

$$\frac{18}{6} = \frac{24}{NZ} \Rightarrow \frac{3}{1} = \frac{24}{NZ} \Rightarrow NZ = \frac{24 \times 1}{3} = 8 \text{ units.}$$

Thus, the length of $NZ$ is 8 units.

## 7.8 Similarity Transformations

Transformations that maintain the shape of a figure while possibly altering its size, position, or orientation are known as similarity transformations. These changes can affect the figure's dimensions and placement, but its fundamental shape remains unchanged.

**Key Point**

Similarity transformations include translations, reflections, rotations, and dilations, and they preserve the angles of a shape while changing side lengths in a proportional manner.

### Types of Similarity Transformations

**Translation:** A translation moves every point of a figure the same distance in the same direction. Here, the shape, size, and orientation of the figure remain unchanged.

**Reflection:** A reflection flips a figure over a line known as the line of reflection. The figure and its image are symmetrical with respect to this line.

**Rotation:** During a rotation, a figure is turned about a fixed point, the center of rotation, by a certain angle in a specified direction.

**Dilation:** A dilation scales a figure larger or smaller with respect to a fixed point, the center of dilation. The angles stay the same, but side lengths are scaled by a factor.

Similarity transformations are characterized by the fact that they produce an image that, although it may differ in size, will be similar to the original figure:

**Key Point**

Corresponding angles between the original figure and the image remain congruent.

**Key Point**

Corresponding side lengths are proportional; they are altered by the same scale factor in a dilation (whereas translations, reflections, and rotations preserve side lengths).

A triangle with vertices $A(2,3)$, $B(5,7)$, and $C(6,2)$ is translated 3 units to the right and 2 units up. Determine the coordinates of the image triangle $A'B'C'$.

**Solution:** By moving each vertex 3 units right and 2 units up, we apply a translation to the original triangle:

$$A'(2+3,3+2) = A'(5,5),$$
$$B'(5+3,7+2) = B'(8,9),$$
$$C'(6+3,2+2) = C'(9,4).$$

Hence, the vertices of the image triangle $A'B'C'$ are $A'(5,5)$, $B'(8,9)$, and $C'(9,4)$.

## 7.9 Partitioning a Line Segment

Partitioning a line segment is a practical application that enables the division of a segment into parts with specific proportional relationships. This process is essential for various geometric constructions and proofs.

**Key Point**

Partitioning a line segment involves finding a point along the segment that divides it into two segments with lengths in a given ratio.

To partition a line segment $AB$ with endpoints $A(x_1,y_1)$ and $B(x_2,y_2)$ and a ratio of $m:n$, the coordinates $(x,y)$ of the point $P$ that partitions the segment in the given ratio are calculated using the formula:

$$x = \frac{mx_2+nx_1}{m+n}, \text{ and } y = \frac{my_2+ny_1}{m+n}.$$

The steps to find this point are straightforward:

1. Identify the endpoints of the line segment and their coordinates.
2. Determine the desired ratio for partitioning the segment.
3. Substitute the known values into the formula to find the coordinates of the partition point.

Divide the line segment with endpoints $A(2,5)$ and $B(10,13)$ in the ratio $3:4$. What are the coordinates of the point that partitions the segment?

**Solution:** Using the formula, the $x$-coordinate is calculated as:

$$x = \frac{3(10)+4(2)}{3+4} = \frac{30+8}{7} = \frac{38}{7}.$$

And the $y$-coordinate is:

$$y = \frac{3(13)+4(5)}{3+4} = \frac{39+20}{7} = \frac{59}{7}.$$

Hence, the partition point is $P(\frac{38}{7}, \frac{59}{7})$.

## 7.10 Similar Polygons

Similarity in polygons is established when there is congruence between their corresponding angles and equality in the ratios of the lengths of their corresponding sides.

**Key Point**

Similar polygons have two important properties:

- Corresponding angles in similar polygons are congruent.
- The lengths of corresponding sides in similar polygons are proportional.

These properties ensure that not only do similar polygons look the same in shape, but all their corresponding elements hold a consistent ratio or equivalency. Similarity is denoted with the symbol "$\sim$". For instance, if two quadrilaterals $ABCD$ and $WXYZ$ are similar, this is written as $ABCD \sim WXYZ$.

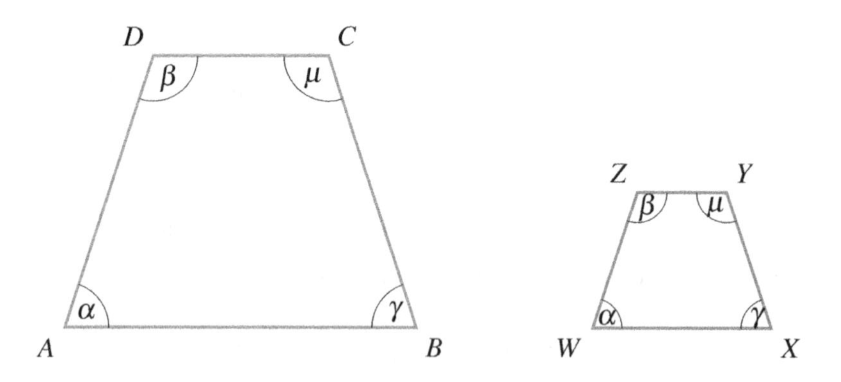


Let us detail the conditions required for two polygons to be similar:

1. They must have the same number of sides.
2. Their corresponding angles must be congruent.
3. The lengths of their corresponding sides must be proportional.

Let us delve into examples to understand the application of these properties.

**Example** Quadrilateral $ABCD$ is similar to quadrilateral $EFGH$. If $AB = 15$ cm and $EF = 10$ cm, what is the ratio of their corresponding sides and the length of side $BC$ if $FG = 14$ cm?

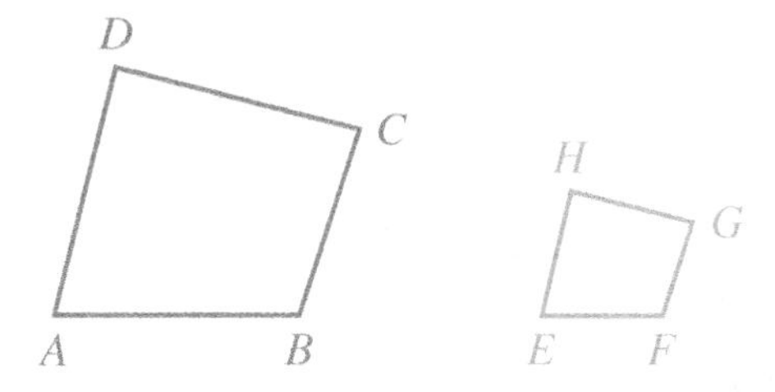


**Solution:** Since quadrilateral $ABCD \sim EFGH$, the sides are proportional. The ratio of the corresponding sides $AB$ to $EF$ is given by $\frac{AB}{EF} = \frac{15}{10} = \frac{3}{2}$. This constant ratio applies to all other corresponding sides. For the side $BC$ to correspond to $FG$, $\frac{BC}{FG} = \frac{3}{2}$ as well. Thus, $BC = \frac{3}{2}FG = \frac{3}{2}(14) = 21$ cm. So, side $BC$ is 21 cm long.

**Example** Triangle $ABC$ is similar to triangle $DEF$, with a side $AC$ measuring 18 cm and side $DF$ measuring 12 cm. Find the lengths of the sides $AB$ when $DE = 10$.

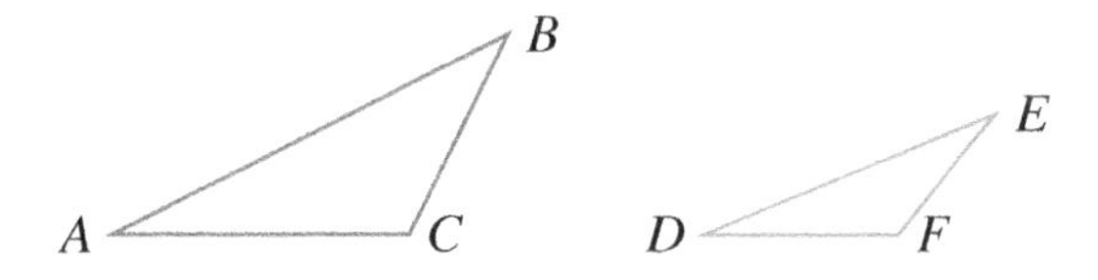


**Solution:** Since triangles $ABC$ and $DEF$ are similar, the corresponding sides are proportional. Therefore, we have: $\frac{AC}{DF} = \frac{AB}{DE}$. Substituting the given values: $\frac{18}{12} = \frac{AB}{10}$. Solving for $AB$: $AB = \frac{18}{12} \times 10 = 15$ cm. Thus, the length of side $AB$ is 15 cm.

## 7.11 Right Triangles and Similarity

When an altitude is drawn from the right angle to the hypotenuse of a right triangle, it creates two smaller right triangles inside the original triangle. Remarkably, these smaller triangles are similar to each other and to the original triangle. This similarity arises from the fact that all three triangles share the same acute angles.

**Key Point**

The drawing of an altitude from the right angle to the hypotenuse in a right triangle creates three similar triangles.

**Key Point**

The lengths of the hypotenuse segments created by the altitude are proportional to the lengths of the corresponding legs of the parent right triangle.

Consider a right triangle *ABC*, where *AB* is perpendicular to *AC*. Upon drawing an altitude *AD* from angle *A*, we observe that:

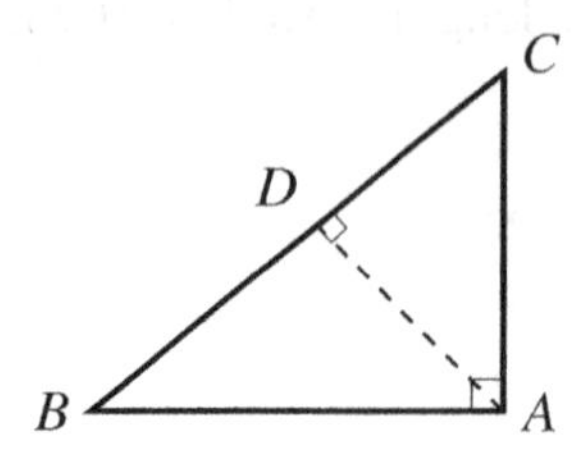


$$ABD \sim CAD \Rightarrow \frac{AD}{CD} = \frac{BD}{AD} \Rightarrow AD^2 = BD \times CD.$$

$$ABD \sim CBA \Rightarrow \frac{BD}{BA} = \frac{AB}{CB} \Rightarrow AB^2 = CB \times BD.$$

$$ACD \sim CBA \Rightarrow \frac{AC}{CB} = \frac{AD}{CA} \Rightarrow AC^2 = CB \times AD.$$

**Example** For a right triangle with hypotenuse segments of 5 *cm* and 20 *cm* formed by an altitude, determine the length of the altitude using the geometric mean relationship.

**Solution:** Using the geometric mean relationship:

$$h^2 = 5 \times 20 \Rightarrow h^2 = 100 \Rightarrow h = 10 \text{ cm}.$$

## 7.12 Similar Solids

When dealing with solids in geometry, understanding the concept of similarity is crucial. Similar solids have corresponding angles that are equal and corresponding linear dimensions that are proportional. This proportionality is the cornerstone in solving various geometric problems related to dimensions and volumes of these solids.

The concept of similarity can be summarized in the following steps when determining whether two solids are similar:

1. Identify the corresponding dimensions between the two solids.
2. Calculate the ratio of the lengths of corresponding dimensions.
3. Verify that all calculated ratios are equal; if they are, the solids are similar.
4. Obtain the scale factor; the ratio by which one solid is larger or smaller than the other.

**Key Point**

The volumes of similar solids scale with the cube of the linear scale factor, while areas scale with its square.

Let us see this principle in action with the following example.

**Example** Two cylindrical jars are similar. The smaller jar has a radius of 3 cm and a height of 8 cm. The larger jar has a height of 20 cm. Find the radius of the larger jar and its volume.

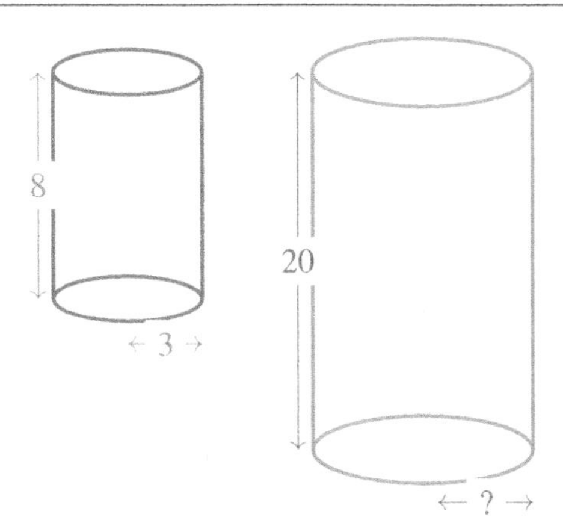


**Solution:** To find the radius of the larger jar, we set up the ratio of their corresponding heights:

$$\frac{\text{Height of larger jar}}{\text{Height of smaller jar}} = \frac{20 \text{ cm}}{8 \text{ cm}}.$$

The scale factor for the height is $\frac{20}{8} = 2.5$, and since the jars are similar, the scale factor for the radius must be the same.

$$R = 3 \text{ cm} \times 2.5 = 7.5 \text{ cm}.$$

To find the volume of the larger jar, we use the volume formula for cylinders:

$$V = \pi R^2 h = \pi \times (7.5 \text{ cm})^2 \times 20 \text{ cm} = \pi \times 56.25 \text{ cm}^2 \times 20 \text{ cm} = 1125\pi \text{ cm}^3.$$

To reinforce the understanding of similar solids, let us look at another example that incorporates this geometric concept.

**Example** The two cones are similar. The smaller cone has a radius of 4 units and a volume of $48\pi$ cubic units. If the larger cone has a radius of 10 units, find the volume of the larger cone.

**Solution:** Since the cones are similar, the scale factor between their radii is:

$$\text{Scale factor} = \frac{\text{Radius of larger cone}}{\text{Radius of smaller cone}} = \frac{10}{4}.$$

The volume of similar solids scales with the cube of the scale factor. Thus, the volume of the larger cone is:

$$V_{\text{larger}} = 48\pi \times (\frac{10}{4})^3 = 750\pi \text{ cubic units.}$$

## 7.13 Practices

**1)** Solve.

1-1) A rectangle has a length of 8 units and a width of 6 units. If the rectangle is dilated by a scale factor of 3, what will be the length and width of the resulting rectangle?

1-2) A triangle has sides measuring 7 units, 10 units, and 12 units. If the triangle is dilated by a scale factor of 0.5, what will be the length of each side of the resulting triangle?

1-3) The diagram shows two concentric circles with radii labeled as follows: What is the scale factor of Circle $B$ relative to Circle $A$?

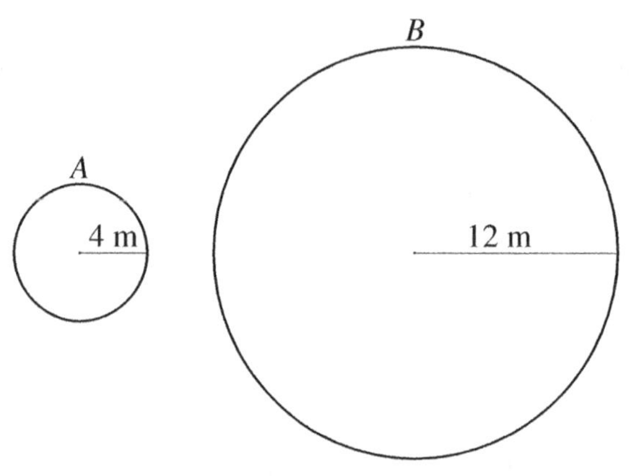


1-4) A hexagon has an area of 120 square centimeters. If it undergoes a dilation with a scale factor of 0.25, what will be the area of the dilated hexagon?

1-5) A rhombus has angles measuring $60^\circ$ and $120^\circ$. If it undergoes a dilation with a scale factor of 2, what will be the angles of the dilated rhombus? Also, if the original rhombus had a side length of 5 cm, what will be the side length of the dilated rhombus?

1-6) A trapezoid has angles measuring $60^\circ$, $120^\circ$, $60^\circ$, and $120^\circ$. If it undergoes a dilation with a scale factor of 1.5, what will be the angles of the dilated trapezoid?

**2)** State if the polygons are similar.

2-1)

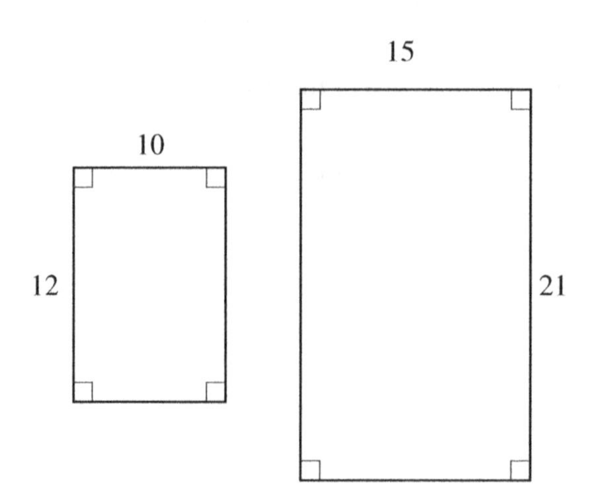

2-2)

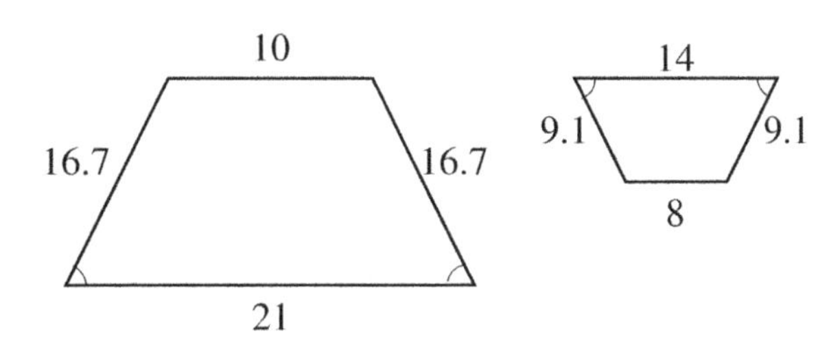


2-3)

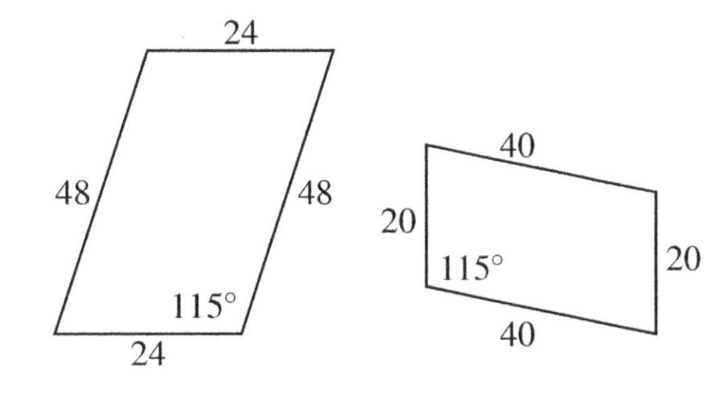


2-4)

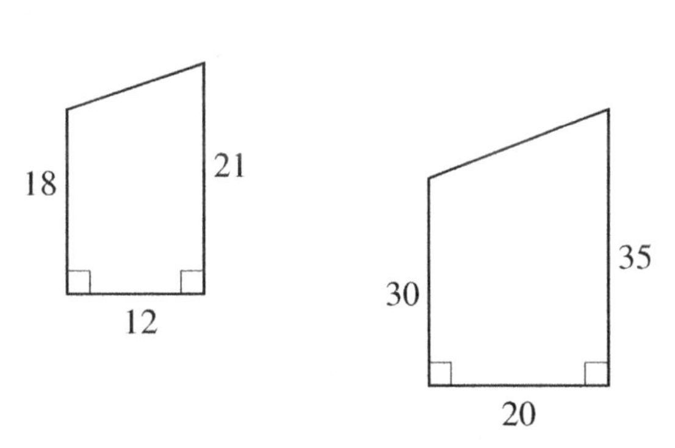


**3)** Solve.

3-1) Two similar triangles have a ratio of their corresponding sides as 3:5. If the shortest side of the larger triangle is 15 *cm*, find the length of the shortest side of the smaller triangle. Also, find the length of the longest side of the larger triangle if it is twice the length of its shortest side.

3-2) In two similar pentagons, the ratio of their corresponding sides is 2:3. If the perimeter of the smaller pentagon is 30 *cm*, find the perimeter of the larger pentagon. Additionally, if the longest side of the smaller pentagon is 8 *cm*, what is the length of the longest side of the larger pentagon?

3-3) Two similar circles have a ratio of their radii as 5:8. If the circumference of the smaller circle is $10\pi$ *cm*, find the circumference of the larger circle. Also, if the area of the smaller circle is $25\pi\ cm^2$, what is the area of the larger circle?

**4)** State if the two triangles are congruent. If they are, state how you know.

4-1)

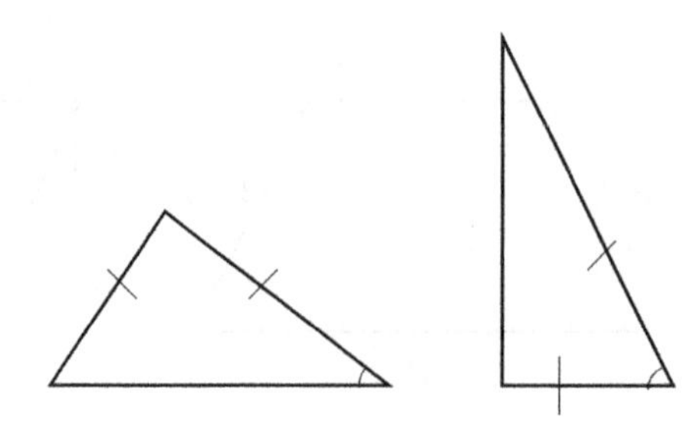

4-2)

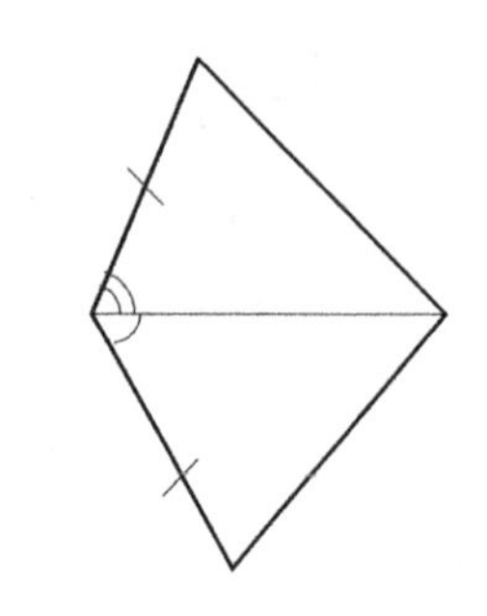

4-3)

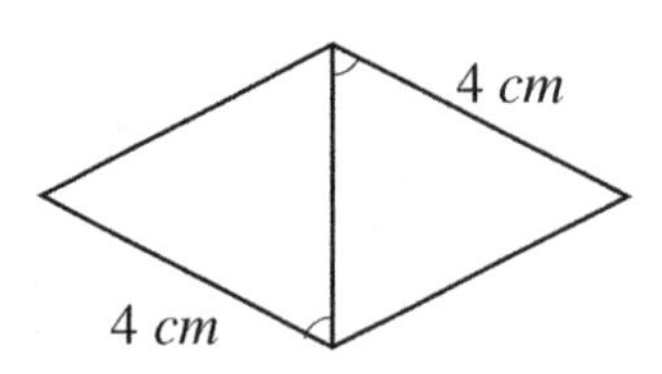


4-4)

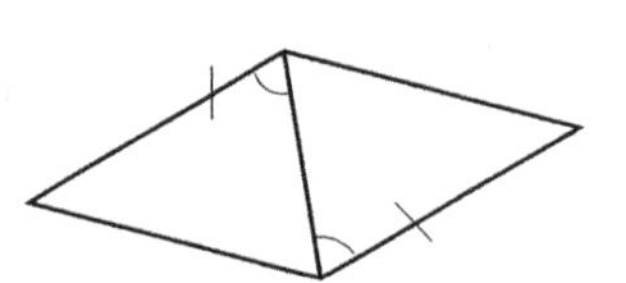

4-5)

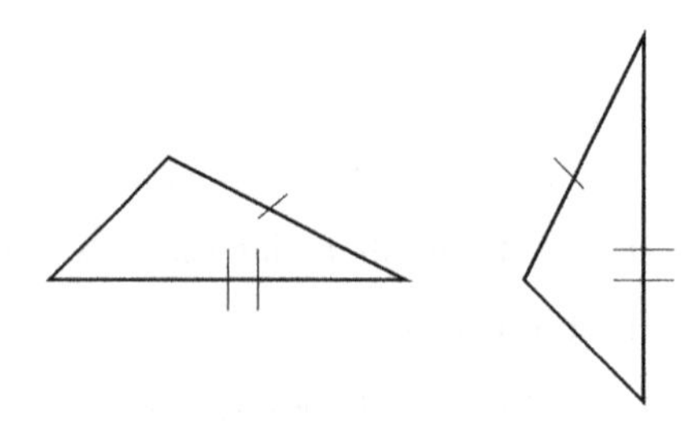

4-6)

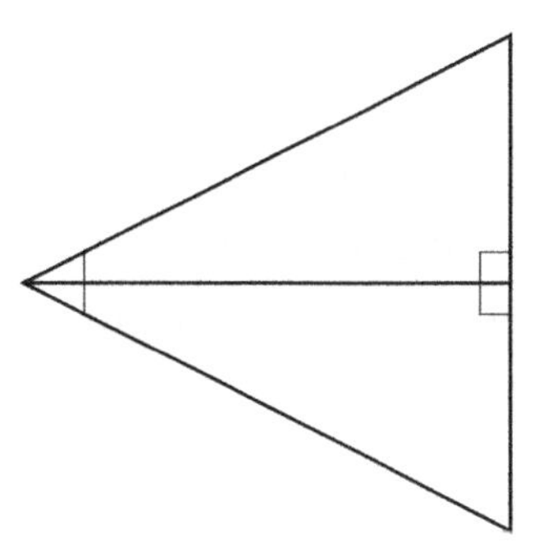

4-7)

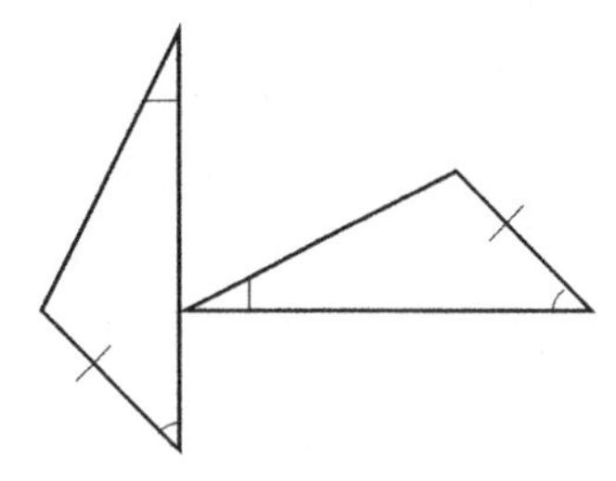

4-8)

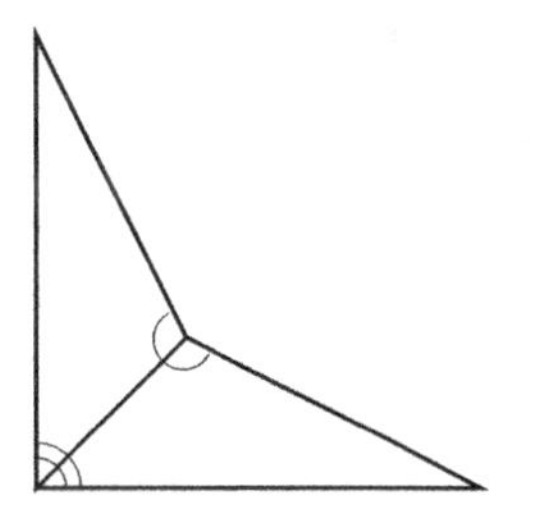

**5)** Solve.

5-1) Are the two following triangles congruent?

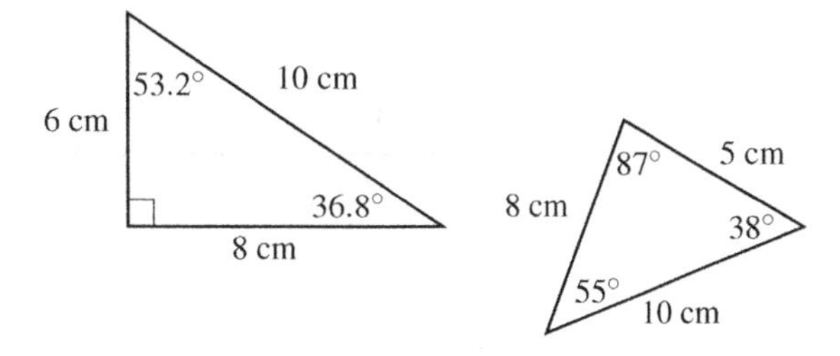


5-2) Are the two following Quadrilateral similar?

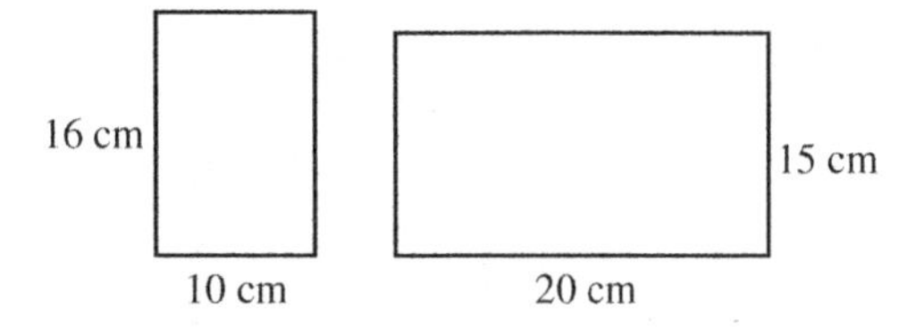

5-3) A square has a perimeter of 20 units. If the length and width are both increased by a factor of 5, what is the new perimeter and area of the square?

5-4) The length of a rectangular garden is 10 meters and the width is 6 meters. If the dimensions are increased by a factor of 2, what is the new perimeter, and area of the garden?

5-5) A cylindrical tank has a radius of 3 meters and a height of 10 meters. If these dimensions increase by a factor of 2, what will be the new surface area, and volume of the tank?

5-6) A circular swimming pool has a diameter of 10 meters. If the diameter increases by a factor of 2, what will be the new diameter, circumference, and area of the pool?

5-7) A model car has a scale of 1:24, which means that every inch on the model car represents 24 inches in the actual car. If the model car is 6 inches long, what is the length of the actual car in feet?

5-8) A map of a city block shows a width of 2 inches and a length of 3 inches. If the map is drawn with a scale factor of $\frac{1}{4}$ inch $=100$ feet, what are the actual dimensions of the city block in feet?

5-9) A blueprint for a garden shows a rectangular plot with a length of 8 centimeters and a width of 6 centimeters. If the blueprint is a scale drawing with a scale factor of $\frac{1}{8}$ centimeter $=1$ meter, what are the actual dimensions of the garden in meters?

5-10) A blueprint for a park shows a playground with a length of 20 meters and a width of 15 meters. If the blueprint is a scale drawing with a scale factor of $\frac{1}{5}$ meter $=1$ yard, what is the actual area of the playground?

**6)** Find the missing length indicated.

6-1)

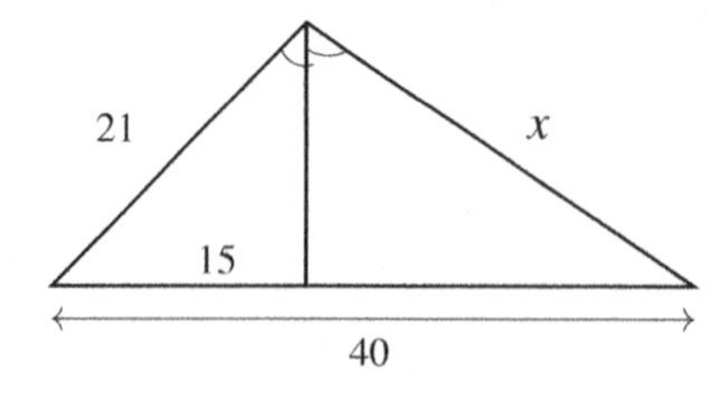


6-2)

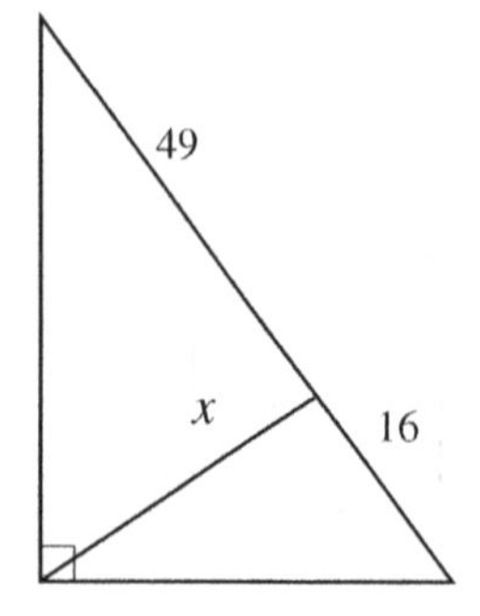

6-3)

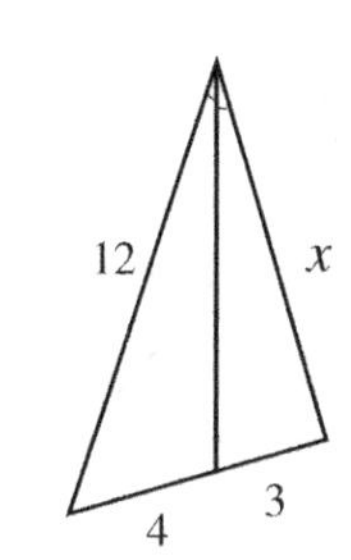


6-4)

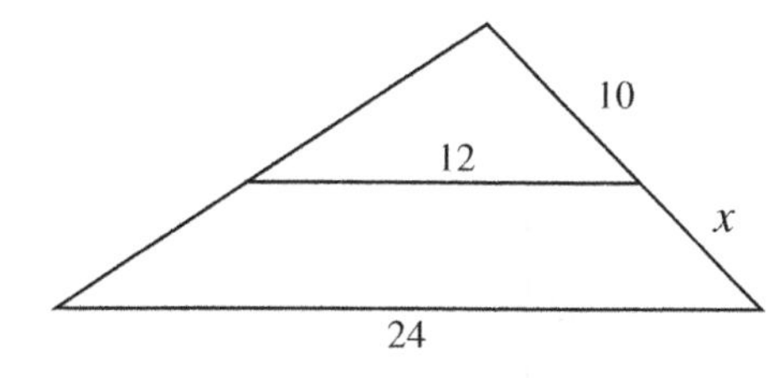


6-5)

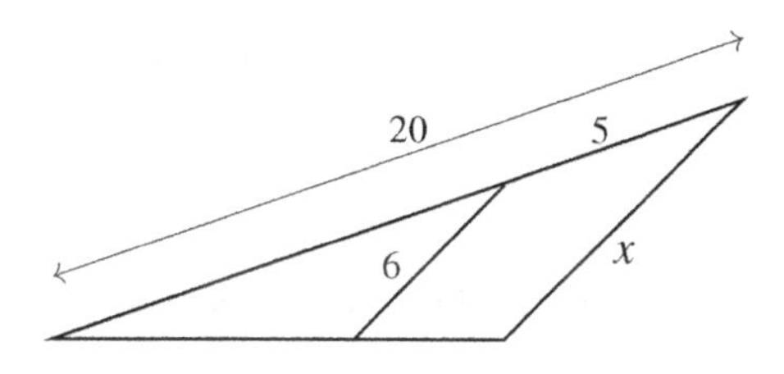


6-6)

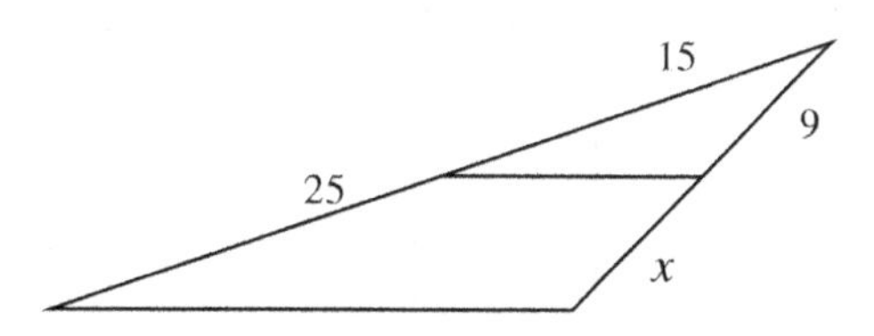


6-7)

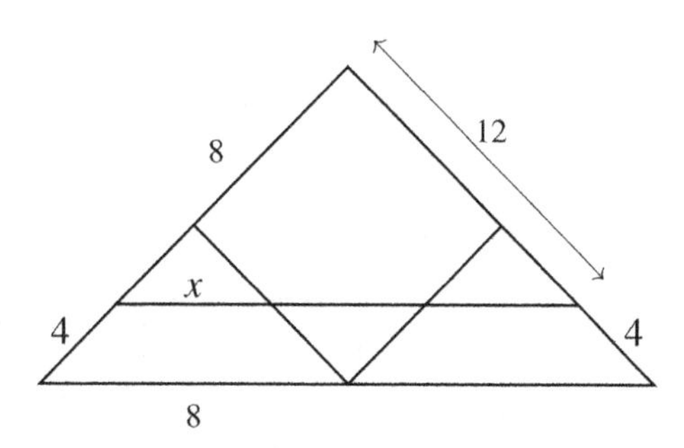


**7)** Find the scale factor of dilation of the triangle $MNP$ to the triangle $QRS$.

7-1)

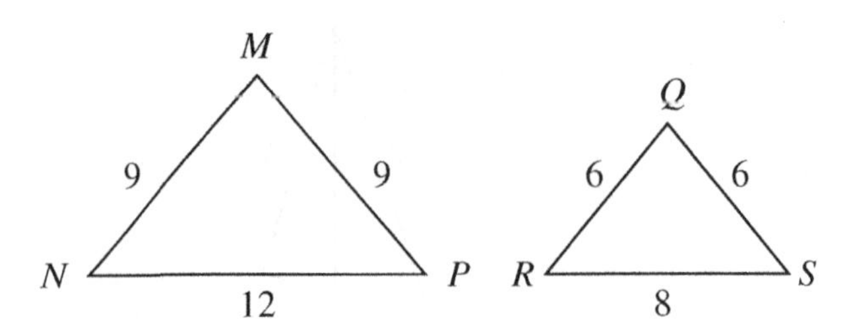


7-2)

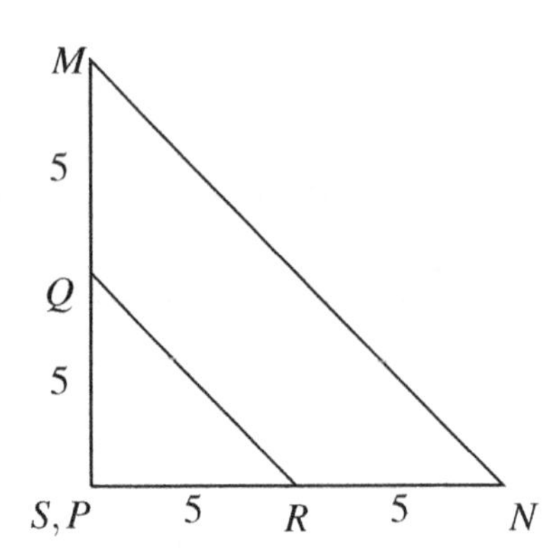


**8)** Find the scale factor of dilation of the rectangle $ABCD$ to the rectangle $EFGH$.

8-1)

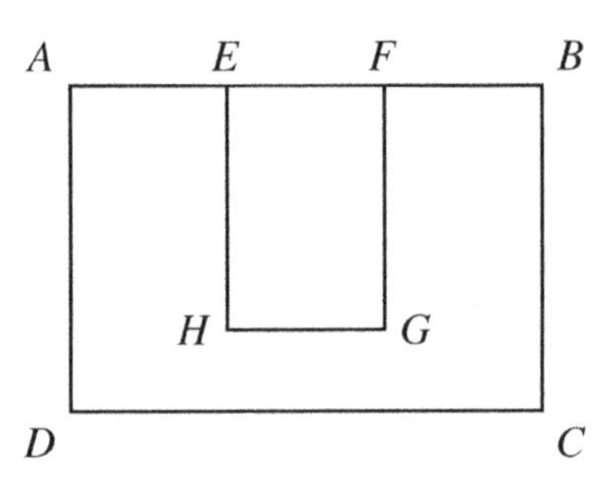


8-2)

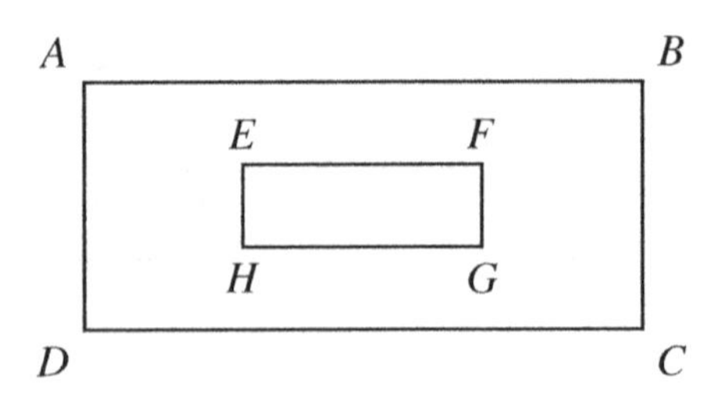


**9)** Solve.

9-1) A line segment has endpoints $A(-2,4)$ and $B(6,-3)$. Find the coordinates of the point that divides AB in the ratio 2:3.

9-2) A line segment has endpoints $M(3,-1)$ and $N(-5,7)$. Find the coordinates of the point that divides $MN$ in the ratio 3:1.

9-3) A line segment has endpoints $P(1,2)$ and $Q(7,-4)$. Find the coordinates of the point that divides $PQ$ in the ratio 4:1.

**10)** State if the polygons are similar.

10-1)

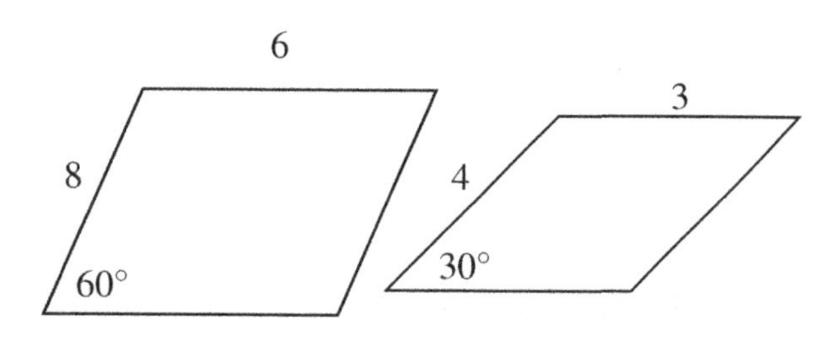


10-2)

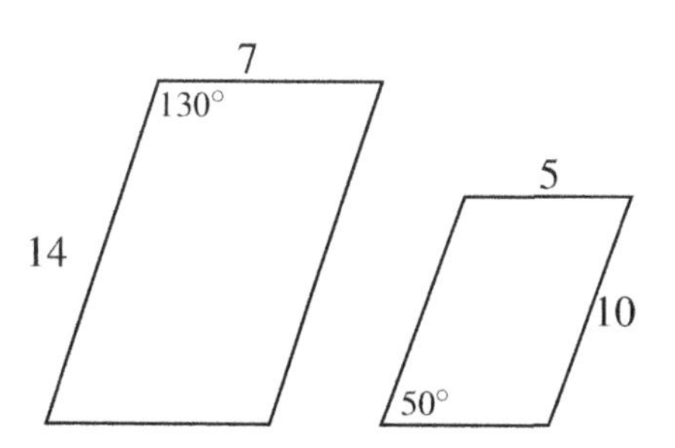


**11)** The polygons in each pair are similar. Find the scale factor of the smaller figure to the larger figure.

11-1)

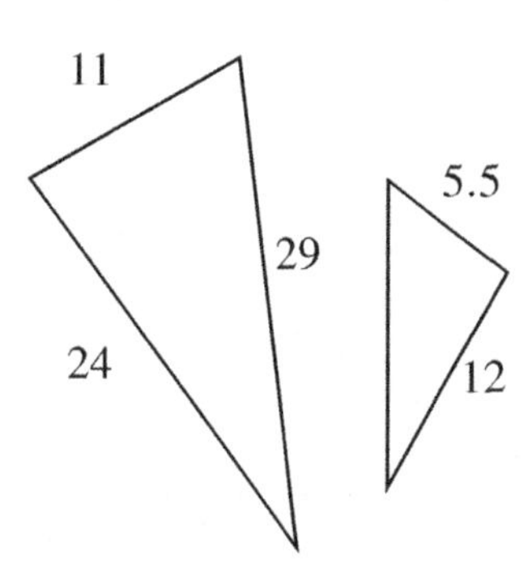

11-2)

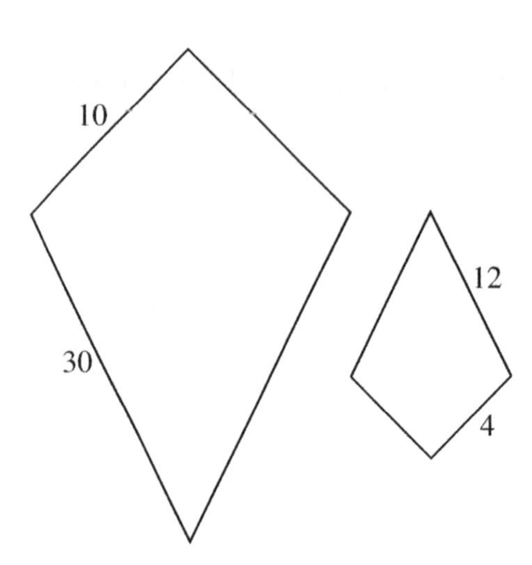


**12)** Find the missing length indicated. Leave your answer in simplest radical form.

12-1)

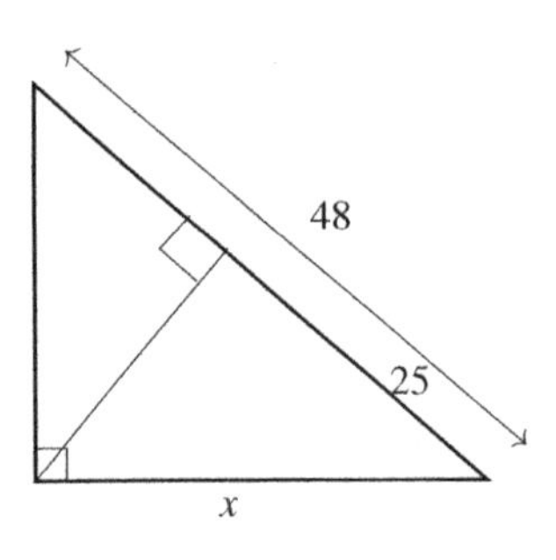


12-2)

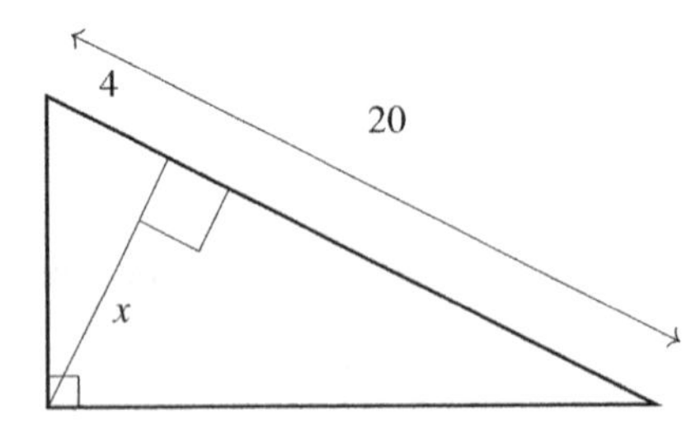


12-3)

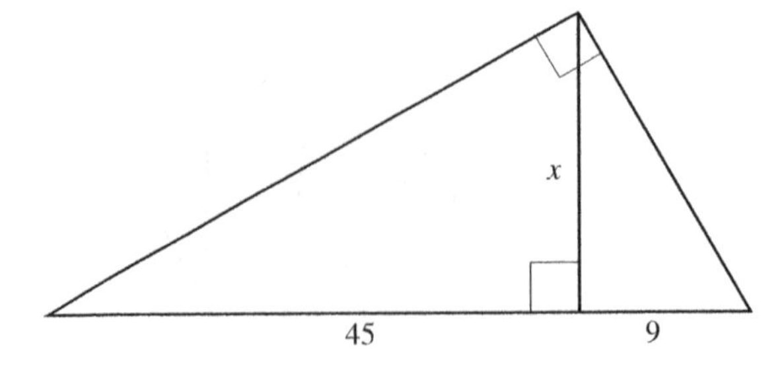

12-4)

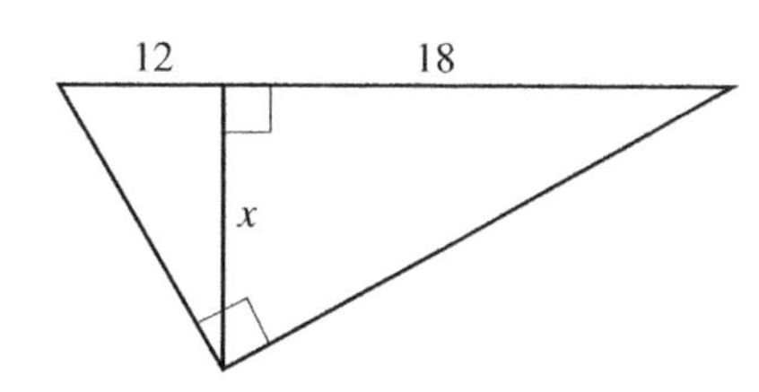


12-5)

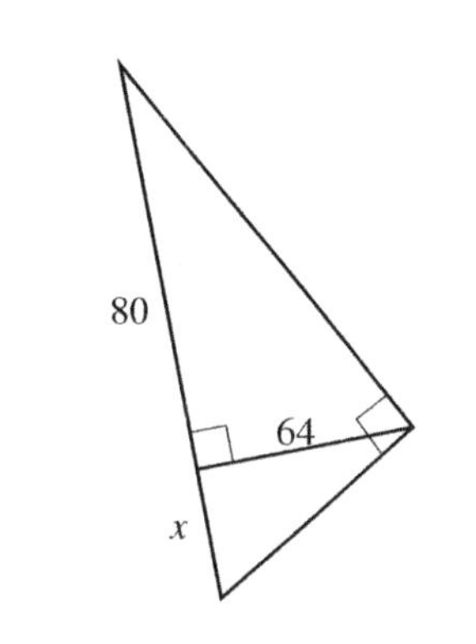


12-6)

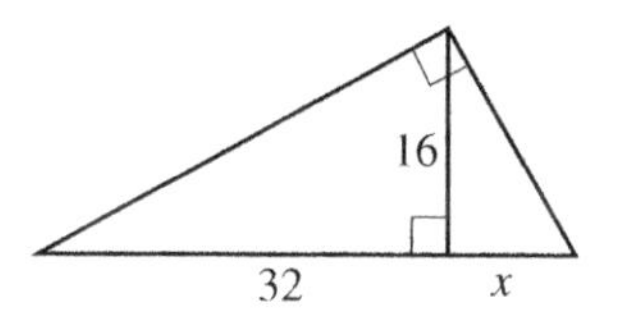


**13)** Solve.

13-1) Two rectangular prisms are similar. What is the volume of the larger prisms?

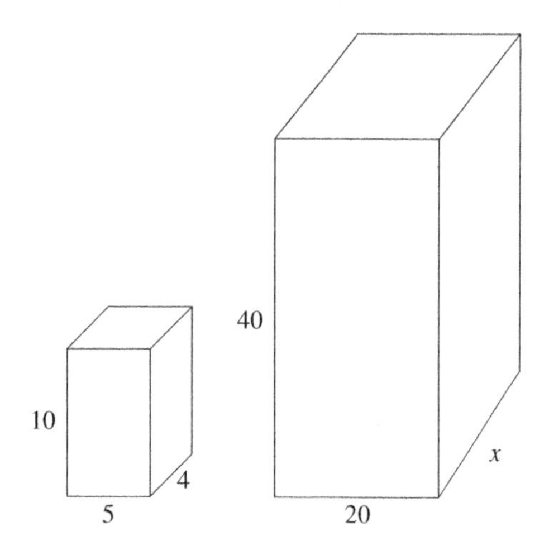

13-2) Two rectangular prisms are similar. What is the volume of the smaller prisms?

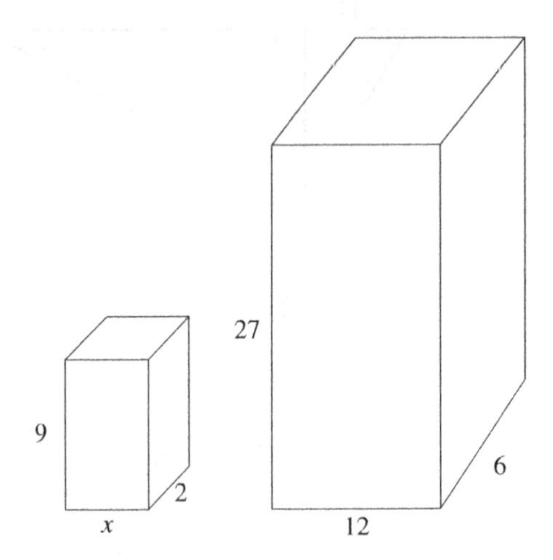

## 7.14 Answers

**1)**

1-1) $L = 24, W = 18$

1-2) 3.5, 5, 6

1-3) 3

1-4) 7.5

1-5) Angels unchanged, $L = 10$ cm

1-6) Angels unchanged

**2)**

2-1) Not Similar

2-2) Not Similar

2-3) Similar

2-4) Similar

**3)**

3-1) 9 $cm$, 30 $cm$

3-2) 45 $cm$, 12 $cm$

3-3) $16\pi, 64\pi$

**4)**

4-1) Not congruent

4-2) Not congruent

4-3) SAS

4-4) SAS

4-5) Not congruent

4-6) ASA

4-7) AAS

4-8) ASA

**5)**

5-1) Two triangles are not congruent

5-2) Two quadrilaterals are not similar

5-3) Perimeter=100 units, area=625 square units

5-4) Perimeter=64 meters, area=240 square meters

5-5) surface area=979.68 square meters, volume=2260.8 cubic meters

5-6) 20 meters, $20\pi$ meters, $100\pi$ square meters

5-7) 12 ft

5-8) 800 ft, 1,200 ft

5-9) 48 m, 64m

5-10) 7,500 yd

**6)**

6-1) 35

6-2) 28

6-3) 9

6-4) 10

6-5) 8

6-6) 15

6-7) 4

**7)**

7-1) 3:2

7-2) 2:1

**8)**

8-1) 2:1

8-2) 3:1

**9)**

9-1) $(\frac{6}{5}, \frac{6}{5})$

9-2) $(-3, 5)$

9-3) $(\frac{29}{5}, -\frac{14}{5})$

**10)**

10-1) Not similar

10-2) Similar

**11)**

11-1) 1:2

11-2) 2:5

**12)**

12-1) $20\sqrt{3}$

12-2) 8

12-3) $9\sqrt{5}$

12-4) $6\sqrt{6}$

12-5) 51.2

12-6) 8

**13)**

13-1) 12800

13-2) 72

# 8. Trigonometry

## 8.1 Pythagorean Identities

The Pythagorean identities consist of a series of equalities originating from the Pythagorean theorem. These identities establish relationships between the squared values of the sine, cosine, and tangent functions, connecting them either to the value 1 or to other trigonometric functions.

The most fundamental Pythagorean identity is:

$$\sin^2\theta + \cos^2\theta = 1.$$

This identity can be directly derived from the Pythagorean theorem when considering a right-angled triangle where the hypotenuse has a length of 1. From this primary identity, we can derive two essential related identities:

$$1 + \tan^2\theta = \sec^2\theta, \text{ and } \cot^2\theta + 1 = \csc^2\theta.$$

These are obtained by dividing the primary Pythagorean identity by $\cos^2\theta$ and $\sin^2\theta$, respectively.

**Key Point**

The Pythagorean identities are relations involving trigonometric functions that are true for any angle $\theta$.

Let us look at some examples to see how these identities work in practice.

Verify the identity $\sin^2\theta + \cos^2\theta = 1$ for $\theta = \frac{\pi}{6}$.

**Solution:** Start by evaluating the sine and cosine:

$$\sin\frac{\pi}{6} = \frac{1}{2}, \text{ and } \cos\frac{\pi}{6} = \frac{\sqrt{3}}{2}.$$

Now, square both values and sum them:

$$\sin^2\frac{\pi}{6}+\cos^2\frac{\pi}{6}=\left(\frac{1}{2}\right)^2+\left(\frac{\sqrt{3}}{2}\right)^2=\frac{1}{4}+\frac{3}{4}=1.$$

As expected, the identity holds for $\theta=\frac{\pi}{6}$.

Prove that $1+\tan^2\theta=\sec^2\theta$ using the primary Pythagorean identity.

**Solution:** Starting with the primary identity $\sin^2\theta+\cos^2\theta=1$, we divide both sides by $\cos^2\theta$:

$$\frac{\sin^2\theta}{\cos^2\theta}+\frac{\cos^2\theta}{\cos^2\theta}=\frac{1}{\cos^2\theta}.$$

This simplifies to:

$$\tan^2\theta+1=\sec^2\theta.$$

Thus, we have proven the identity.

## 8.2 Special Right Triangles

Here, we examine special right triangles, which apply these identities for swift side length calculations, circumventing trigonometric functions or the Pythagorean Theorem. Key types include the 30-60-90 and 45-45-90 triangles.

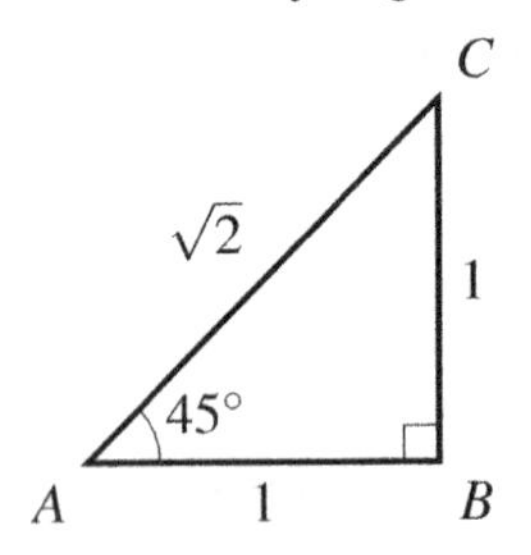


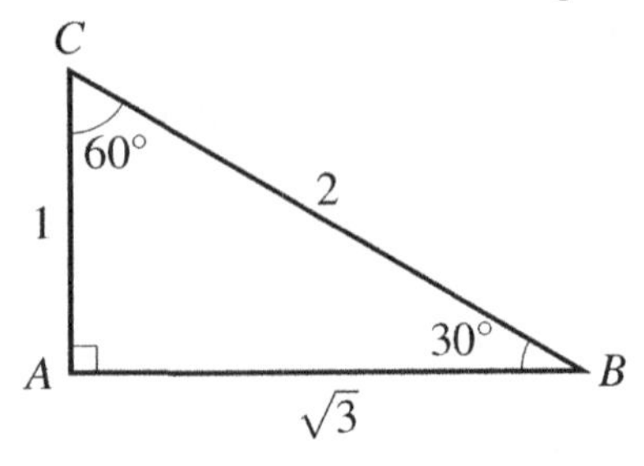


**Key Point**

A $45°-45°-90°$ or isosceles right triangle has sides in a $1:1:\sqrt{2}$ ratio.

**Key Point**

A $30°-60°-90°$ right triangle has sides in a $1:\sqrt{3}:2$ ratio, where the smallest side is opposite the $30°$ angle.

Find the length of the hypotenuse of a right triangle if the length of the other two sides are both 4 inches.

**Solution:** Since the two sides are equal, the triangle is a $45^\circ - 45^\circ - 90^\circ$ triangle. By the ratio $1:1:\sqrt{2}$, let the side length be $x$. Then, the hypotenuse $y$ must satisfy $x:x:y = 1:1:\sqrt{2}$, so $y = x\sqrt{2}$. With $x = 4$, we have $y = 4\sqrt{2}$ inches.

## 8.3 Trigonometric Ratios

Trigonometry has six fundamental functions, and together they encompass the different aspects of relating the sides of a triangle with its angles. These are sine (sin), cosine (cos), tangent (tan), cosecant (csc), secant (sec), and cotangent (cot).

Trigonometric ratios are unique to each angle in a right-angled triangle, enabling us to compare side lengths effectively.

**Definitions of the Trigonometric Ratios**

The six trigonometric ratios for an angle $\theta$ are defined as follows:

- $\sin(\theta) = \frac{\text{opposite side}}{\text{hypotenuse}}$
- $\cos(\theta) = \frac{\text{adjacent side}}{\text{hypotenuse}}$
- $\tan(\theta) = \frac{\text{opposite side}}{\text{adjacent side}}$
- $\csc(\theta) = \frac{1}{\sin(\theta)} = \frac{\text{hypotenuse}}{\text{opposite side}}$
- $\sec(\theta) = \frac{1}{\cos(\theta)} = \frac{\text{hypotenuse}}{\text{adjacent side}}$
- $\cot(\theta) = \frac{1}{\tan(\theta)} = \frac{\text{adjacent side}}{\text{opposite side}}$

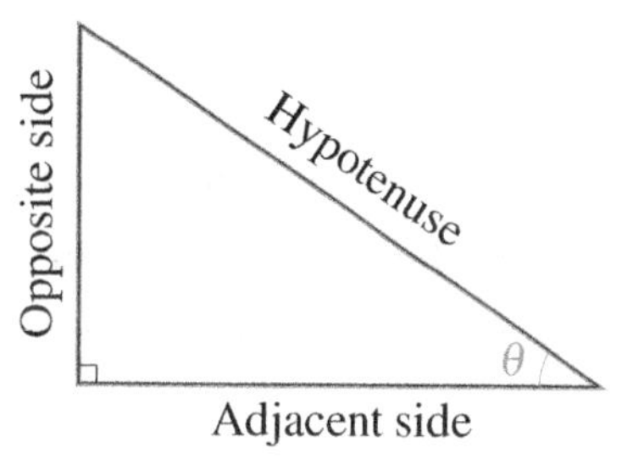


Using the definitions of these ratios, we can solve various problems involving right-angled triangles.

**Example** A right-angled triangle has a hypotenuse length of 13 units. The side opposite to angle $\theta$ is 5 units. Find $\sin(\theta)$ and $\cos(\theta)$.

**Solution:** To find $\sin(\theta) = \frac{\text{opposite}}{\text{hypotenuse}}$, we simply substitute the given lengths: $\sin(\theta) = \frac{5}{13}$. For $\cos(\theta)$, we first need to calculate the length of the adjacent side using the Pythagorean theorem:

$$\text{Adjacent} = \sqrt{\text{hypotenuse}^2 - \text{opposite}^2} = \sqrt{13^2 - 5^2} = \sqrt{169 - 25} = 12 \text{ units}.$$

Now we can find $\cos(\theta) = \frac{\text{adjacent}}{\text{hypotenuse}} = \frac{12}{13}$.

**Example** Calculate the length of the opposite side in a right-angled triangle where the angle $\theta$ is $30^\circ$ and the hypotenuse is 10 units long.

**Solution:** Using the sin function:

$$\sin\theta = \frac{\text{opposite}}{\text{hypotenuse}} \Rightarrow \sin 30^\circ = \frac{\text{opposite}}{10},$$

since $\sin 30^\circ = \frac{1}{2}$, we can solve for the opposite:

$$\frac{1}{2} = \frac{\text{opposite}}{10} \Rightarrow \text{opposite} = \frac{10}{2} = 5 \text{ units}.$$

## 8.4 Trigonometric Ratios for General Angles

Trigonometry extends beyond the scope of standard angles frequently encountered in mathematical tables. To understand and solve a broader array of problems, it is essential to explore general angles, which may not be explicitly listed in common trigonometric tables.

**Standard Trigonometric Values**. The following table summarizes the trigonometric ratios for some standard angles:

| $\theta$ | $0^\circ$ | $30^\circ$ | $45^\circ$ | $60^\circ$ | $90^\circ$ |
|---|---|---|---|---|---|
| $\sin\theta$ | 0 | $\frac{1}{2}$ | $\frac{\sqrt{2}}{2}$ | $\frac{\sqrt{3}}{2}$ | 1 |
| $\cos\theta$ | 1 | $\frac{\sqrt{3}}{2}$ | $\frac{\sqrt{2}}{2}$ | $\frac{1}{2}$ | 0 |
| $\tan\theta$ | 0 | $\frac{\sqrt{3}}{3}$ | 1 | $\sqrt{3}$ | Undefined |

**Understanding Sign Conventions in Trigonometry**. Trigonometric functions take on specific signs depending on the quadrant of the angle in the coordinate plane. Angles are measured from the positive $x$-axis, with positive angles going counterclockwise and negative angles going clockwise. The sign conventions for sine, cosine, and tangent in each quadrant are as follows:

- **Quadrant I** ($0^\circ$ to $90^\circ$): All trigonometric functions are positive.
- **Quadrant II** ($90^\circ$ to $180^\circ$): Sine is positive; cosine and tangent are negative.
- **Quadrant III** ($180^\circ$ to $270^\circ$): Tangent is positive; sine and cosine are negative.
- **Quadrant IV** ($270^\circ$ to $360^\circ$): Cosine is positive; sine and tangent are negative.

With these conventions, determining the trigonometric ratios for angles in any quadrant becomes straightforward.

**Key Trigonometric Relationships**. These relationships are pivotal for extending trigonometric calculations beyond the first quadrant:

- $\cos(x) = \sin(90^\circ - x)$, a reflection of the complementary nature of sine and cosine.
- $\sin(x) = \cos(90^\circ - x)$, indicating the interdependency between sine and cosine.
- $\cos(-x) = \cos(x)$, showing that cosine is an even function.
- $\sin(-x) = -\sin(x)$, indicating that sine is an odd function.

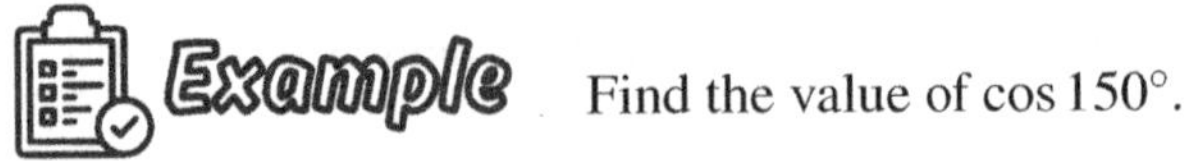

Find the value of $\cos 150^\circ$.

**Solution:** Using the identity $\cos(x) = \sin(90^\circ - x)$, we rewrite $\cos 150^\circ$ as $\sin(90^\circ - 150^\circ)$:

$$\cos 150^\circ = \sin(90^\circ - 150^\circ) = \sin(-60^\circ) = -\sin(60^\circ) = -\frac{\sqrt{3}}{2}.$$

Therefore, $\cos 150^\circ = -\frac{\sqrt{3}}{2}$.

Evaluate $\sin 120°$.

**Solution:** Using the identity $\sin(x) = \cos(90° - x)$, we rewrite $\sin 120°$ as $\cos(90° - 120°)$:

$$\sin 120° = \cos(90° - 120°) = \cos(-30°) = \cos(30°) = \frac{\sqrt{3}}{2}.$$

Therefore, $\sin 120° = \frac{\sqrt{3}}{2}$.

## 8.5 Trigonometry and the Calculator

Calculators allow us to take the theoretical understanding of trigonometric functions and apply them to real-world problems with speed and precision.

Always set your calculator to the correct angle measurement mode–degrees or radians–before calculating trigonometric values.

Calculators typically have separate modes for degrees and radians because they are two different ways of measuring angles:

- Degrees (°) are based on dividing a circle into 360 equal parts.
- Radians are based on the radius of the circle, where the arc length equal to the radius is 1 radian. There are $2\pi$ radians in a full circle.

Key Point

Degrees and radians are interconvertible, with 360° equal to $2\pi$ radians, implying $1° = \frac{\pi}{180}$ radians.

To calculate the value of a trigonometric function using a scientific calculator, follow these steps:

1. Press the MODE button on your calculator.
2. Select the correct angle measurement: DEGREE or RADIAN.
3. Press the trigonometric function button you need (sin, cos, or tan).
4. Enter the angle measurement.
5. Press 'Enter' to display the result.

Find $\cos(3.142)$ in radian mode to four decimal places.

**Solution:** First, ensure your calculator is in radian mode, as the angle is given in radians. Enter $\cos(3.142)$ and press 'Enter'. You should get: $\cos(3.142) \approx -1.0000$, which is almost the value of $\cos(\pi)$ since 3.142 is an approximation for $\pi$.

 **Example** Find $\tan(45^\circ)$ to four decimal places.

**Solution:** Make sure the calculator is in degree mode as the angle is given in degrees. Enter $\tan(45)$ and press 'Enter'. The result is: $\tan(45^\circ) = 1.0000$.

 Calculate the value of $\sin(225^\circ)$ to four decimal places.

**Solution:** Change the calculator to degree mode, then calculate $\sin(225^\circ)$. The expected result is: $\sin(225^\circ) = -0.7071$, this result corresponds to the sine of an angle in the third quadrant where sine values are negative.

## 8.6 Inverse Trigonometric Ratios

In the context of right-angled triangles, when determining an angle, one should select the appropriate trigonometric ratio corresponding to the known side lengths and apply the inverse function to ascertain the angle.

**Key Point**

Inverse trigonometric functions, namely arcsine ($\sin^{-1}$), arccosine ($\cos^{-1}$), and arctangent ($\tan^{-1}$), are pivotal in determining angles when the lengths of sides in a triangle are known.

Let us revisit the concepts we have covered:

- For arcsine, given that $\sin\theta = a$, it follows that $\theta = \sin^{-1} a$.
- In the case of arccosine, if $\cos\theta = a$, then the angle $\theta$ is obtained by $\theta = \cos^{-1} a$.
- Concerning arctangent, if $\tan\theta = a$, then the angle $\theta$ is derived as $\theta = \tan^{-1} a$.

Each of these functions returns an angle within its principal range:

- The function $\sin^{-1}$ yields values within the range $[-90^\circ, 90^\circ]$ or $[-\frac{\pi}{2}, \frac{\pi}{2}]$ in radians.
- The function $\cos^{-1}$ provides angles in the range $[0^\circ, 180^\circ]$ or $[0, \pi]$ in radians.
- The function $\tan^{-1}$ produces results in the interval $(-90^\circ, 90^\circ)$ or $(-\frac{\pi}{2}, \frac{\pi}{2})$ in radians.

**Example** Given a right triangle where the opposite side to our angle is 7 units and the hypotenuse is 25 units, calculate the angle.

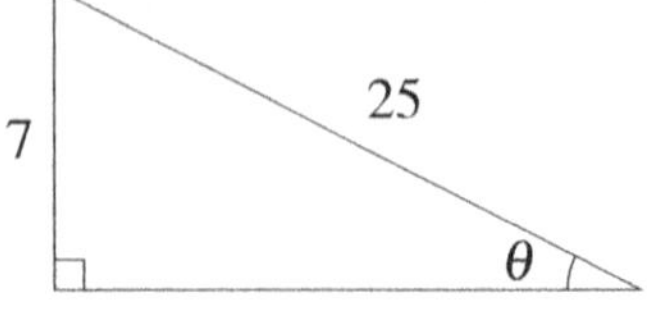

**Solution:** For the angle $\theta$, the opposite side is 7 and the hypotenuse is 25, so we use the arcsine ratio.

$$\sin\theta = \frac{7}{25} \Rightarrow \theta = \sin^{-1}\left(\frac{7}{25}\right).$$

To find the angle $\theta$, we can use a calculator:

$$\theta = \sin^{-1}\left(\frac{7}{25}\right) \approx 16.26^\circ.$$

 A ladder leans against a wall, forming an angle with the ground. If the bottom of the ladder is 5 feet from the wall and it reaches a height of 12 feet, find the angle the ladder makes with the ground.

**Solution:** Use the tangent function, as the ground and wall form a right angle. The opposite side is 12, the adjacent side is 5.

$$\tan\theta = \frac{12}{5} \Rightarrow \theta = \tan^{-1}\left(\frac{12}{5}\right).$$

Using a calculator,

$$\theta = \tan^{-1}\left(\frac{12}{5}\right) \approx 67.38^\circ.$$

## 8.7 Solving Right Triangles

A right triangle consists of the adjacent, opposite, and hypotenuse sides relative to a particular angle. Specifically:

- The **adjacent** side is next to the angle of interest.
- The **opposite** side is across from the angle of interest.
- The **hypotenuse** is the longest side, opposite the right angle.

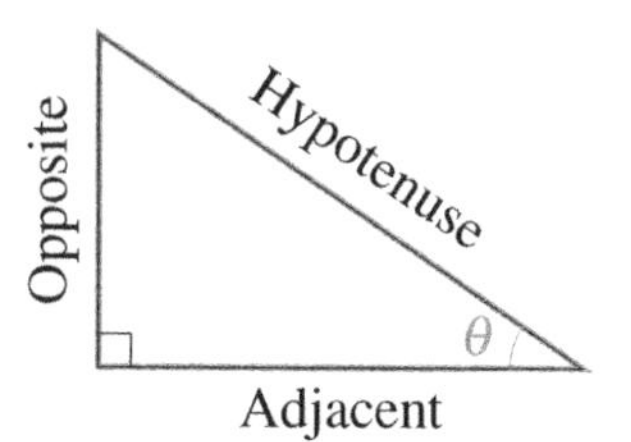


We apply the mnemonic SOH–CAH–TOA to remember that:

- $\sin\theta = \frac{\text{opposite}}{\text{hypotenuse}}$
- $\cos\theta = \frac{\text{adjacent}}{\text{hypotenuse}}$
- $\tan\theta = \frac{\text{opposite}}{\text{adjacent}}$

The angle of interest in SOH–CAH–TOA is always the one which is not the right angle.

 Find $AC$ in the right triangle where $\angle A = 30^\circ$ and $BC = 10$ units.

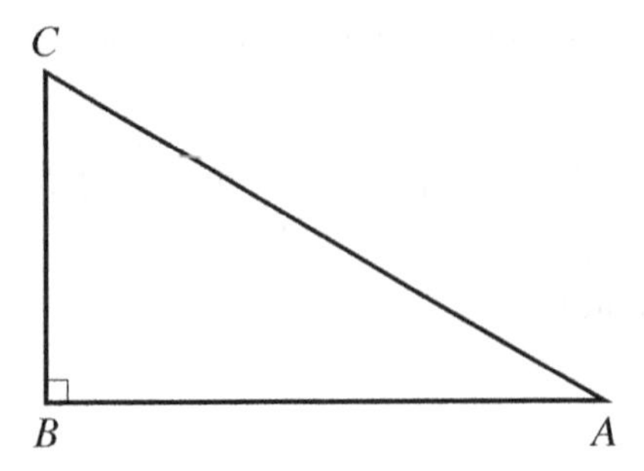


**Solution:** We will use the trigonometric function that relates the angle to the opposite side and hypotenuse, which is sine:

$$\sin(30^\circ) = \frac{BC}{AC} \Rightarrow \frac{1}{2} = \frac{10}{AC} \Rightarrow AC = 20.$$

Note that $\sin(30^\circ) = \frac{1}{2}$.

## 8.8 Trigonometry and Area of Triangles

Trigonometry provides a special way to calculate the area of a triangle using two sides and the angle between them. This approach is particularly helpful when the triangle's height is not readily available or challenging to figure out.

The area '$A$' of a triangle given two sides $a$ and $b$ and the included angle $C$ is given by the formula $A = \frac{1}{2}ab\sin C$.

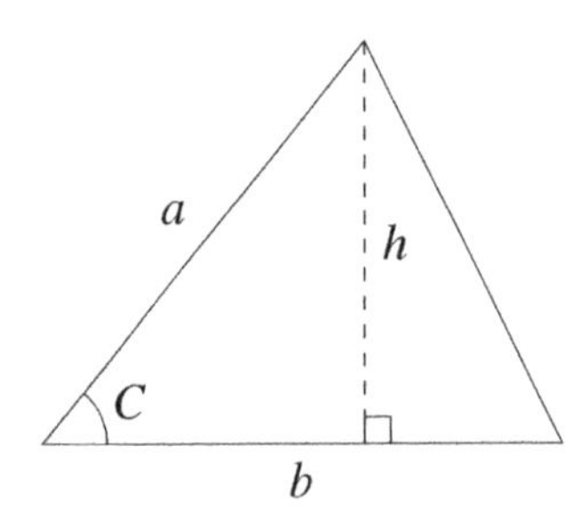


This formula is a direct result of understanding the relationship between the sides of a triangle and its altitudes. More precisely, when you drop an altitude from the vertex opposite the included angle $C$, you effectively form two right triangles. These right triangles allow us to use trigonometric functions to find the height $h$. In mathematical terms, $h = a\sin C$. Subsequently, the area of the triangle can be calculated using the base $b$ and the height $h$, leading us back to the formula: $A = \frac{1}{2}ab\sin C$.

Find the area of a triangle with sides of lengths 5 $cm$ and 7 $cm$ and an included angle of $60^\circ$.

**Solution:** Implementing our formula, we calculate:

$$A = \frac{1}{2} \times 5 \times 7 \times \sin 60^\circ = \frac{1}{2} \times 5 \times 7 \times \frac{\sqrt{3}}{2}$$
$$= 35\frac{\sqrt{3}}{4} \approx 15.2\, cm^2.$$

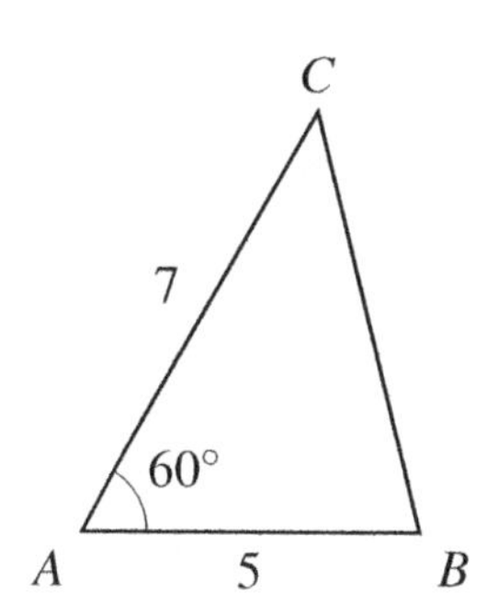


## 8.9 Law of Sines

In trigonometry, the Law of Sines is invaluable when dealing with non-right triangles. It allows us to find unknown angles and sides when we know at least one side and its opposite angle (ASA or AAS situations).

The Law of Sines states that for any triangle $\triangle ABC$, the ratio of the length of a side to the sine of its opposite angle is the same for all three sides of the triangle:

$$\frac{a}{\sin A} = \frac{b}{\sin B} = \frac{c}{\sin C}.$$

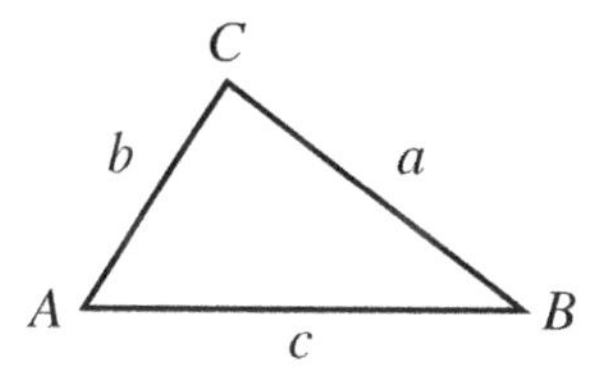


**Key Point**

The Law of Sines describes a relationship between the sides and angles of a triangle.

Given a triangle with $a = 12$ cm, $b = 15$ cm, and angle $B = 62^\circ$. Calculate the measure of angle $A$.

**Solution:** Using the Law of Sines, we start by finding the ratio for given side $b$ and its opposite angle $B$:

$$\frac{a}{\sin A} = \frac{b}{\sin B} \Rightarrow \frac{12}{\sin A} = \frac{15}{\sin 62^\circ} \Rightarrow \sin A = \frac{12}{15} \sin 62^\circ \approx 0.71.$$

Therefore, $A \approx \arcsin(0.71)$ or $A \approx 45^\circ$.

## 8.10 Law of Cosines

The Law of Cosines helps in determining the length of the third side of a triangle, given the lengths of two sides and the angle between them. Furthermore, the Law of Cosines is useful for calculating the angles of a triangle when the lengths of all three sides are known.

Key Point

The Law of Cosines is pivotal in finding an unknown side or angle in a triangle when given two sides and the included angle, or when all sides are known.

Consider a triangle $ABC$ with sides $a$, $b$, and $c$, opposite to angles $A$, $B$, and $C$, respectively. The Law of Cosines can be represented in three equivalent forms:

$$a^2 = b^2 + c^2 - 2bc\cos A \Rightarrow \cos A = \frac{b^2+c^2-a^2}{2bc}.$$

$$b^2 = a^2 + c^2 - 2ac\cos B \Rightarrow \cos B = \frac{a^2+c^2-b^2}{2ac}.$$

$$c^2 = a^2 + b^2 - 2ab\cos C \Rightarrow \cos C = \frac{a^2+b^2-c^2}{2ab}.$$

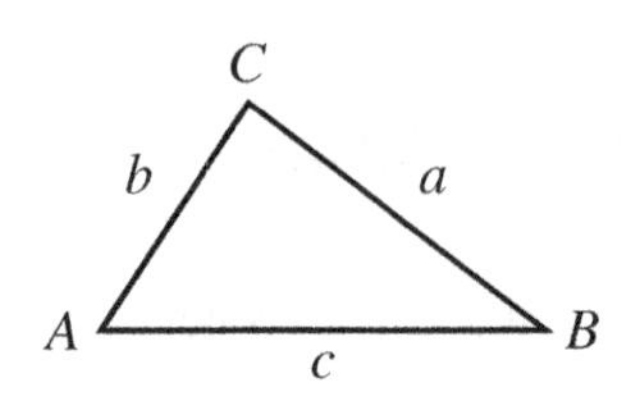


Let us proceed with illustrative examples.

Find angle $B$ in the triangle $ABC$ where $a = 15$, $b = 13$, and $c = 9$.

**Solution:** To determine angle $B$, use the Law of Cosines:

$$\cos B = \frac{a^2+c^2-b^2}{2ac} = \frac{15^2+9^2-13^2}{2\times 15\times 9} = \frac{137}{270} \approx 0.51.$$

Thus, $B = \cos^{-1}(0.51) \approx 59.51^\circ$.

Determine the length of side $c$ in the triangle $ABC$ where sides $a$ and $b$ have lengths 7 and 11, respectively, and angle $C$ measures $65^\circ$.

**Solution:** Using the Law of Cosines:

$$\begin{aligned} c^2 = a^2 + b^2 - 2ab\cos C &= 7^2 + 11^2 - 2(7)(11)\cos 65^\circ \\ &= 170 - 154\cos 65^\circ \\ &\approx 170 - 65.08 = 104.92. \end{aligned}$$

Thus, side $c$ is approximately $\sqrt{104.92} \approx 10.24$ units long.

## 8.11 Trigonometric Applications

We continue our exploration of trigonometry by studying its real-world applications. Trigonometry is not just confined to theoretical mathematics but also extends into many practical fields such as physics, engineering, astronomy, architecture, and navigation.

**Key Point**

The basic trigonometric functions—the sine (sin), cosine (cos), and tangent (tan)—describe the ratios of sides in right triangles related to specific angles.

**Key Point**

The core trigonometric functions can be used to model physical phenomena, such as wave oscillations in Physics and roof inclinations in Architecture.

-In Physics, trigonometry is indispensable in understanding complex waveforms and vibrational patterns. For example, the sine and cosine functions are the backbone of harmonic motion equations and the analysis of sound and light waves.
-Engineers rely on trigonometry to calculate forces, mechanical stress, and angles within constructed structures. It is fundamental in both static designs, where structures must withstand certain loads, and in dynamic systems involving rotating components.
-Astronomy harnesses the power of trigonometry to interpret the cosmos, measuring distances to stars and mapping their positions. The principles learned from triangles on Earth apply to vast celestial distances when angles are measured with precision.
-In Architecture, the aesthetic and structural integrity of buildings often depend on trigonometric calculations. Roof pitches and the angles of supports are all influenced by architectural trigonometry.
-For Navigation, the direction, and distance between two points can be solved using trigonometry, which is crucial for maritime and aerial travel as well as in surveying and mapmaking.

**Example** A ladder leaning against a wall forms a $60^\circ$ angle with the horizontal. The base of the ladder is 2.5 meters from the wall. How far up the wall does the ladder reach?

**Solution:** We represent the scenario with a right-angled triangle where the ladder serves as the hypotenuse, and we seek the opposite side (the height up the wall) to the $60^\circ$ angle. Using the tangent function:

$$\tan(60^\circ) = \frac{\text{opposite}}{\text{adjacent}}.$$

We can now solve for the height ($h$):

$$h = \text{adjacent} \times \tan(60^\circ) = 2.5 \times \sqrt{3} \approx 4.33 \text{ meters}.$$

The ladder reaches approximately 4.33 meters up the wall.

## 8.12 Practices

**1)** Simplify each trigonometric expression using Pythagorean identities.

1-1) $(\sin x + \cos x)^2$

1-2) $(1 + \cot^2 x)\sin^2 x$

1-3) $\csc^2 x - \cot^2 x$

1-4) $2\sin^2 x + \cos^2 x$

**2)** Find the value of $x$ and $y$ in the following special right triangles.

2-1)

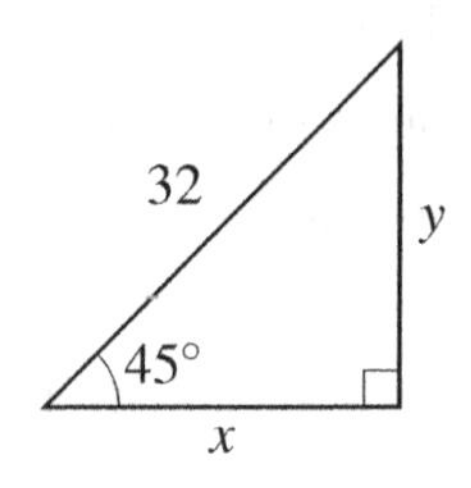


2-3)

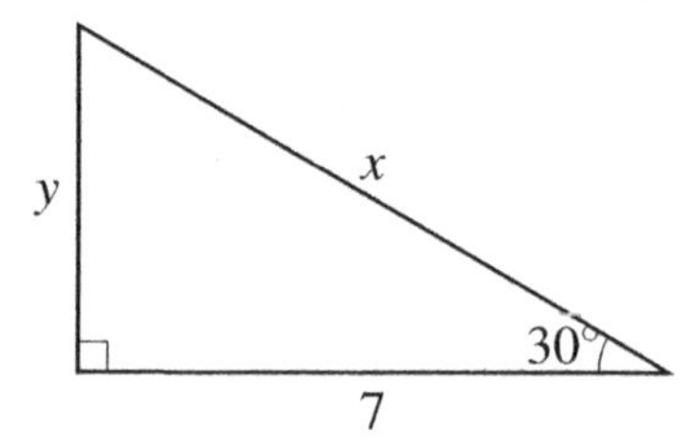


2-2)

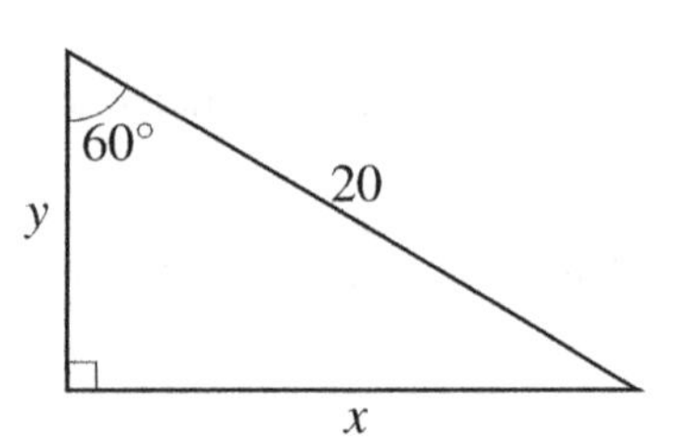


2-4)

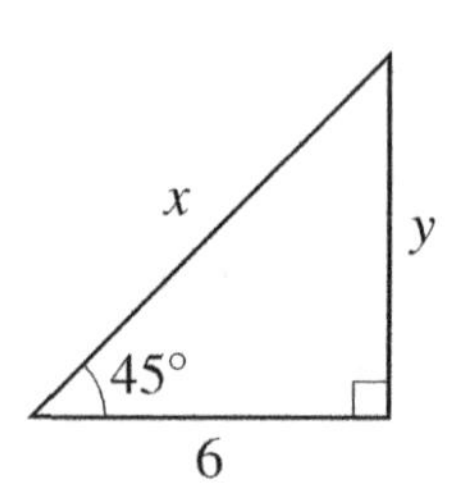


**3)** Find the given trigonometric ratio.

3-1) $\tan O$

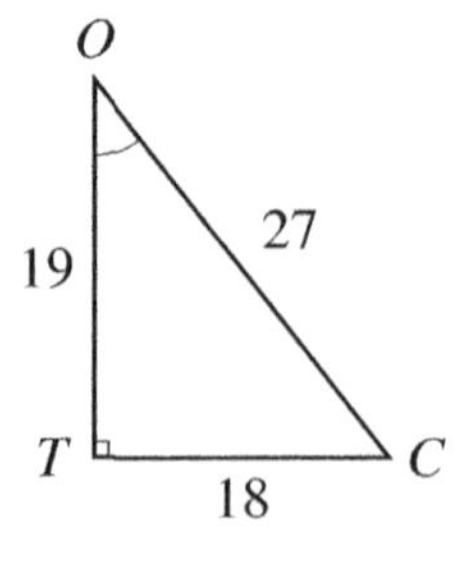


3-2) $\sin X$

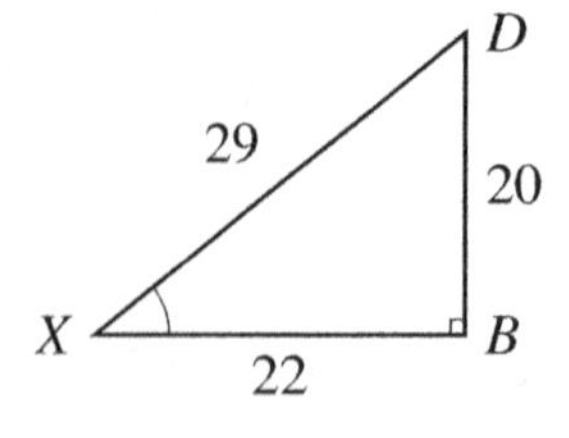


3-3) $\cos X$

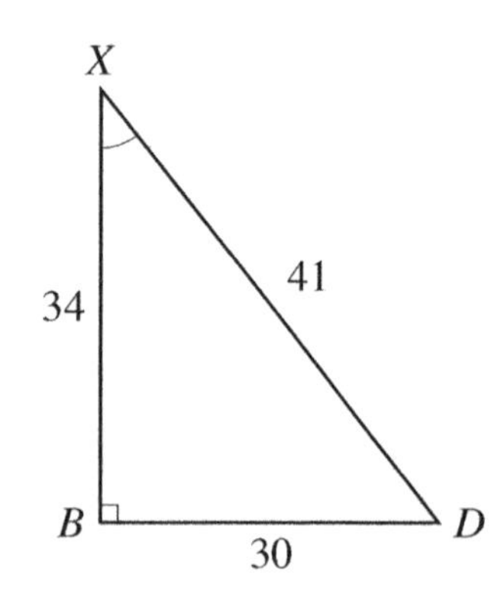


**4)** Evaluate.

| | | | |
|---|---|---|---|
| 4-1) $\sin 120^\circ$ | 4-3) $\tan(-90^\circ)$ | 4-5) $\cos(-90^\circ)$ | 4-7) $\csc 480^\circ$ |
| 4-2) $\sin(-330^\circ)$ | 4-4) $\cot 90^\circ$ | 4-6) $\sec 60^\circ$ | 4-8) $\cot(-135^\circ)$ |

**5)** Find the measure of each side indicated. round to the nearest tent.

5-1)

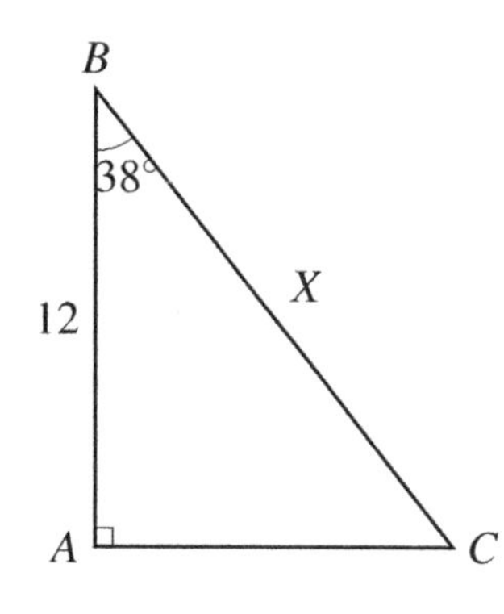


5-2)

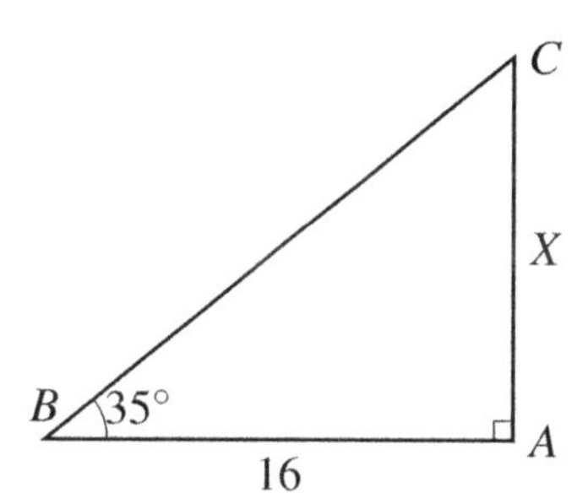


5-3)

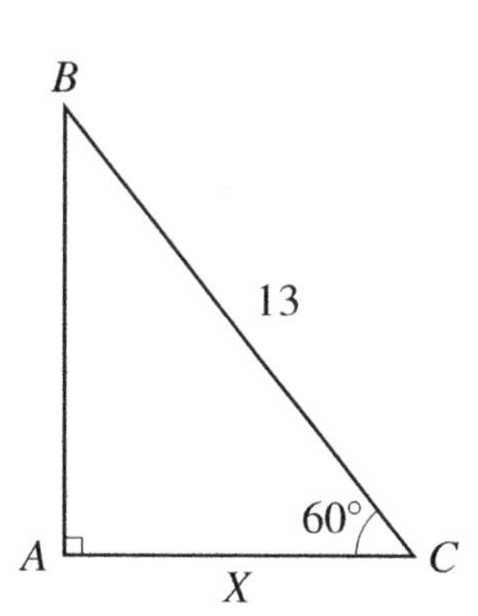


**6)** Find angles to four decimal places.

6-1) $\csc 66^\circ$

6-2) $\sec 56^\circ$

6-3) $\cos 120^\circ$

6-4) $\tan 44^\circ$

**7)** Find each angle measure to the nearest degree.

7-1) $\cos A = 0.4226$

7-2) $\tan W = 1$

7-3) $\sin A = 0.9659$

7-4) $\tan W = 1.1106$

7-5) $\sin A = 0.3907$

7-6) $\tan W = 0.5773$

7-7) $\sin A = 0.5000$

7-8) $\cos B = 0.1392$

**8)** Find the measure of each angle indicated.

8-1)

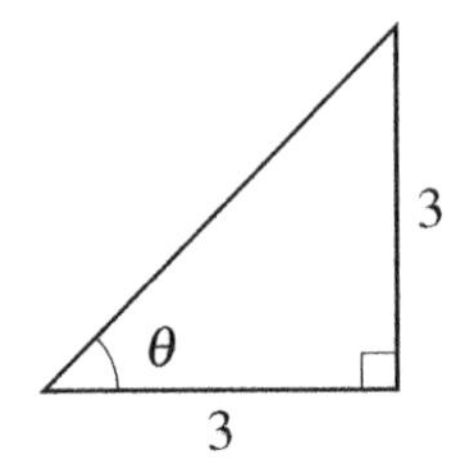


8-2)

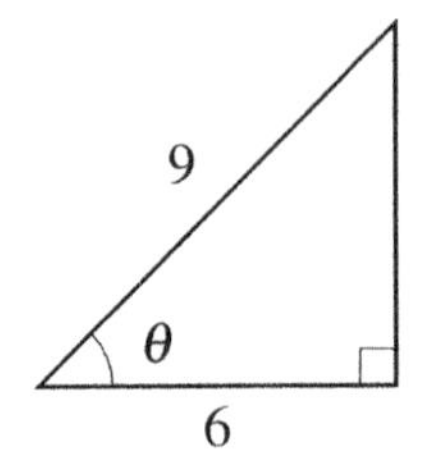


**9)** Find the missing sides. Round answers to the nearest tenth.

9-1)

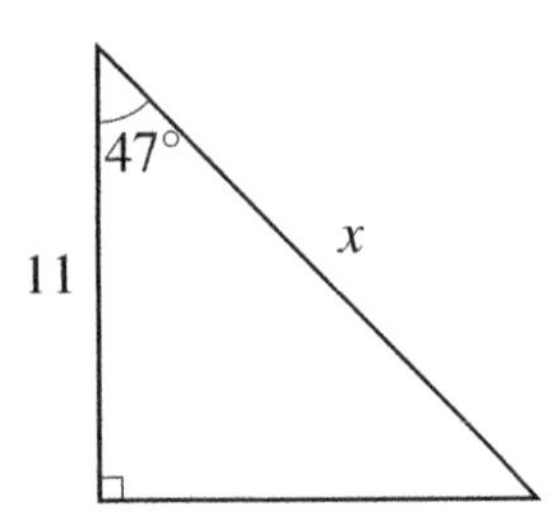


9-2)

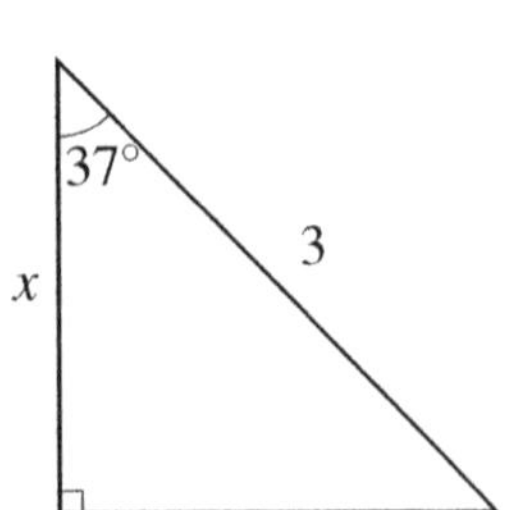


9-3)

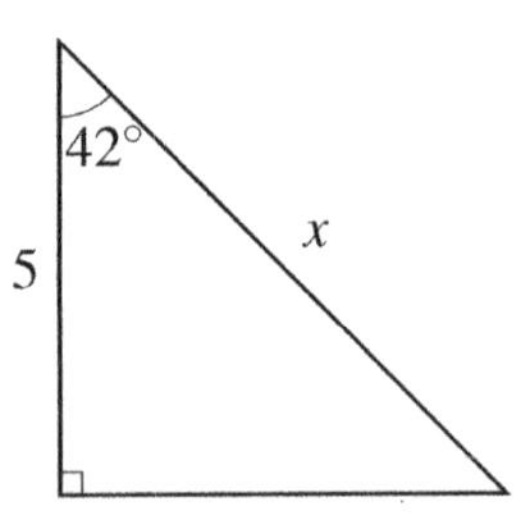


9-4)

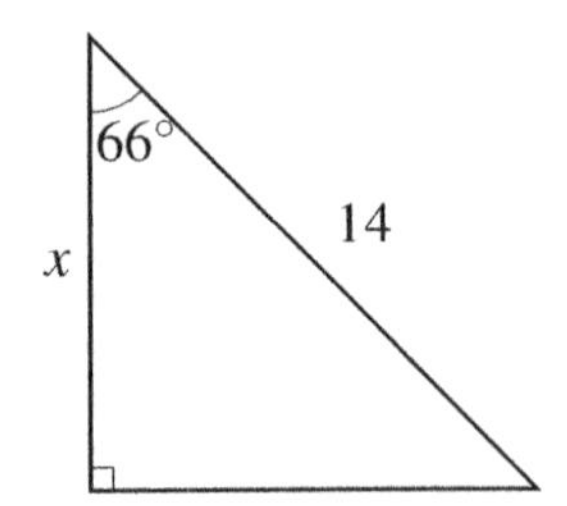

**10)** Find the area of each regular polygon. Round your answer to the nearest tenth.

10-1) Perimeter $= 27\ mi$

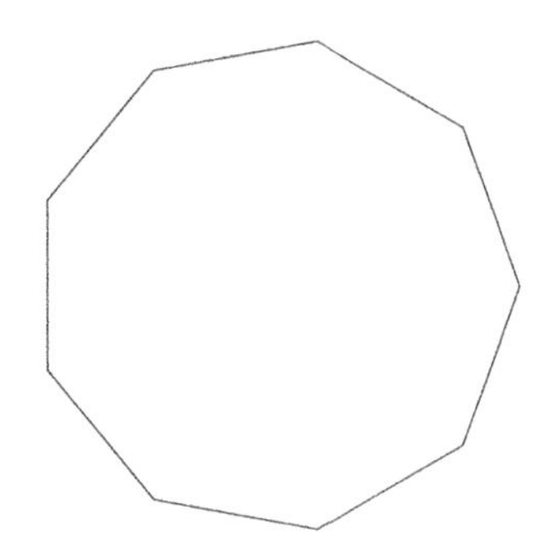

10-2)

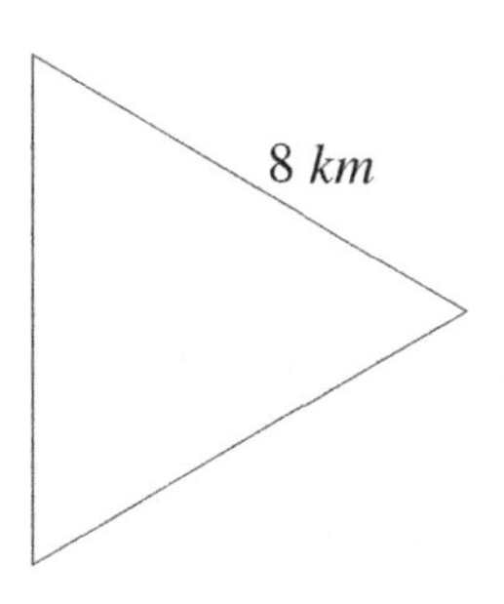


10-3) Perimeter $= 60\ cm$

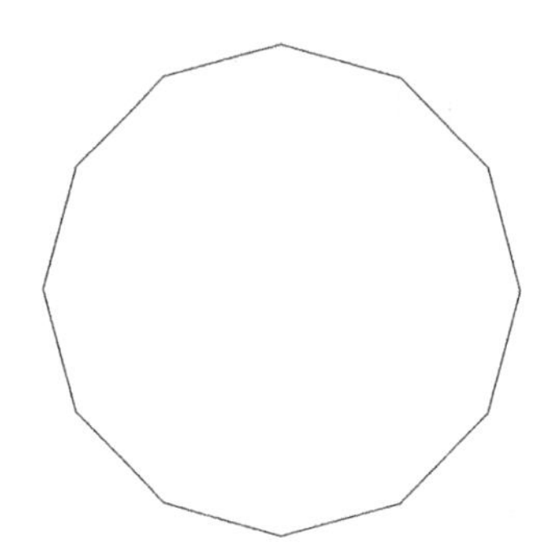

10-4)

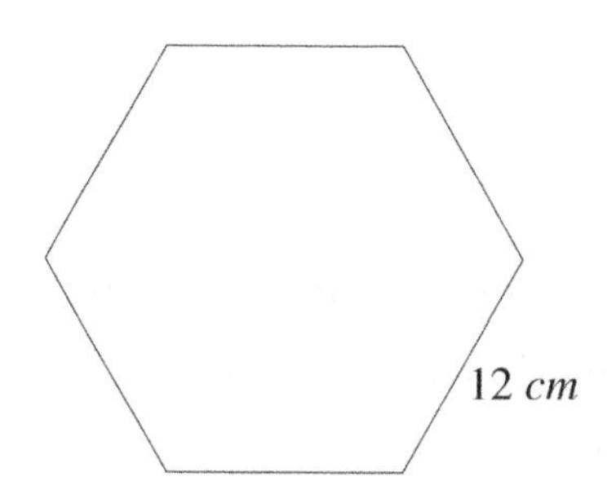


10-5) A regular hexagon with a perimeter of 60 $yd$.

10-6) A regular pentagon 8 $ft$ on each side.

**11)** Find each measurement indicated. Round your answers to the nearest tenth.

11-1)

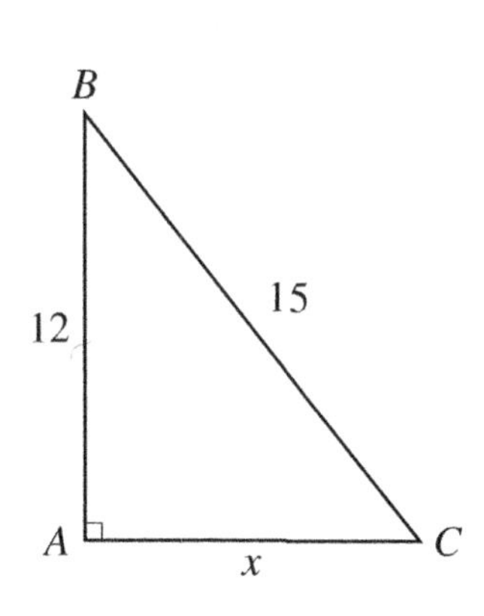


11-2)

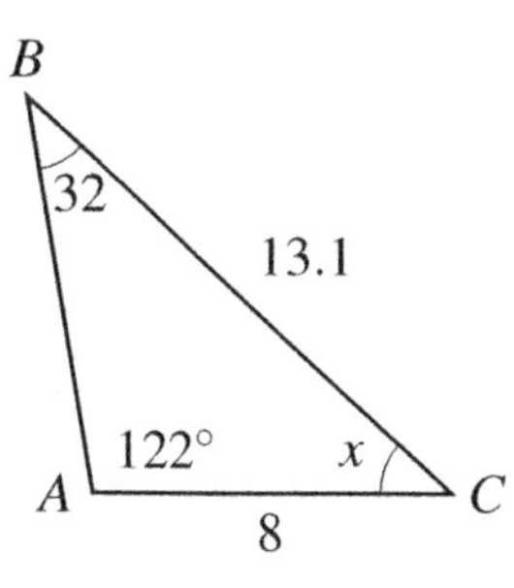


11-3)

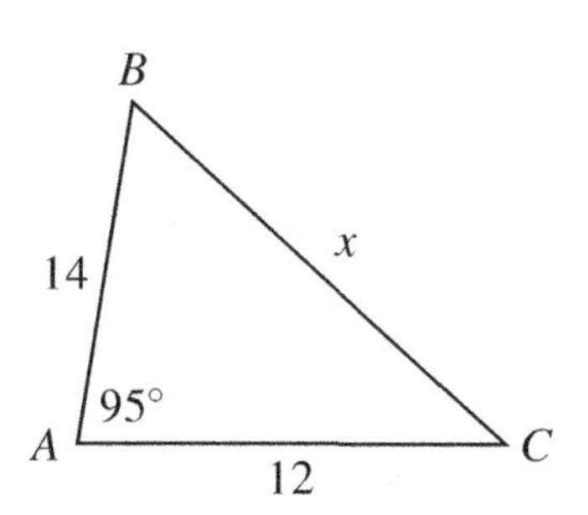

11-4)

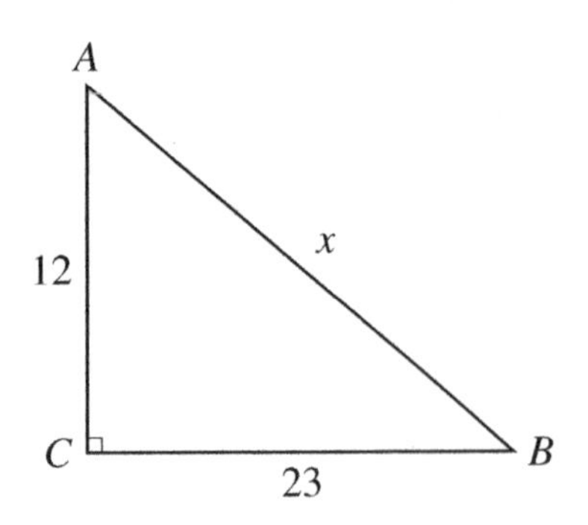


11-5) In $\triangle ABC$, $a = 15\ cm$, $b = 10\ cm$, $c = 7\ cm$

11-6) In $\triangle ABC$, $a = 18\ cm$, $b = 15\ cm$, $c = 11\ cm$

**12)** Solve.

12-1) A string of a kite is 10 meters long and the inclination of the string with the ground is $30^\circ$. Find the height of the kite.

12-2) From the top of the tree 90 $m$ height a man is observing the base of a tree at an angle of depression measuring $30^\circ$. Find the distance between the two trees.

12-3) A ladder is leaning against a vertical wall makes an angle of $45^\circ$ with the ground. The foot of the ladder is 5 $m$ from the wall. Find the length of ladder.

12-4) To find the height of a building, a boy measure the length of its shadow as 12 $m$. From the end of the shadow on the ground to the highest height of the building, He observes that an angle of $34^\circ$. Calculate the height of the building.

## 8.13 Answers

**1)**

1-1) $1+2\sin x\cos x$

1-2) 1

1-3) 1

1-4) $1+\sin^2 x$

**2)**

2-1) $x=y=16\sqrt{2}$

2-2) $x=10\sqrt{3}, y=10$

2-3) $x=\frac{14\sqrt{3}}{3}, y=\frac{7\sqrt{3}}{3}$

2-4) $x=6\sqrt{2}, y=6$

**3)**

3-1) $\frac{18}{19}$

3-2) $\frac{20}{29}$

3-3) $\frac{34}{41}$

**4)**

4-1) $\frac{\sqrt{3}}{2}$

4-2) $\frac{1}{2}$

4-3) Undefined

4-4) 0

4-5) 0

4-6) 2

4-7) $\frac{2\sqrt{3}}{3}$

4-8) 1

**5)**

5-1) 15.2

5-2) 11.2

5-3) 6.5

**6)**

6-1) 1.0946

6-2) 1.7882

6-3) $-0.5$

6-4) 0.9656

**7)**

7-1) $65^\circ$

7-2) $45^\circ$

7-3) $75^\circ$

7-4) $48^\circ$

7-5) $23^\circ$

7-6) $30^\circ$

7-7) $30^\circ$

7-8) $82^\circ$

**8)**

8-1) $45^\circ$

8-2) $48.19^\circ$

**9)**

9-1) 16.1

9-2) 2.4

9-3) 6.7

9-4) 5.7

**10)**

10-1) 55.6

10-2) 27.7

10-3) 279.9

10-4) 374.1

10-5) 259.8

10-6) 110.1

**11)**

11-1) $x=9$

11-2) $x=26^\circ$

11-3) $x=19.2$

11-4) $x=26$

11-5) $\angle A=122.9^\circ, \angle B=34.1^\circ, \angle C=23^\circ$

11-6) $\angle A=86.1^\circ, \angle B=56.3^\circ, \angle C=37.6^\circ$

**12)**

12-1) $5\ m$

12-2) $90\sqrt{3}\ m$

12-3) $5\sqrt{2}\ m$

12-4) $8.04\ m$

# 9. Circle Geometry

## 9.1 The Unit Circle

The unit circle is a fundamental concept in geometry and trigonometry. It refers to a circle with a radius of 1, centered at the origin of a coordinate system. The unit circle is not just a geometric shape; it serves as a bridge to trigonometric functions and their relationship with angles.

**Key Point**

The equation for the unit circle in the Cartesian coordinate system is $x^2 + y^2 = 1$.

When we place an angle in standard position (its vertex at the origin and one side along the positive $x$-axis), the terminal side of the angle will intersect the unit circle at a point $P$. If this angle is $\theta$ and the coordinates of $P$ are $(x, y)$, then we have a direct connection between angles and points on the circle.

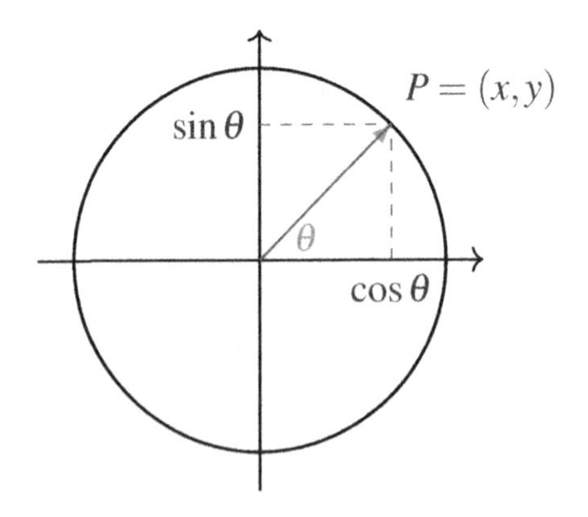


**Key Point**

The $x$-coordinate of point $P$ on the unit circle represents the cosine of angle $\theta$, written as $\cos(\theta) = x$.

**Key Point**

The $y$-coordinate of point $P$ on the unit circle represents the sine of angle $\theta$, written as $\sin(\theta) = y$.

These relationships give us the tools to convert between geometric representations of angles and their trigonometric functions.

If point $P\left(\frac{\sqrt{3}}{2}, -\frac{1}{2}\right)$ is on the unit circle and corresponds to an angle $\theta$ in standard position, find $\sin(\theta)$ and $\cos(\theta)$.

**Solution:** Given the coordinates of $P$, we identify $\cos(\theta) = \frac{\sqrt{3}}{2}$ and $\sin(\theta) = -\frac{1}{2}$.

## 9.2 Arc Length and Sector Area

The arc length of a sector and the area of a sector are two important measures in circle geometry. Both of these quantities can be determined using the central angle of the sector.

**Sector Area**. To find the area of a sector of a circle, you have to know its radius ($r$) and the central angle ($\theta$) in degrees.

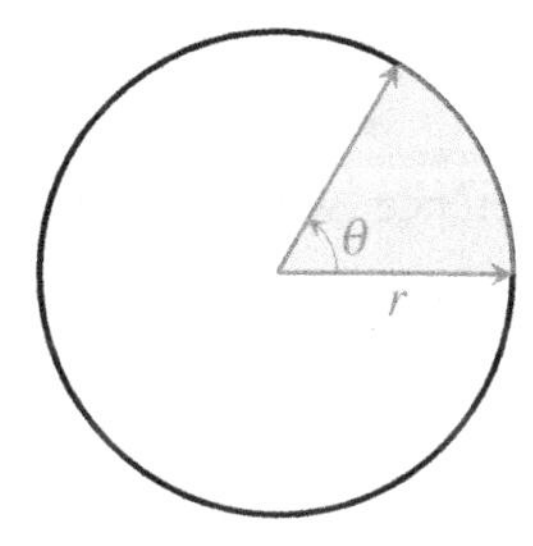


Use the following formula:

$$\text{Area of a sector} = \pi r^2 \left(\frac{\theta}{360}\right).$$

This formula gives us the portion of the circle's area that the sector occupies.

**Key Point**

The formula for the area of a sector is derived from the fact that $\frac{\theta}{360}$ is the fraction of the full circle that the sector represents.

**Arc Length**. Similarly, to find the length of the arc of a sector, we use the radius and the central angle. The formula is given by:

$$\text{Arc of a sector} = \left(\frac{\theta}{180}\right)\pi r.$$

The arc length is a measure of the distance along the edge of the circle that makes up the sector.

**Key Point**

Remember to always express the central angle in degrees when using the arc length of a sector and the area of sector formulas.

 **Example** Given a circle of radius 10 $cm$, find the length of an arc and the area of the corresponding sector for a 90° central angle. ($\pi \approx 3.14$)

**Solution:** For the arc length, use the formula:

$$\text{Length of arc} = \left(\frac{90}{180}\right)\pi(10) = \frac{1}{2}\pi(10) = 5\pi\ cm \approx 15.7\ cm.$$

For the sector area:

$$\text{Area of sector} = \pi(10)^2\left(\frac{90}{360}\right) = \pi(100)\left(\frac{1}{4}\right) = 25\pi\ cm^2 \approx 78.5\ cm^2.$$

The approximate length of the arc is 15.7 $cm$ and the area of the sector is approximately 78.5 $cm^2$.

## 9.3 Arcs and Central Angles

An arc represents a portion of the circle's circumference, while a central angle is the angle from the center of the circle to the endpoints of the arc on the circumference. The central angle's measure determines the size of the corresponding arc.

**Key Point**

The measure of an arc in degrees is equivalent to its central angle's measure in degrees.

A complete circle has 360°, and a circle's arc is a fraction of the circle's circumference. The relationship between the arc length and central angle is expressed through the formula:

$$\text{Arc Length} = \left(\frac{\theta}{360}\right)2\pi r$$

where $\theta$ is the central angle in degrees, and $r$ is the radius of the circle.

**Key Point**

To find an arc length, multiply the fraction of the central angle over 360° by the circumference of the entire circle.

This understanding is an extension from our previous exploration of arc length and sector area, completing our grasp of how central angles are at the heart of circle geometry.

 **Example** A circle with a radius of 8 units has a central angle that measures 45°. Calculate the length of the arc.

**Solution:** Using the arc length formula:

$$L = \left(\frac{\theta}{360}\right) 2\pi r = \left(\frac{45}{360}\right) 2\pi(8) = \frac{1}{8}(2\pi)(8) = 2\pi.$$

Hence, the length of the arc is $2\pi$ or approximately 6.28 units.

## 9.4 Arcs and Chords

The way arcs and chords work together is an important part of learning about circles in geometry. Arcs and chords show interesting patterns and links inside a circle, which lead to different rules and uses.

**Key Point**

A chord is a straight-line segment with endpoints on the circle's circumference. The diameter, a special type of chord, spans the circle through its center and is the longest possible chord.

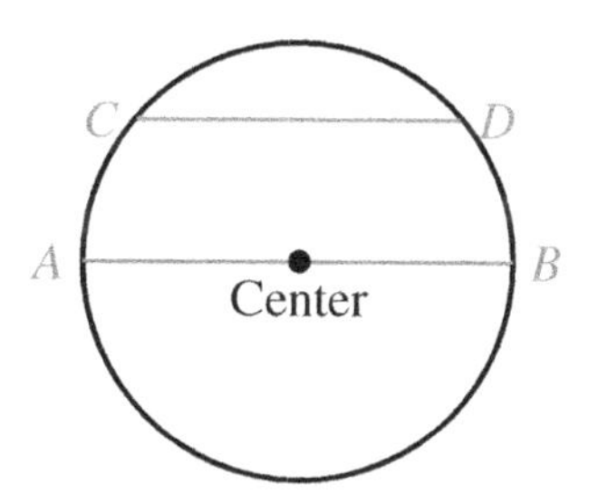


Diameter $AB$ and Chord $CD$

**Key Point**

Equal chords in a circle correspond to arcs of equal length. This reveals the chord's property of equidistance from the circle's center when lengths are consistent.

**Key Point**

The perpendicular bisector of a chord passes through the circle's center, dividing the chord into two equal segments. This is a fundamental attribute in the congruence of chords and arcs.

**Key Point**

Chords that yield arcs of equal length lie an equal distance from the circle's center. This equidistance is pivotal in determining the lengths of unknown chords by comparing them to known ones.

Chords and arcs showcase various properties significant to circle geometry. For example, if two chords are of equal length, the arcs they subtend will also be of equal length. This relationship can be used to solve problems involving distances and measures in a circle.

**Example** In a circle with radius 10 units, a chord is drawn 8 units from the center. Calculate the length of the chord.

**Solution:** Using the Pythagorean theorem, let $d$ be the distance from the center to the chord, and $l$ be the half-length of the chord. Hence, $l^2 + d^2 = r^2$. Substituting the given values: $l^2 + 8^2 = 10^2$, we find $l = \sqrt{10^2 - 8^2} = 6$. Therefore, the length of the chord is $2l$, which computes to $2(6) = 12$ units.

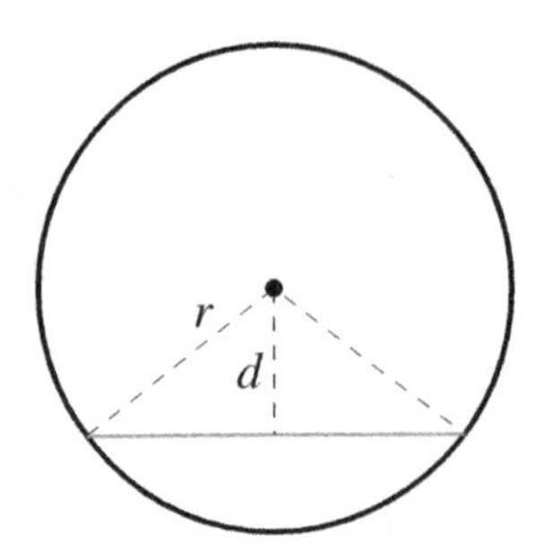


**Example** In a circle, two chords, $AB$ and $CD$, are parallel and equidistant from the center. If $AB$ is 8 units in length, find the length of $CD$.

**Solution:** Since $AB$ and $CD$ are equidistant and parallel, they are essentially the same length due to the circle's symmetry; therefore, $CD$ is also 8 units long.

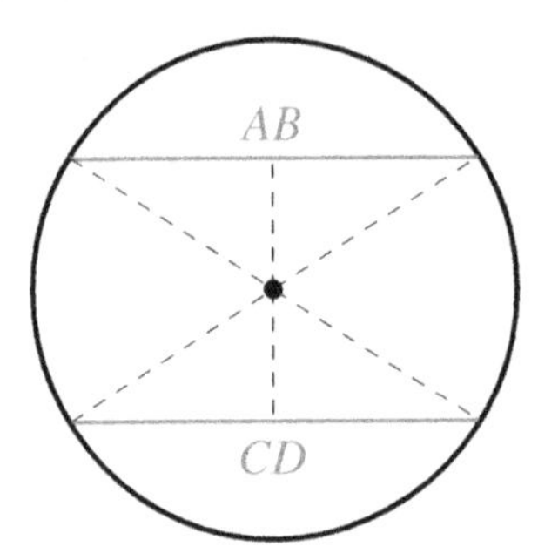


## 9.5 Inscribed Angles

An inscribed angle is created by two chords in a circle that meet at a single point on the circumference.

**Key Point**

The measure of an inscribed angle is half the measure of its intercepted arc. This foundational property is known as the Inscribed Angle Theorem.

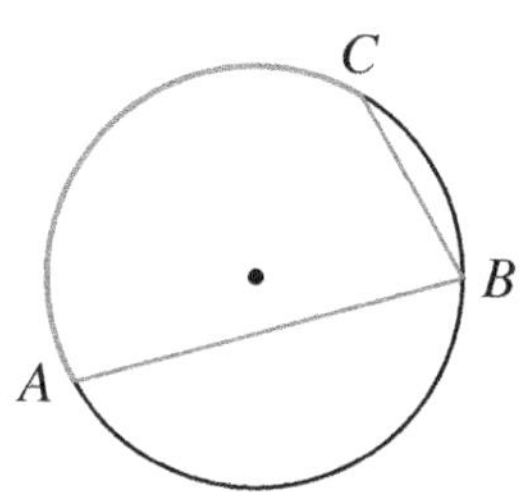


Inscribed angle $\measuredangle ABC$ and its intercepted arc $\overset{\frown}{AC}$; $\measuredangle ABC = \frac{1}{2}\overset{\frown}{AC}$.

We can also note an angle inscribed in a semicircle always forms a right angle, as due to its intercepted arc being a diameter, the phenomenon holds true.

**Key Point**

Angles inscribed in the same arc, or subtended by the same chords, are congruent.

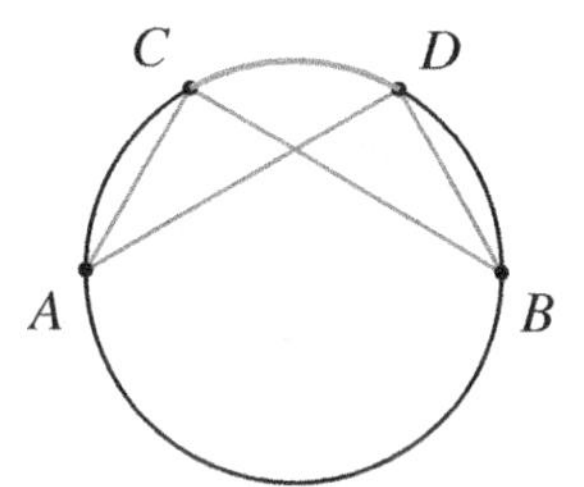


Angles $\angle DAC$ and $\angle CBD$ are congruent and intercept the same arc $CD$.

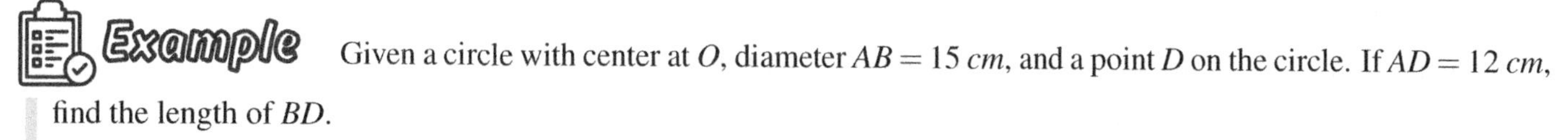

**Example** Given a circle with center at $O$, diameter $AB = 15$ $cm$, and a point $D$ on the circle. If $AD = 12$ $cm$, find the length of $BD$.

**Solution:** The angle $ADB$ is inscribed in a semicircle with diameter $AB$, hence it is a right angle $90°$.

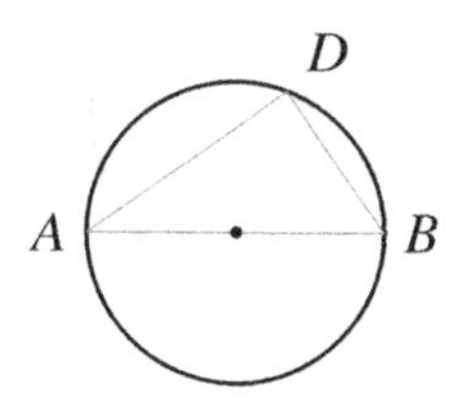


Using the Pythagorean theorem on triangle $ADB$, we have:

$$12^2 + BD^2 = 15^2 \Rightarrow BD^2 = 225 - 144 = 81 \Rightarrow BD = \sqrt{81} = 9.$$

Therefore, the length of $BD$ is 9 $cm$.

## 9.6 Tangents to Circles

A tangent to a circle is a straight line that touches the circle at exactly one point, known as the *point of tangency*. Circle tangents provide a wealth of geometric properties that can be applied to solve various problems. Let us explore some of these key properties.

**Key Point**

The radius drawn to the point of tangency is perpendicular to the tangent.

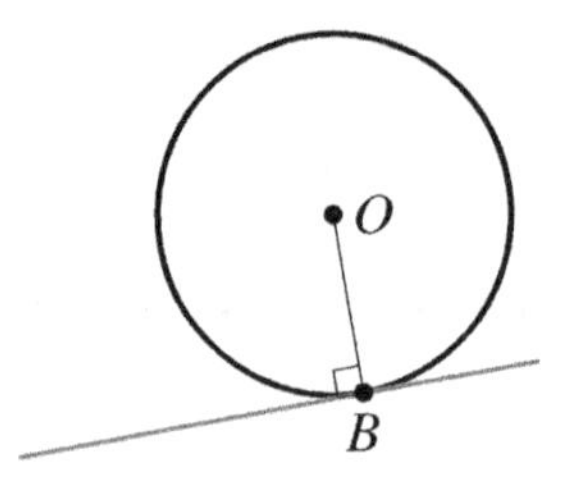


**Key Point**

If two tangents are drawn from a single external point to a circle, they will be of equal length.

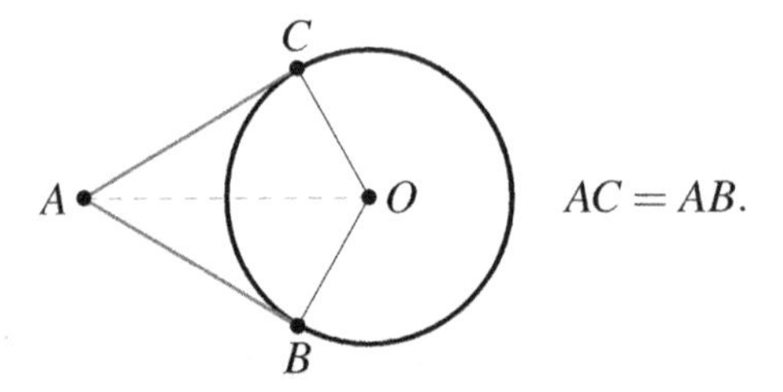


$AC = AB.$

**Key Point**

The angle formed by a tangent and a chord is equal to half the arc intercepted by the chord.

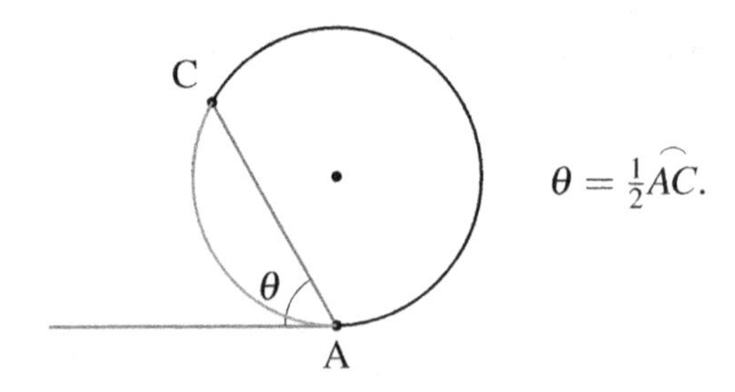


$\theta = \frac{1}{2}\overset{\frown}{AC}.$

**Example** A circle has a radius of 7 *cm* and a tangent is drawn from a point 24 *cm* away from the center of the circle. Find the length of the tangent.

**Solution:** By creating a right triangle with the radius, the tangent, and the line from the circle's center to the external point, we can apply the Pythagorean theorem. Hence, the tangent length $AB$ is:

$$AB = \sqrt{24^2 - 7^2} = \sqrt{576 - 49} = \sqrt{527} \approx 23\ cm.$$

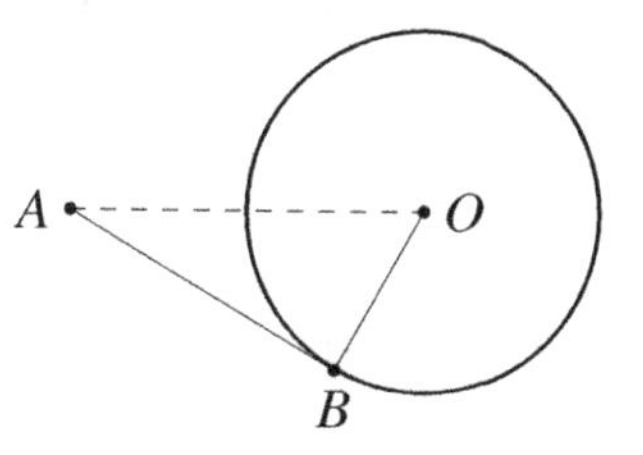


## 9.7 Secant Angles

We now explore secant angles. Secants, intersecting a circle at two points, reveal intriguing properties and predictable angles. When two secants extend from an external point, they form an angle outside the circle known as the secant-secant angle.

The measure of an angle formed by two intersecting secants from an outside point is half the difference of the intercepted arcs.

The intercepted arcs are the sections of the circle's circumference that lie inside the angle formed by the secants. This property can be summarized by the formula:

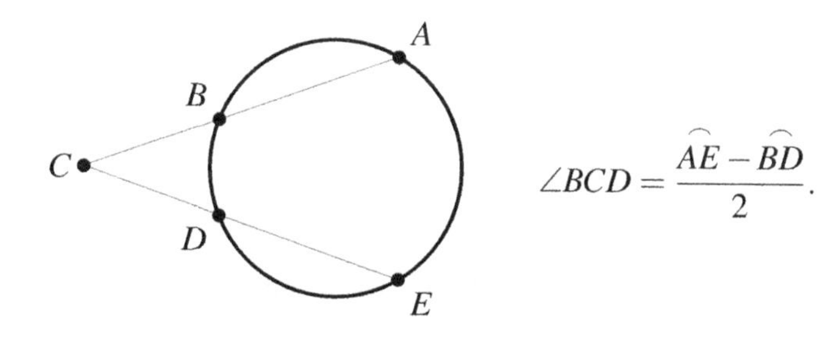


$$\angle BCD = \frac{\overset{\frown}{AE} - \overset{\frown}{BD}}{2}.$$

Let us illustrate these concepts with an example.

**Example** Two secant lines $ABC$ and $ADE$ intersect outside a circle at point $A$. Arc $CE$ measures $140^\circ$ and Arc $BD$ measures $80^\circ$. What is the measure of $\angle A$?

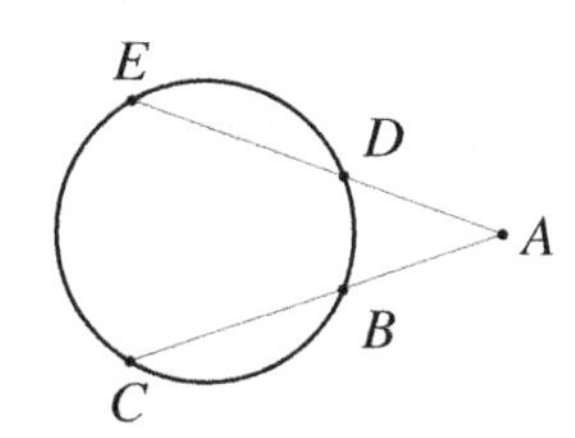


**Solution:** According to the formula for the angles between two secants intersecting outside a circle, the measure of $\angle A$ is half the difference of the measures of the intercepted arcs, i.e.,

$$\angle A = \frac{1}{2}\left(\overset{\frown}{CE} - \overset{\frown}{BD}\right) = \frac{1}{2}(140^\circ - 80^\circ) = 30^\circ.$$

So, $\angle A$ measures 30°.

## 9.8 Secant-tangent and Tangent-tangent Angles

In circle geometry, the interactions between secants and tangents are pivotal, leading to two important angle types: the secant-tangent angle and the tangent-tangent angle.

**Key Point**

[Secant-Tangent Angle] When a secant and a tangent intersect at a common external point, the angle formed is half the difference of the measures of the intercepted arcs.

To graphically represent a secant-tangent angle in a circle:

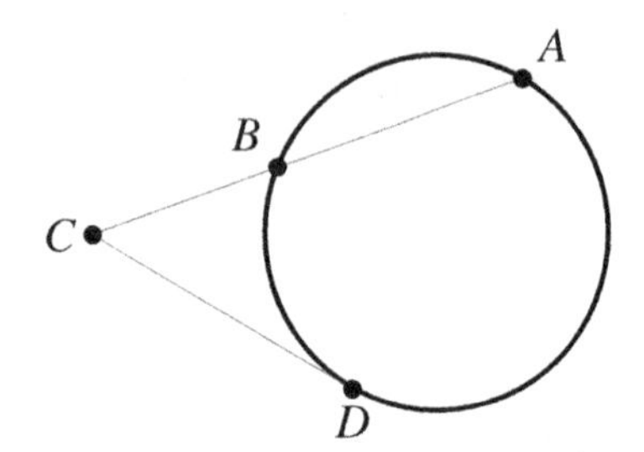


Secant-Tangent Angle:

$$\angle BCD = \frac{\overset{\frown}{AD} - \overset{\frown}{BD}}{2}.$$

**Key Point**

[Tangent-Tangent Angle] The angle between two tangents drawn from the same external point to a circle is half the difference of the intercepted arcs.

To illustrate the tangent-tangent angle in a circle:

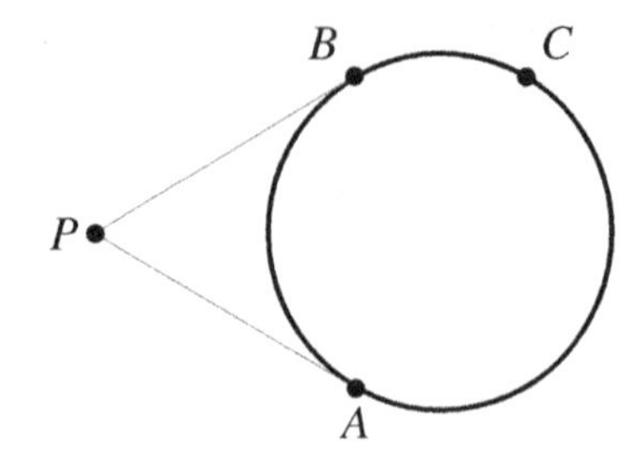


Tangent-Tangent Angle:

$$\angle BPA = \frac{\overset{\frown}{BCA} - \overset{\frown}{BA}}{2}.$$

**Example** Consider the following figure, where line segment DC is a diameter of the circle. The ratio of arc BC to arc DB is 2:1. What is the measure of $\angle BAD$?

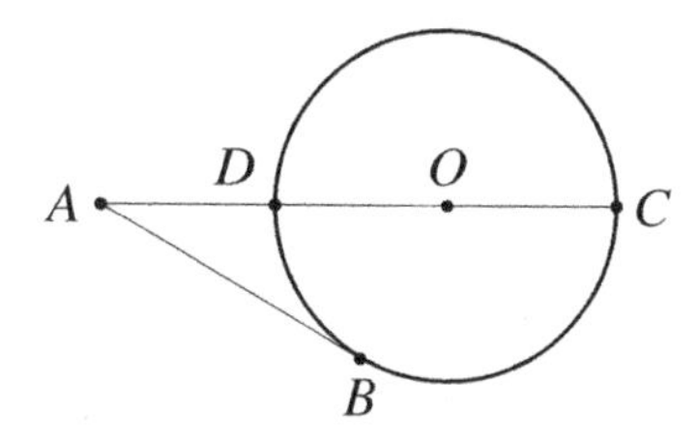


**Solution:** Since DC is a diameter, arc CBD forms a semicircle and measures 180°. Given the ratio of arc BC to arc DB is 2:1, let the measure of arc DB be $x$. Therefore, the measure of arc BC is $2x$, making the total measure of arc CBD (a semicircle) $3x$. This gives us the equation $3x = 180°$, so $x = 60°$.

According to the Secant-Tangent Angle Theorem, the measure of $\angle BAD$ is half the difference of the measures of arcs BC and DB, which is:

$$\frac{1}{2} \times (2x - x) = \frac{1}{2} \times (x) = \frac{1}{2} \times 60° = 30°.$$

Therefore, $\angle BAD$ measures 30°.

## 9.9 Segment Lengths in Circle

Segments within a circle are formed by chords, tangents, and secants. Their lengths are often interconnected, and understanding these connections is crucial in solving various geometric problems.

**Key Point**

The Chord-Chord Power Theorem states that if two chords intersect inside a circle, the product of the lengths of the segments of one chord equals the product of the lengths of the segments of the other chord.

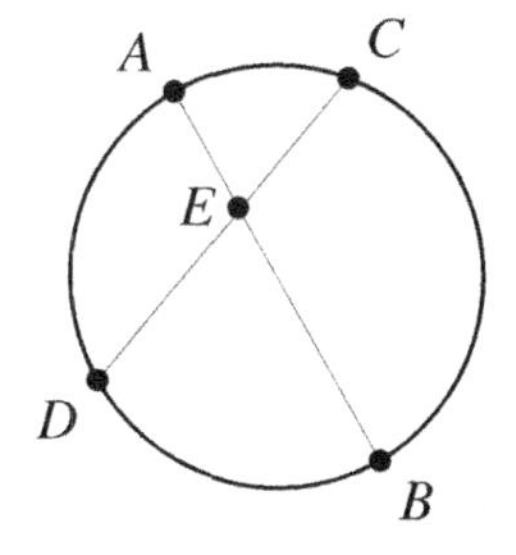


Chord-Chord Power Theorem:

$AE \times EB = CE \times ED$

**Key Point**

The Tangent-Secant Power Theorem reveals that if a tangent and a secant extend from a common external point, the square of the tangent's length is equal to the product of the secant's entire length and its external segment length.

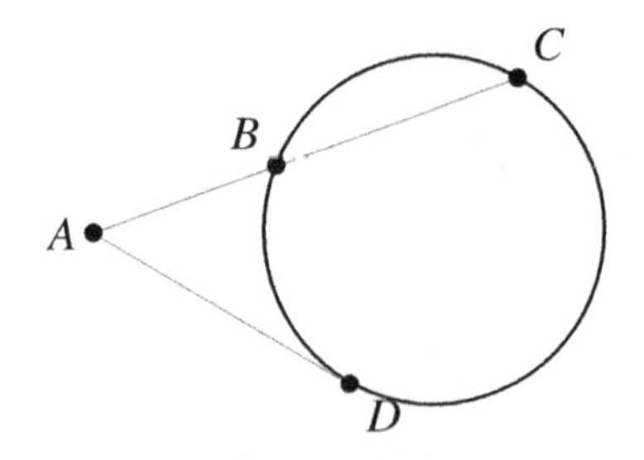


Tangent-Secant Power Theorem:

$AD^2 = AC \times AB$

**Key Point**

The Secant-Secant Power Theorem describes that if two secants intersect outside the circle, the product of the entire length of one secant and its external segment equals the product of the entire length of the second secant and its external segment.

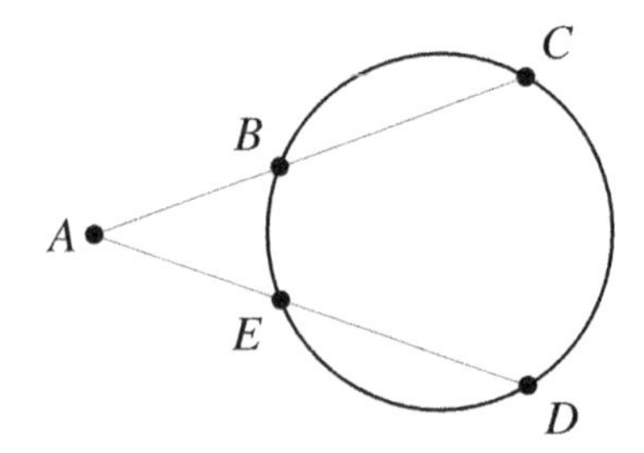


Secant-Secant Power Theorem:

$AD \times AE = AC \times AB$

Let us explore each theorem further with examples.

A chord $AB$ intersects chord $CD$ inside a circle at point $E$, with $AE = 6$, $EB = 3$, $CE = 4$, and $ED$ unknown. Find the length of $ED$.

**Solution:** Applying the Chord-Chord Power Theorem:

$$AE \times EB = CE \times ED.$$

Substituting the given values:

$$6 \times 3 = 4 \times ED \Rightarrow ED = \frac{6 \times 3}{4} = \frac{18}{4}.$$

Therefore, $ED = \frac{18}{4} = 4.5$ units long.

**Example** If a tangent from external point $A$ touches the circle at point $B$ and the secant from $A$ cuts the circle at points $C$ and $D$, find the length of $AC$ given that $AB = 5$ units and $AD = 13$ units.

**Solution:** Using the Tangent-Secant Power Theorem, we have:

$$AB^2 = AC \times AD.$$

With the given values:

$$5^2 = AC \times 13 \Rightarrow AC = \frac{5^2}{13} = \frac{25}{13} \approx 1.92.$$

Thus, $AC$ is approximately 1.92 units long.

## 9.10 Segment Measures

A segment in a circle can be thought of as a "slice" of the circle that lies between a chord and the arc that it subtends. The measure of a segment can refer to different things, such as the length of the chord, the length of the arc, or the area of the segment itself. When discussing segments, we often refer to two types:

- **Minor Segment**: The smaller region between the chord and its arc.
- **Major Segment**: The larger region between the chord and the rest of the circle.

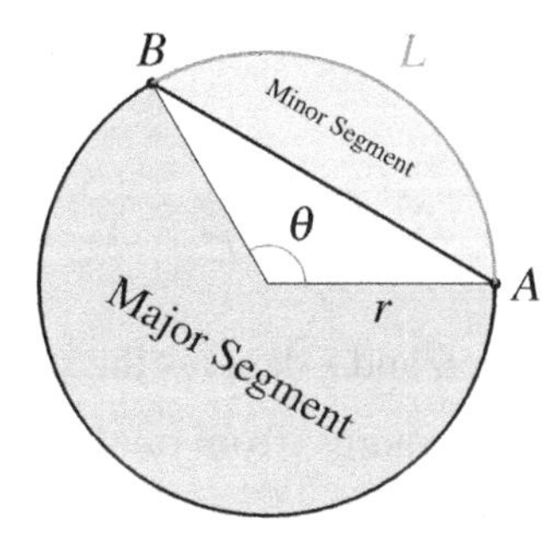


Chord: AB

**Key Point**

The length of an arc can be calculated using the formula

$$L = \left(\frac{\theta}{360}\right) 2\pi r,$$

where $\theta$ is the central angle in degrees and $r$ is the radius.

**Key Point**

The area of a segment is determined by the formula

$$Area = \frac{1}{2} r^2 (\theta(\frac{\pi}{180}) - \sin\theta),$$

with $\theta$ in degrees.

If the radius of a circle is 10 cm and the central angle of an arc is $60^\circ$, what is the length of the arc? ($\pi \approx 3.14$)

**Solution:** We use the arc length formula: $L = \left(\frac{\theta}{360}\right) 2\pi r$. Plugging in the given values:

$$L = \left(\frac{60}{360}\right) 2\pi(10) = \frac{1}{6} \times 20\pi \approx 10.47\ cm.$$

Calculate the area of a segment with a 90° central angle in a circle of radius 10 cm. ($\pi \approx 3.14$)

**Solution:** Apply the area formula for the segment:

$$Area = \frac{1}{2} \times (10)^2 \left(90(\frac{\pi}{180}) - \sin(90)\right) = 50(\frac{\pi}{2} - 1) \approx 28.5\ cm^2.$$

## 9.11 Standard Form of a Circle

Equation of circles in standard form is:

$$(x-h)^2 + (y-k)^2 = r^2,$$

where $(h,k)$ represents the coordinates of the circle's center, and $r$ denotes the radius. This equation simply shows what a circle is: a group of points all the same distance, called the radius, from one center point.
We may also encounter a circle's equation in its general format:

$$ax^2 + by^2 + cx + dy + e = 0.$$

It should be noted that for it to represent a circle, we require $a = b$ and both $a$ and $b$ must be non-zero. This general form is sometimes what we obtain when working through geometric problems, and our goal is often to convert it to the standard form for easier interpretation and use.

Given the general equation of a circle $9x^2 + 9y^2 - 36x + 54y - 63 = 0$, convert it to the standard form.

**Solution:** First, divide the entire equation by 9:

$$x^2 + y^2 - 4x + 6y - 7 = 0.$$

Now, complete the squares for the $x$ and $y$ terms:

$$\begin{aligned}(x^2 - 4x) + (y^2 + 6y) &= 7\\ [(x^2 - 4x + 4) - 4] + [(y^2 + 6y + 9) - 9] &= 7\\ (x-2)^2 + (y+3)^2 &= 20.\end{aligned}$$

Hence, the standard form is $(x-2)^2+(y+3)^2=20$, centered at $(2,-3)$ with radius $\sqrt{20}$.

 **Example** Write the standard form equation of a circle with center $(-9,-12)$ and radius 4.

**Solution:** Using the fundamental equation for a circle $(x-h)^2+(y-k)^2=r^2$, we substitute the given center and radius values:

$$(x-(-9))^2+(y-(-12))^2=4^2 \Rightarrow (x+9)^2+(y+12)^2=16.$$

Thus, the standard form of the circle's equation is $(x+9)^2+(y+12)^2=16$.

## 9.12 Finding the Center and the Radius of Circles

Two key characteristics of a circle are its center and radius.

**Key Point**

The general form of the equation of a circle is $(x-h)^2+(y-k)^2=r^2$, where $(h,k)$ is the center and $r$ is the radius of the circle.

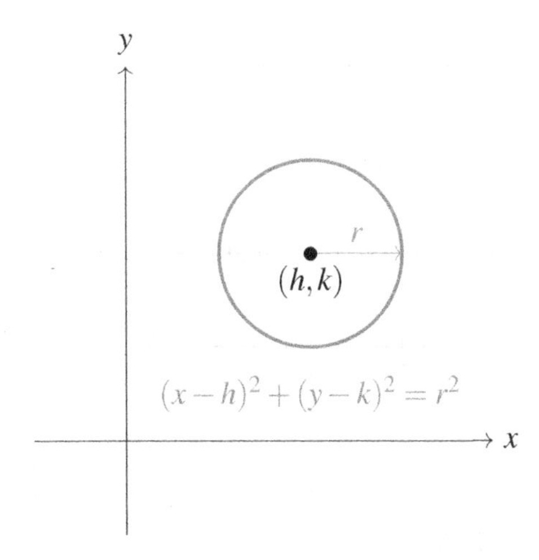


To find the center and radius from a given circle equation, you must often complete the square for the $x$- and $y$-terms. Here is how it is done.

 **Example** Identify the center and radius of the circle given by $x^2+y^2+6x-8y+9=0$.

**Solution:** First, rearrange the equation to group the $x$- and $y$-terms:

$$(x^2+6x)+(y^2-8y)=-9.$$

Complete the square for the $x$-terms and the $y$-terms. For $x$, we add $(\frac{6}{2})^2 = 9$ and for $y$, we add $(\frac{-8}{2})^2 = 16$:

$$(x^2 + 6x + 9) + (y^2 - 8y + 16) = -9 + 9 + 16.$$

This simplifies to:

$$(x+3)^2 + (y-4)^2 = 16.$$

The center of the circle is $(-3, 4)$ and the radius is $\sqrt{16}$ which is 4.

## 9.13 Radian Angle Measure

Understanding angles is crucial for the study of circles, and while degrees are familiar, radians offer a connection directly to the circle's geometry. A radian, the standard unit of angular measure in mathematics, is defined as the angle subtended at the center of a circle by an arc whose length is equal to the radius of the circle.

**Key Point**

The circumference of a circle is $2\pi$ times the radius. There are $2\pi$ radians in a full circle, which is equal to 360 degrees.

To transform degrees into radians, apply the following equation:

$$\text{Radian} = \text{Degrees} \times \frac{\pi}{180}.$$

Conversely, to convert radians into degrees, utilize the equation:

$$\text{Degrees} = \text{Radian} \times \frac{180}{\pi}.$$

Let us look at these translations using examples.

Convert $240^\circ$ to radians.

**Solution:** Using the conversion formula:

$$\text{Radian} = \text{Degrees} \times \frac{\pi}{180}$$

we get:

$$\text{Radian} = 240 \times \frac{\pi}{180} = \frac{240}{180}\pi = \frac{4}{3}\pi \text{ radians}.$$

Convert $\frac{5\pi}{6}$ radians to degrees.

**Solution:** Using the conversion formula:

$$\text{Degrees} = \text{Radian} \times \frac{180}{\pi}$$

we get:

$$\text{Degrees} = \frac{5\pi}{6} \times \frac{180}{\pi} = 5 \times 30 = 150^\circ.$$

## 9.14 Practices

**1)** Solve.

1-1) If $P(-\frac{\sqrt{3}}{2}, \frac{1}{2})$ is a point on the unit circle and the terminal side of an angle in a standard position whose size is $\theta$. Find $\sin\theta$ and $\cos\theta$.

1-2) If $P(\frac{\sqrt{2}}{2}, -\frac{\sqrt{2}}{2})$ is a point on the unit circle and the terminal side of an angle in a standard position whose size is $\theta$. Find $\sin\theta$ and $\cos\theta$.

**2)** Find the length of each arc. Round your answers to the nearest tenth.

2-1) $r = 14\ ft$, $\theta = 45^\circ$

2-2) $r = 18\ m$, $\theta = 60^\circ$

2-3) $r = 26\ m$, $\theta = 90^\circ$

2-4) $r = 20\ m$, $\theta = 120^\circ$

**3)** Find the area of the sector. Round your answers to the nearest tenth.

3-1) $r = 4\ m$, $\theta = 20^\circ$

3-2) $r = 2\ m$, $\theta = 45^\circ$

3-3) $r = 8\ m$, $\theta = 90^\circ$

3-4) $r = 4\ m$, $\theta = 135^\circ$

**4)** Name the central angle of the given arc.

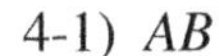

4-1) $AB$

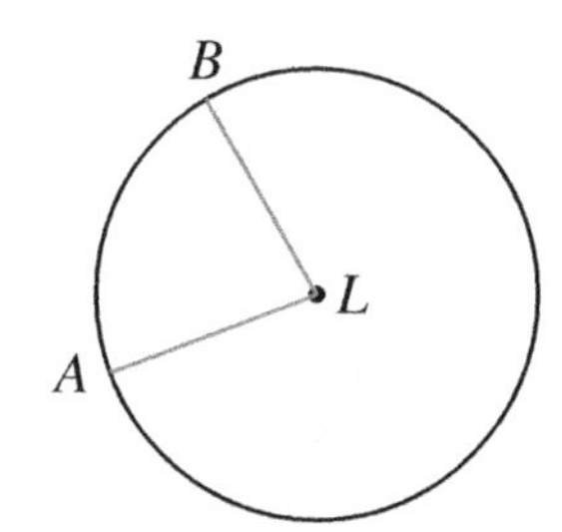


4-2) $AB$

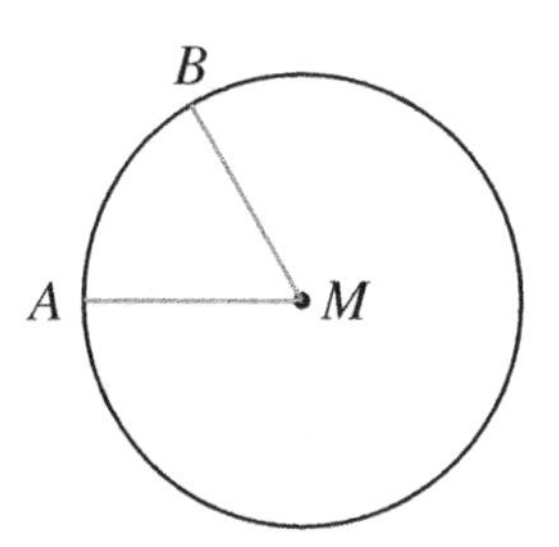


**5)** Solve for $x$. Assume that lines which appear to be diameters are actual diameters.

5-1)

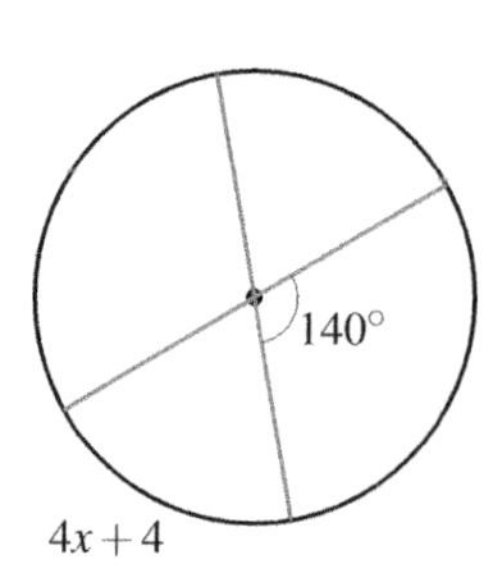


5-2)

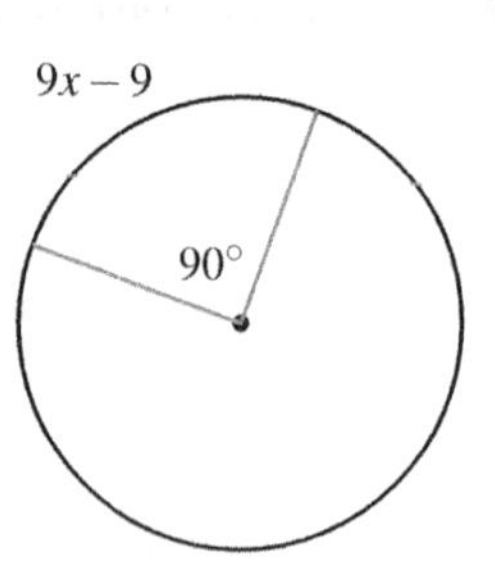


**6)** Find the measure of the arc or central angle indicated. Assume that lines which appear to be diameters are actual diameters.

6-1)

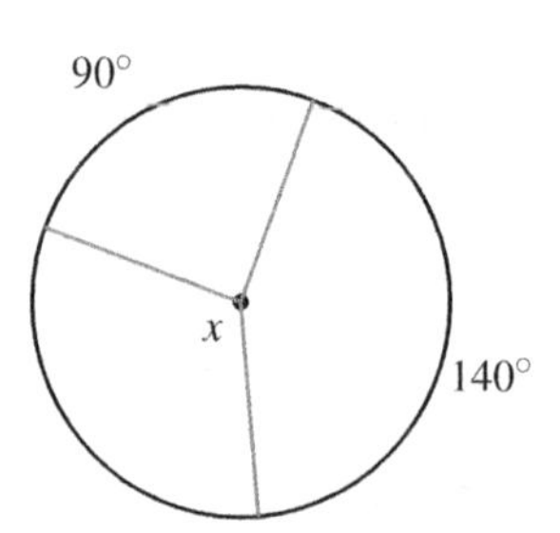


6-2)

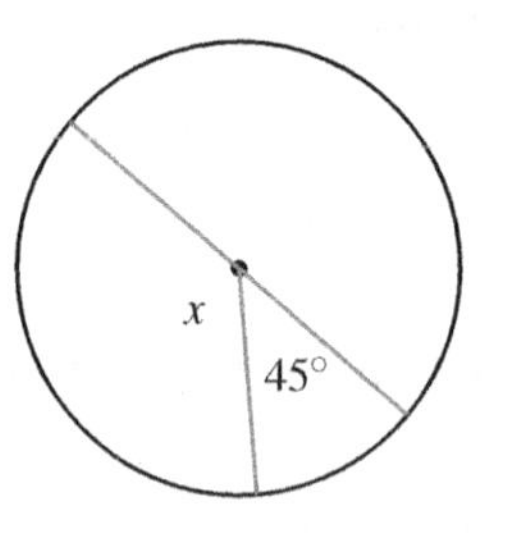


**7)** Find the length of the segment indicated. Round your answer to the nearest tenth if necessary.

7-1)

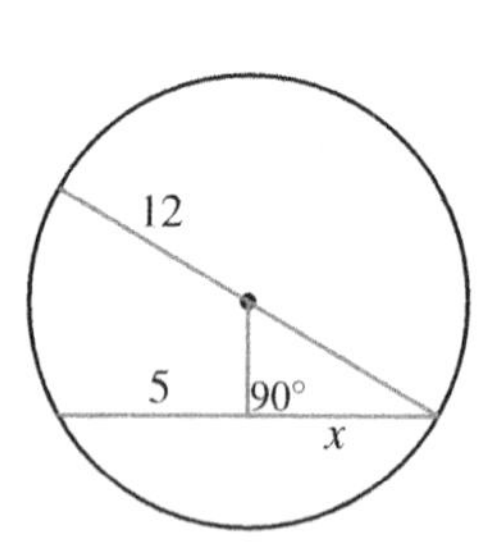


7-3)

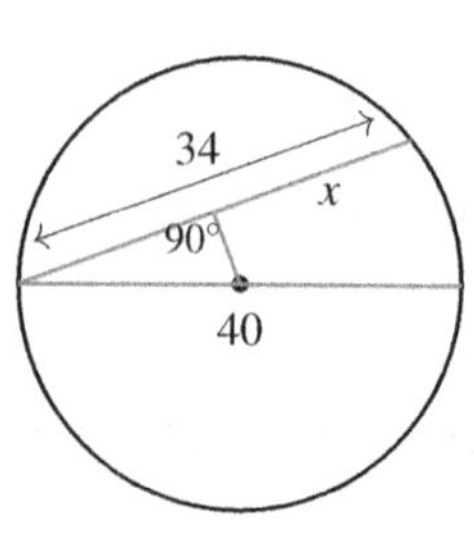


7-2)

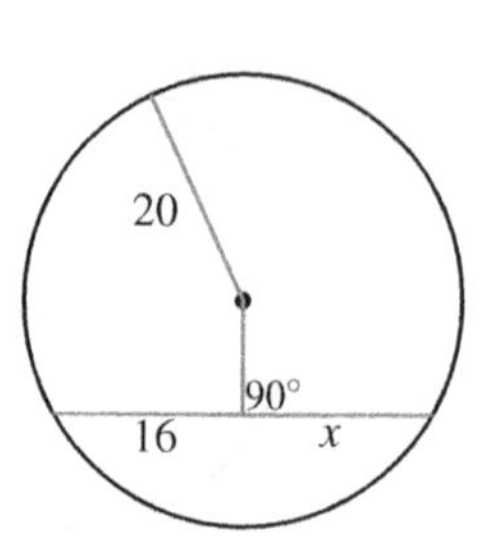


7-4)

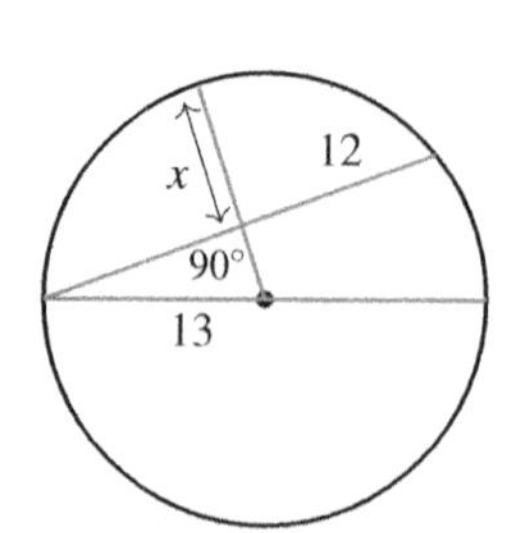

7-5)

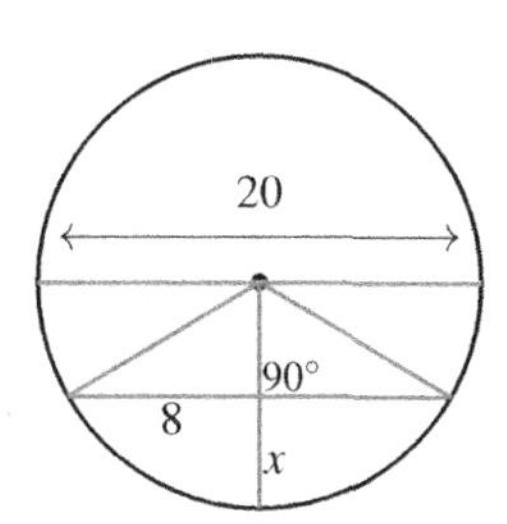


7-6)

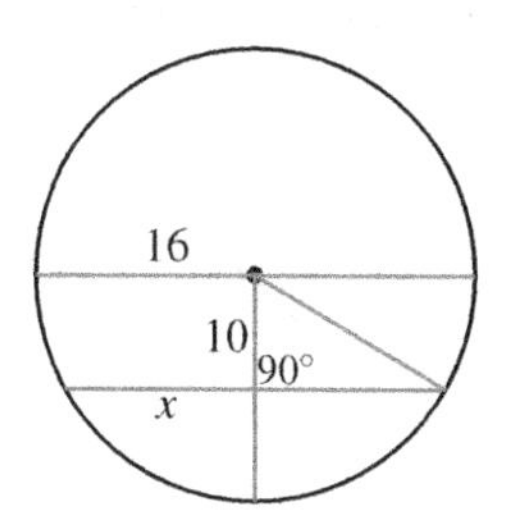


**8)** State if each polygon is an inscribed polygon.

8-1)

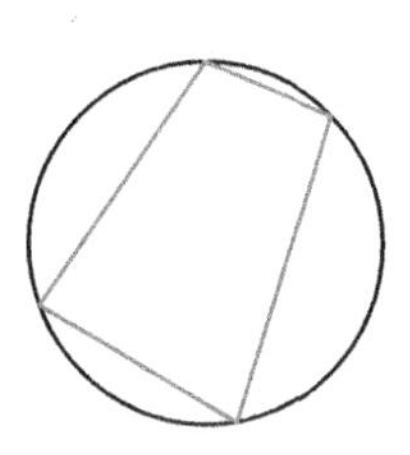

8-3)

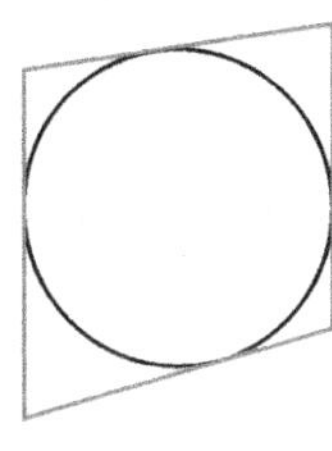

8-2)

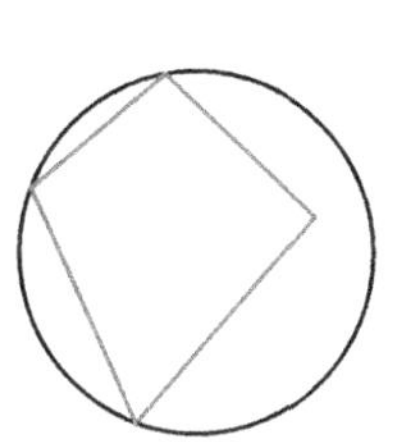

8-4)

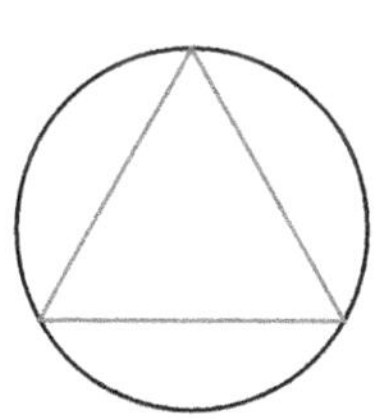

**9)** Find the value of each variable.

9-1)

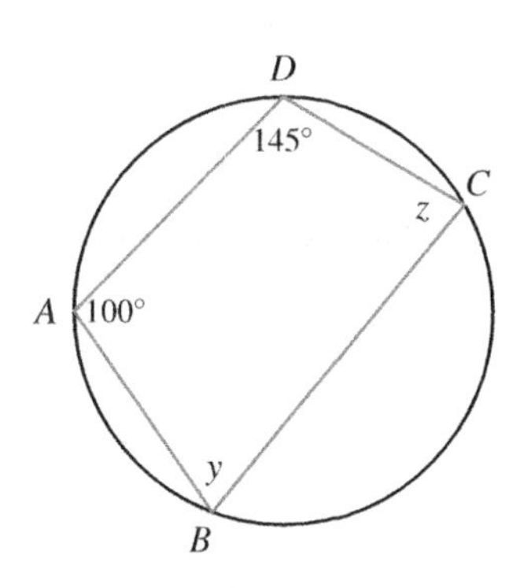


9-2)

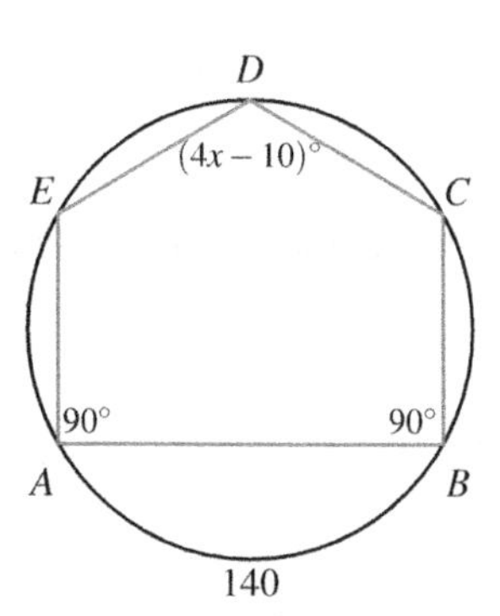

**10)** State if each angle is an inscribed angle.

10-1)

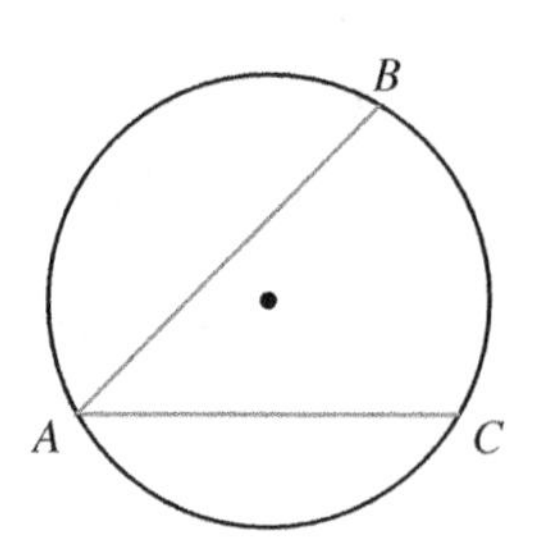


10-3)

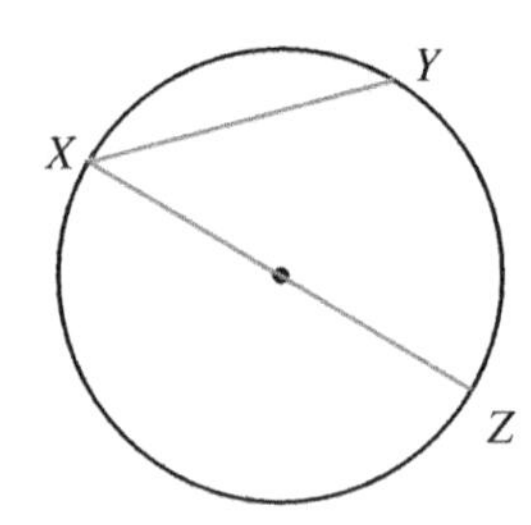


10-2)

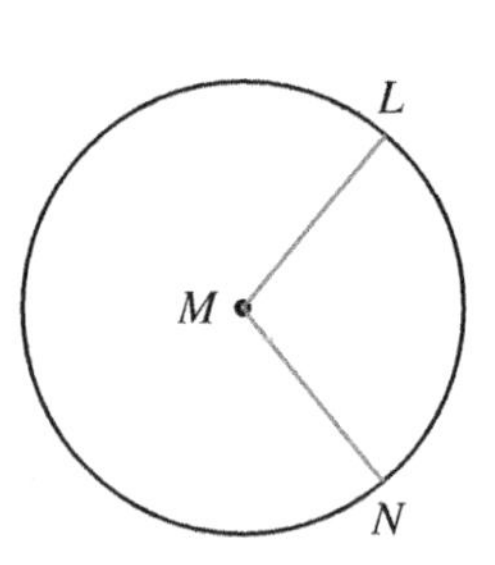


10-4)

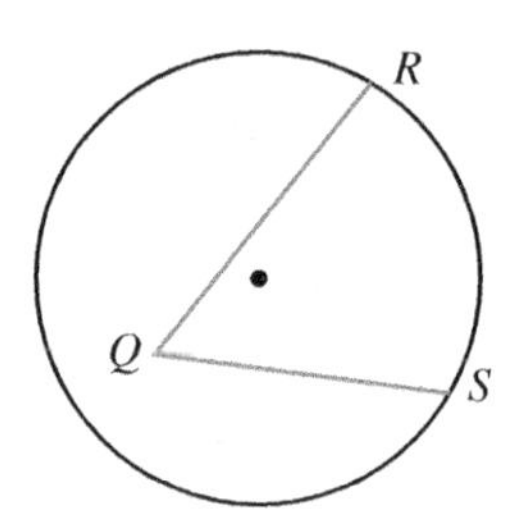


**11)** Find the measure of the angle $\angle BAC$.

11-1)

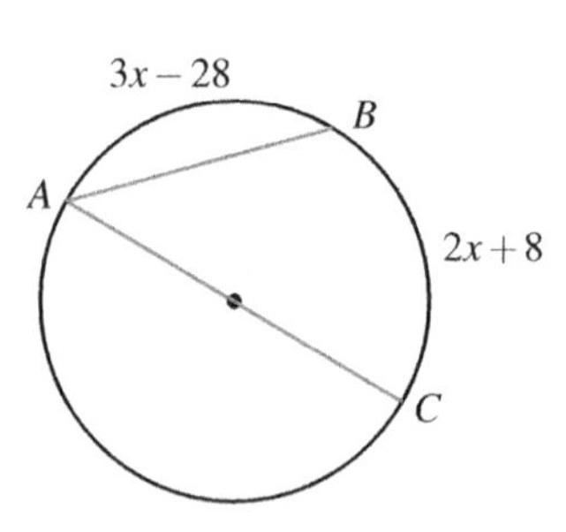


11-2)

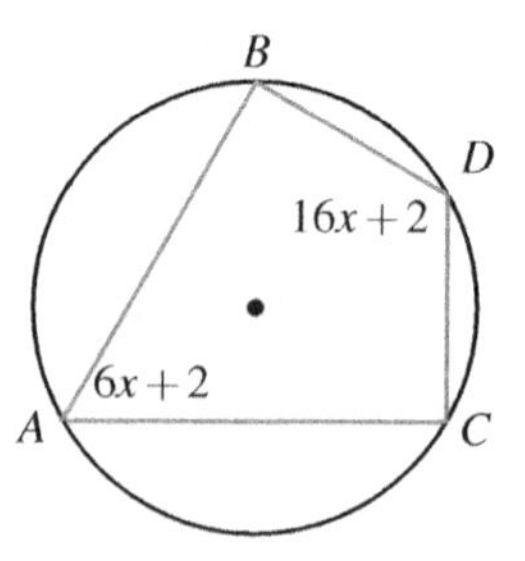


**12)** Find the segment length indicated. Assume that lines which appear to be tangent are tangent.

12-1)

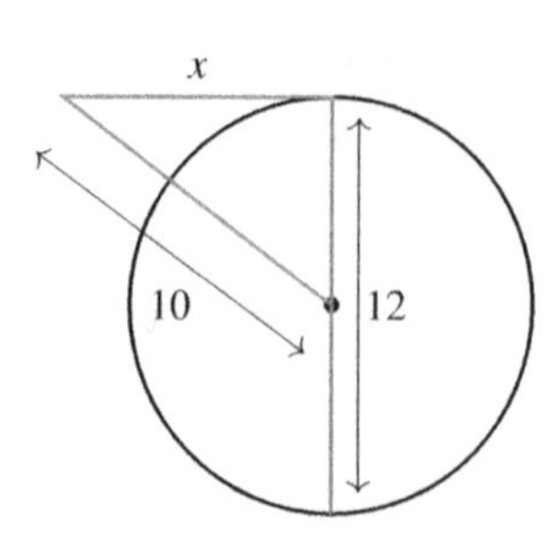


12-2)

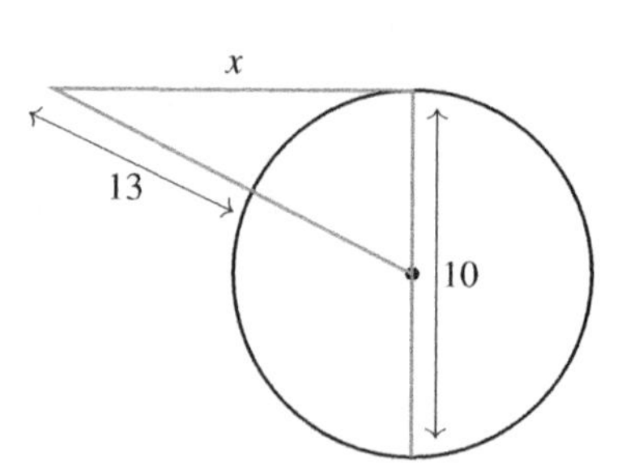

12-3)

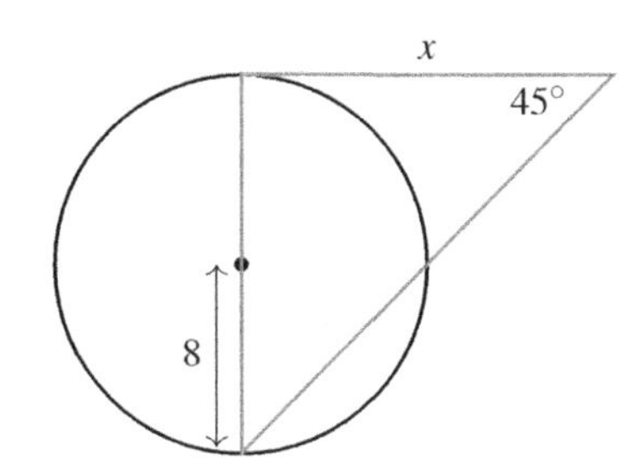


12-4)

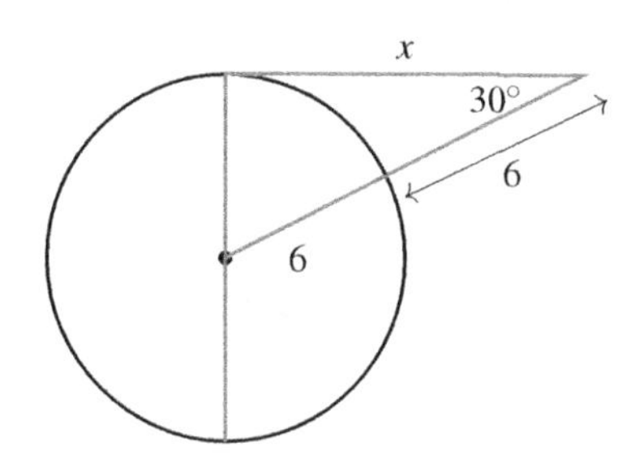


**13)** Find the perimeter of each polygon. Assume that lines which appear to be tangent are tangent.

13-1)

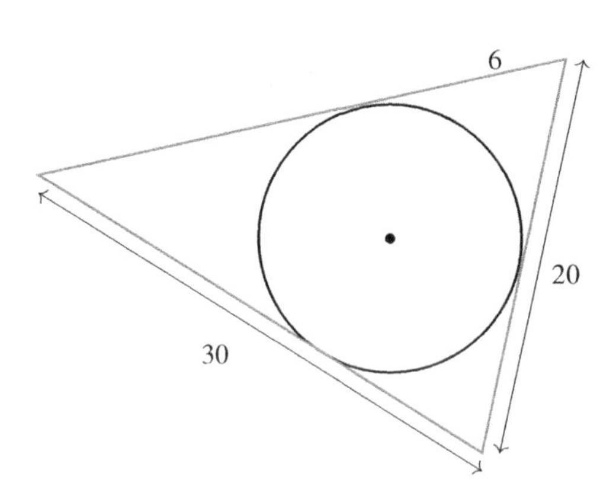


13-2)

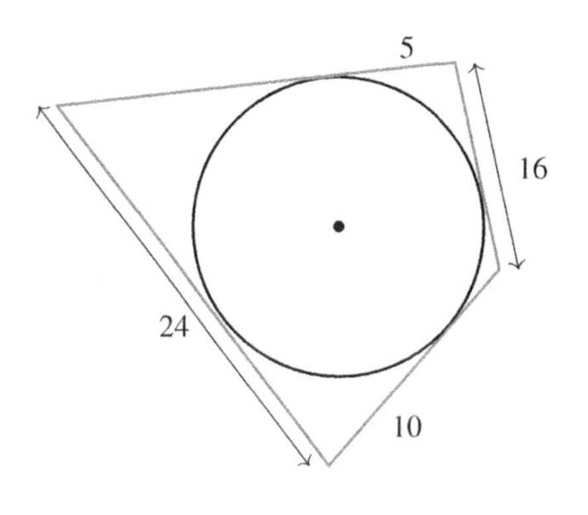


**14)** Find the measure of the arc or angle indicated.

14-1)

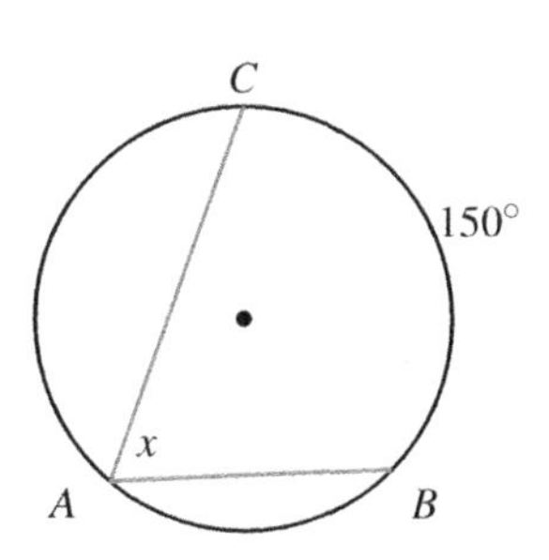


14-3)

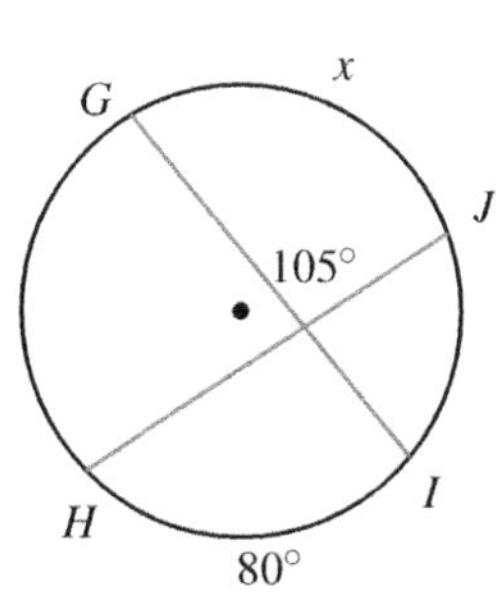


14-2)

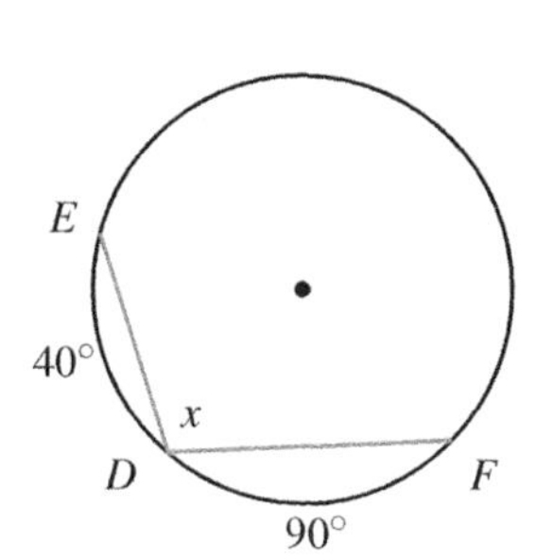


14-4)

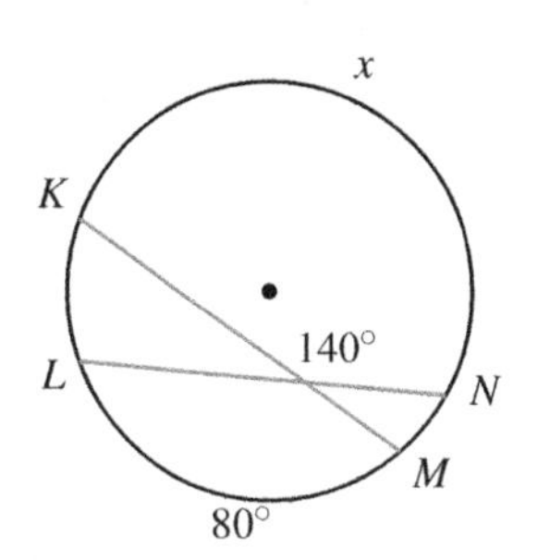

**15)** Solve for $x$.

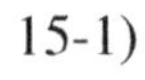


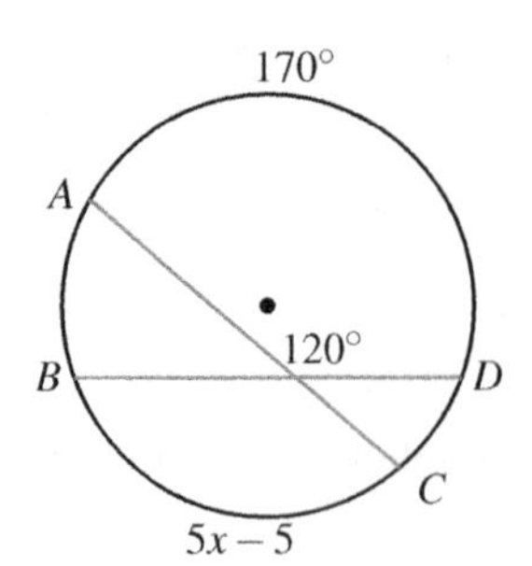


15-2)

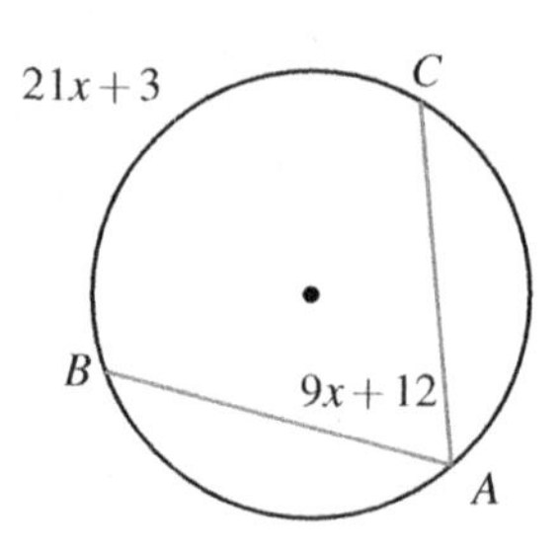


**16)** Solve for $x$. assume that lines which appear tangent are tangent.

16-1)

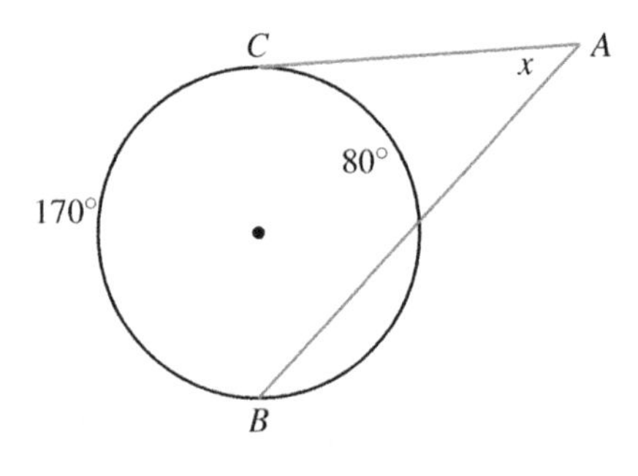


16-3)

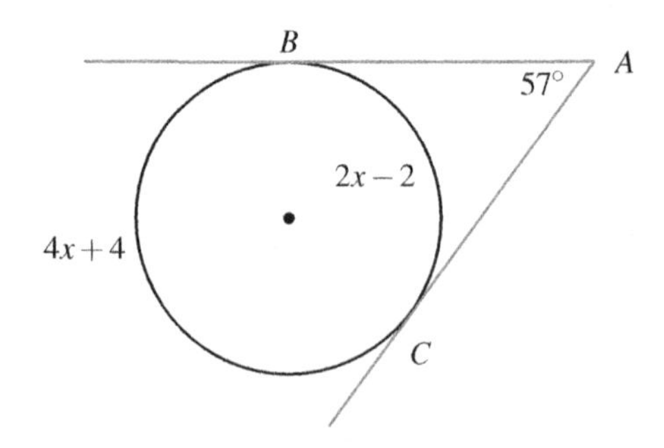


16-2)

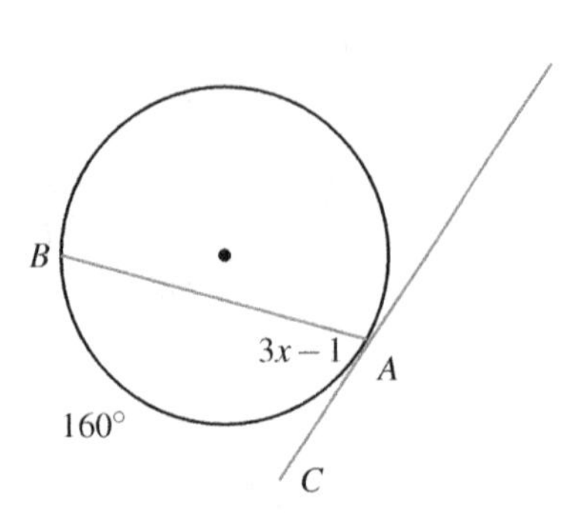


16-4)

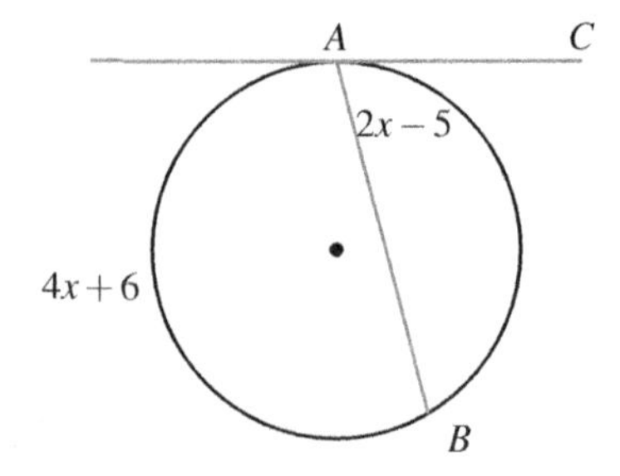


**17)** Find the measure of the arc or angle indicated. Assume that lines which appear tangent are tangent.

17-1) Find measure $\angle A$

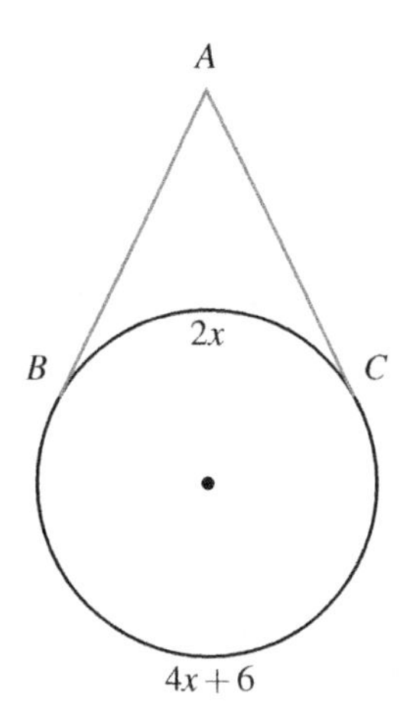


17-2) Find arc $EF$

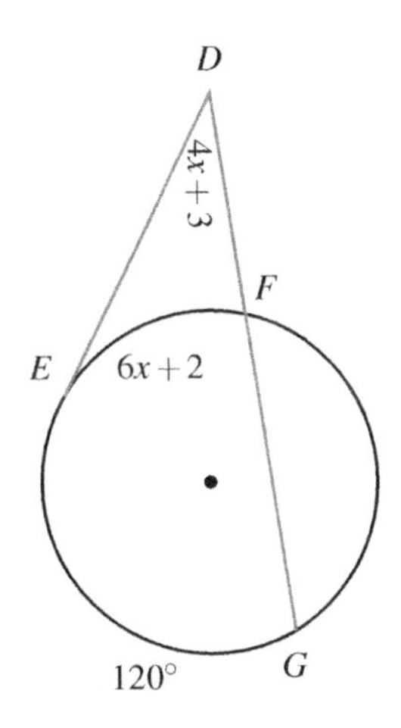


**18)** Solve for $x$. Assume that lines which appear tangent are tangent.

18-1)

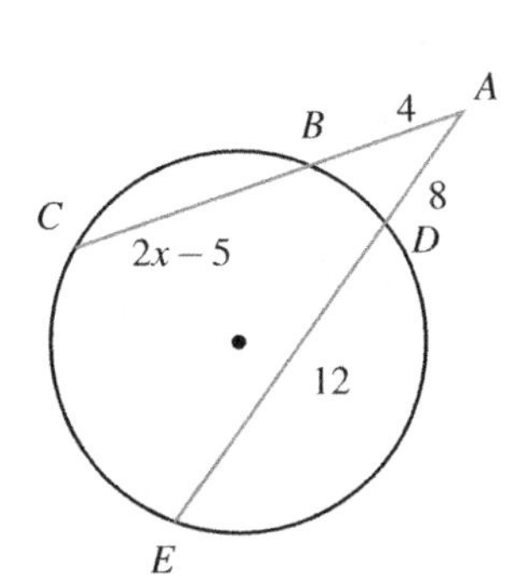


18-2)

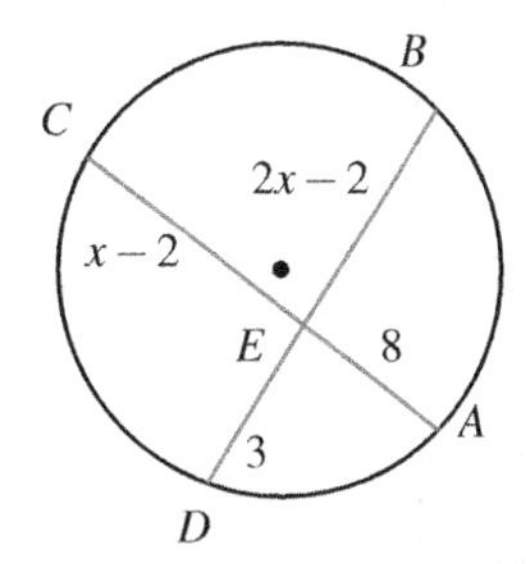


18-3)

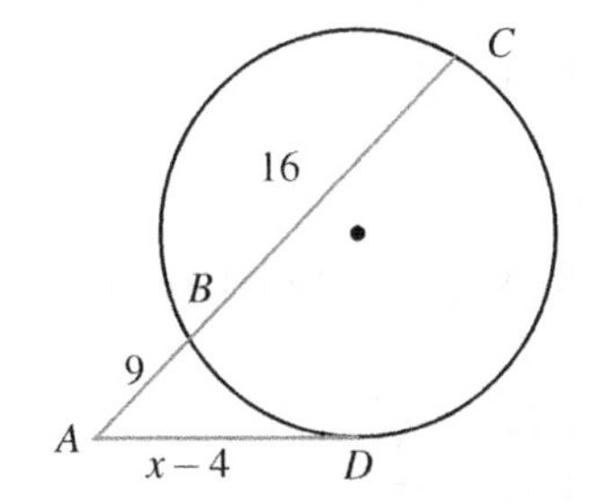

18-4)

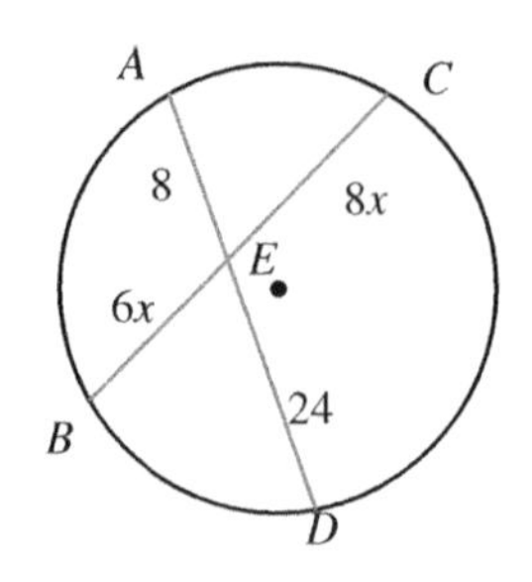


18-5)

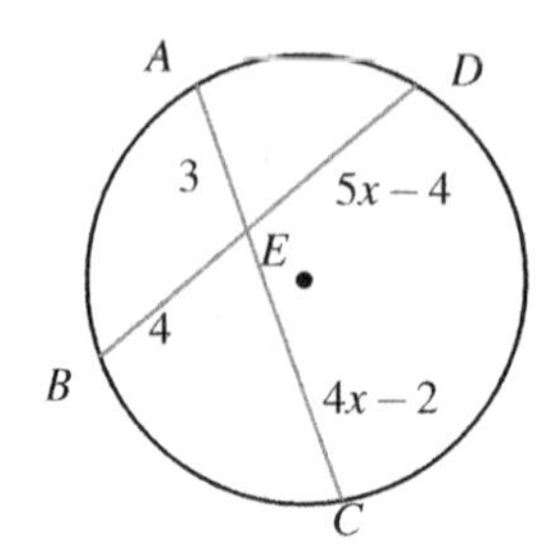


**19)** Write the standard form equation of each circle.

19-1) $y^2+2x+x^2=24y-120$

19-2) $x^2+y^2-2y-15=0$

19-3) $8x+x^2-2y=64-y^2$

19-4) Center: $(-6,-5)$, Radius: 9

19-5) Center: $(-12,-5)$, Area: $4\pi$

19-6) Center: $(-11,-14)$, Area: $16\pi$

19-7) Center: $(-3,2)$, Circumference: $2\pi$

19-8) Center: $(15,14)$, Circumference: $2\pi\sqrt{15}$

**20)** Identify the center and radius of each.

20-1) $(x-2)^2+(y+5)^2=10$

20-2) $x^2+(y-1)^2=4$

20-3) $(x-2)^2+(y+6)^2=9$

20-4) $(x+14)^2+(y-5)^2=16$

**21)** Convert each degree measure into radians.

21-1) $-150°$

21-2) $420°$

21-3) $300°$

21-4) $-60°$

21-5) $315°$

21-6) $600°$

**22)** Convert each radian measure into degrees.

22-1) $-\frac{16\pi}{3}$

22-2) $-\frac{3\pi}{5}$

22-3) $\frac{11\pi}{6}$

22-4) $\frac{5\pi}{9}$

22-5) $-\frac{\pi}{3}$

22-6) $\frac{13\pi}{6}$

## 9.15 Answers

**1)**

1-1) $\sin\theta$ is the $y$-coordinate of $P = \frac{1}{2}$, $\cos\theta$ is the $x$-coordinate of $P = -\frac{\sqrt{3}}{2}$.

1-2) $\sin\theta$ is the $y$-coordinate of $P = -\frac{\sqrt{2}}{2}$, $\cos\theta$ is the $x$-coordinate of $P = \frac{\sqrt{2}}{2}$.

**2)**

2-1) 11.0

2-2) 18.8

2-3) 40.8

2-4) 41.9

**3)**

3-1) 2.8

3-2) 1.6

3-3) 50.3

3-4) 18.8

**4)**

4-1) $\angle ALB$

4-2) $\angle AMB$

**5)**

5-1) 9

5-2) 11

**6)**

6-1) 130°

6-2) 135°

**7)**

7-1) 5

7-2) 16

7-3) 17

7-4) 8

7-5) 4

7-6) 12.5

**8)**

8-1) Yes

8-2) No

8-3) No

8-4) Yes

**9)**

9-1) $y = 35°, z = 80°$

9-2) $x = 20$

**10)**

10-1) Yes

10-2) No

10-3) Yes

10-4) No

**11)**

11-1) $\angle BAC = 44°$

11-2) $\angle BAC = 50°$

**12)**

12-1) 8

12-2) 17.3

12-3) 16

12-4) 10.4

**13)**

13-1) 72

13-2) 80

**14)**

14-1) 75°

14-2) 115°

14-3) 130°

14-4) 200°

**15)**

15-1) 15

15-2) 7

**16)**

16-1) 45

16-2) 27

16-3) 54

16-4) 45.5

**17)**

17-1) 62°

17-2) $50^\circ$

**18)**

18-1) 20.5

18-2) 5

18-3) 19

18-4) 2

18-5) 1.25

**19)**

19-1) $(x+1)^2+(y-12)^2=25$

19-2) $x^2+(y-1)^2=16$

19-3) $(x+4)^2+(y-1)^2=81$

19-4) $(x+6)^2+(y+5)^2=81$

19-5) $(x+12)^2+(y+5)^2=4$

19-6) $(x+11)^2+(y+14)^2=16$

19-7) $(x+3)^2+(y-2)^2=1$

19-8) $(x-15)^2+(y-14)^2=15$

**20)**

20-1) Center: $(2,-5)$, Radius: $\sqrt{10}$

20-2) Center: $(0,1)$, Radius: 2

20-3) Center: $(2,-6)$, Radius: 3

20-4) Center: $(-14,5)$, Radius: 4

**21)**

21-1) $-\frac{5\pi}{6}$

21-2) $\frac{7\pi}{3}$

21-3) $\frac{5\pi}{3}$

21-4) $-\frac{\pi}{3}$

21-5) $\frac{7\pi}{4}$

21-6) $\frac{10\pi}{3}$

**22)**

22-1) $-960^\circ$

22-2) $-108^\circ$

22-3) $330^\circ$

22-4) $100^\circ$

22-5) $-60^\circ$

22-6) $390^\circ$

# 10. Surface Area and Volume

## 10.1 Circumference and Area of a Circle

A circle is a perfect round figure defined by all the points that are the same distance (radius) from a center point. This distance is denoted with the variable $r$, while the diameter, which is twice the radius, is denoted by $d$. The circumference is the distance around the circle, and the area is the space it encloses.

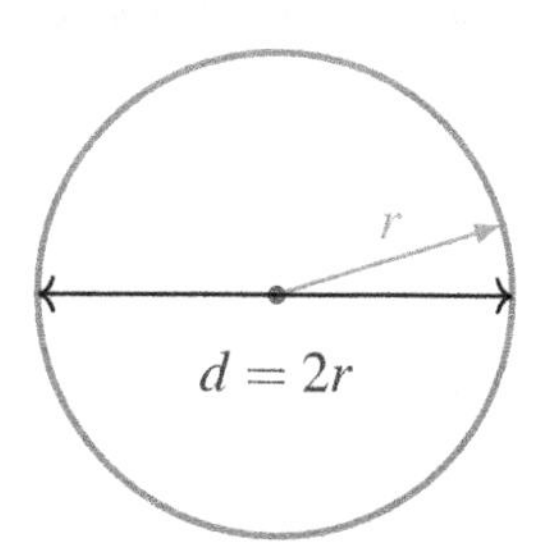


**Key Point**

For any circle, the circumference $C$ is given by $C = 2\pi r$, where $\pi$ is approximately 3.14 and $r$ is the radius.

**Key Point**

The area $A$ of a circle is given by $A = \pi r^2$, demonstrating that the area increases with the square of the radius.

**Example** A circular patio has a diameter of 10 feet. Find the area and circumference of the patio. ($\pi \approx 3.14$)

**Solution:** First, we calculate the radius $r$ which is half of the diameter. So, $r = \frac{d}{2} = \frac{10}{2} = 5$ feet. The circumference $C$ is computed as follows:

$$C = 2\pi r = 2\pi(5) = 10\pi \approx 10 \times 3.14 = 31.4 \text{ feet.}$$

Next, the area $A$ is given by:

$$A = \pi r^2 = \pi(5^2) = 25\pi \approx 25 \times 3.14 = 78.5 \text{ square feet.}$$

**Example** A round flower bed with a radius of 4 meters is planted inside a square lawn. Find the circumference and the area of the flower bed.

**Solution:** Using the radius $r = 4$ meters: The circumference $C$ of the flower bed:

$$C = 2\pi r = 2\pi(4) = 8\pi \approx 8 \times 3.14 = 25.12 \text{ meters.}$$

To find the area $A$:

$$A = \pi r^2 = \pi(4^2) = 16\pi \approx 16 \times 3.14 = 50.24 \text{ square meters.}$$

## 10.2 Area of a Trapezoids

A trapezoid is a special kind of quadrilateral that has at least one pair of parallel sides. These parallel sides are referred to as the *bases* of the trapezoid. The formula to find the area of a trapezoid is useful in numerous real-world contexts, from architecture to engineering.

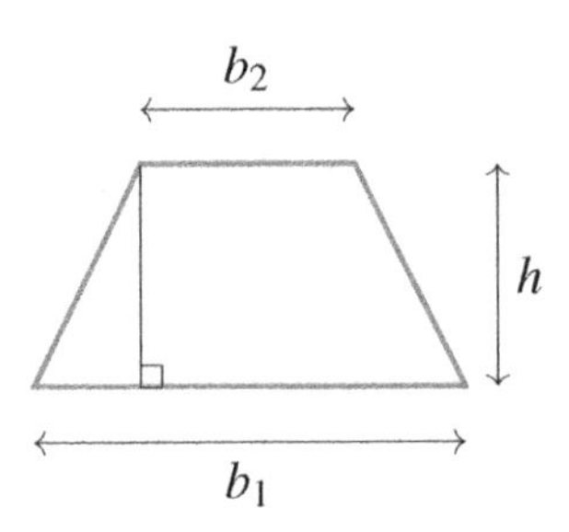


**Key Point**

The area of a trapezoid can be calculated using the formula $A = \frac{1}{2}h(b_1 + b_2)$, where $h$ is the height of the trapezoid (the perpendicular distance between the bases), and $b_1$ and $b_2$ are the lengths of the two bases.

Now, let us see this formula in action through examples.

**Example** Calculate the area of a trapezoid with bases of 6 cm and 10 cm, and a height of 12 cm.

**Solution:** Use the area formula $A = \frac{1}{2}h(b_1 + b_2)$. Here, $b_1 = 6$ cm, $b_2 = 10$ cm and $h = 12$ cm. Substituting these values into the formula gives us:

$$A = \frac{1}{2} \times 12 \times (6 + 10) = 6 \times 16 = 96 \text{ cm}^2.$$

Therefore, the area of the trapezoid is 96 $\text{cm}^2$.

**Example** A trapezoid has a height of 8 cm and one base is 5 cm longer than the other. If the shorter base is 7 cm, find the area of the trapezoid.

**Solution:** Let $b_1$ be the length of the shorter base, so $b_1 = 7$ cm. The longer base, $b_2$, is 5 cm longer than $b_1$, so $b_2 = b_1 + 5 = 7\text{ cm} + 5\text{ cm} = 12$ cm. The height $h$ is given as 8 cm. Using the area formula:

$$A = \frac{1}{2} \times 8 \times (7+12) = 4 \times 19 = 76\text{ cm}^2.$$

Therefore, the area of the trapezoid is 76 $\text{cm}^2$.

## 10.3 Area of Polygons

The study of polygon areas gives us the tools to measure the space these shapes occupy. Polygons are flat, two-dimensional shapes with straight sides that close in a space. Calculating the area of polygons is fundamental in both academic and practical settings, from urban planning to art design.

- **Triangles:** A triangle is a three-sided polygon. Its area can be calculated using the formula:

$$\text{Area} = \frac{1}{2} \times (\text{base}) \times (\text{height}).$$

This formula is used to determine the area when the base length and height are known.

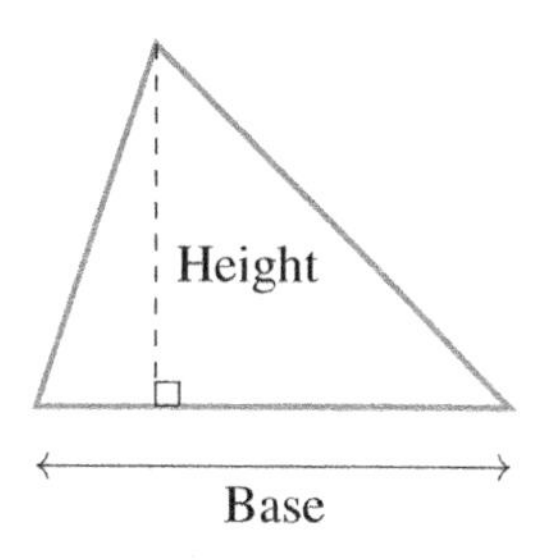


- **Rectangles and Squares:** Both are quadrilaterals. A rectangle's area is calculated by multiplying its length by its width, and a square's area, being a special rectangle with all sides equal, is the square of its side:

$$\textit{Rectangle}: \text{Area} = (\text{length}) \times (\text{width}), \quad \text{and} \quad \textit{Square}: \text{Area} = (\text{side})^2.$$

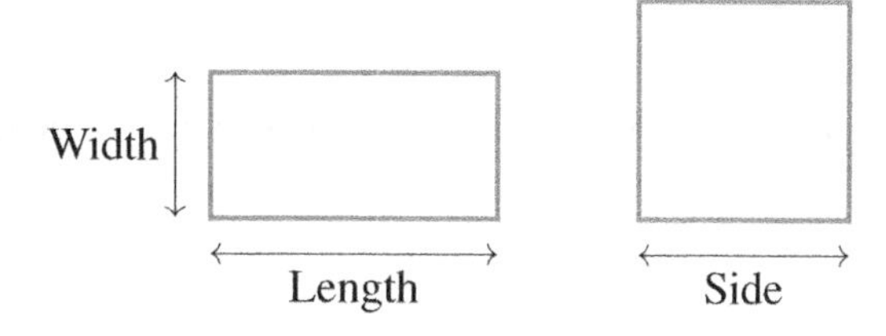

- **Parallelograms:** These are quadrilaterals with opposite sides parallel and equal. The area is found by:

$$\text{Area} = (\text{base}) \times (\text{height}).$$

Height

Base

- **Trapezoids:** A trapezoid is identified by having at least one pair of parallel sides. Its area is calculated as:

$$\text{Area} = \frac{1}{2} \times (\text{base}_1 + \text{base}_2) \times (\text{height}).$$

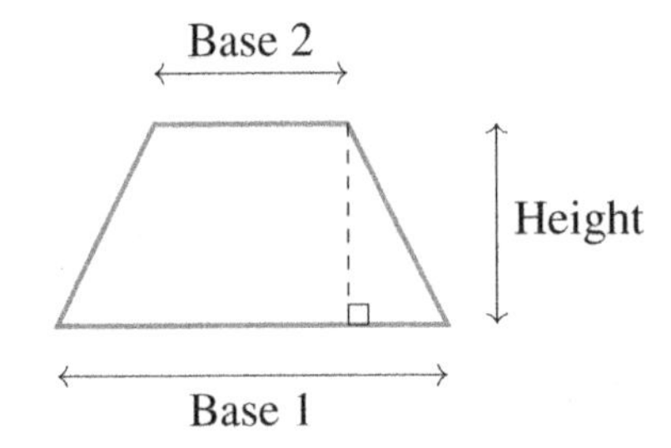


- **Regular Polygons:** These are polygons that are equilateral and equiangular. The area formula is:

$$\text{Area} = \frac{1}{2} \times (\text{perimeter}) \times (\text{apothem}).$$

Where apothem is the perpendicular distance from the center to a side.

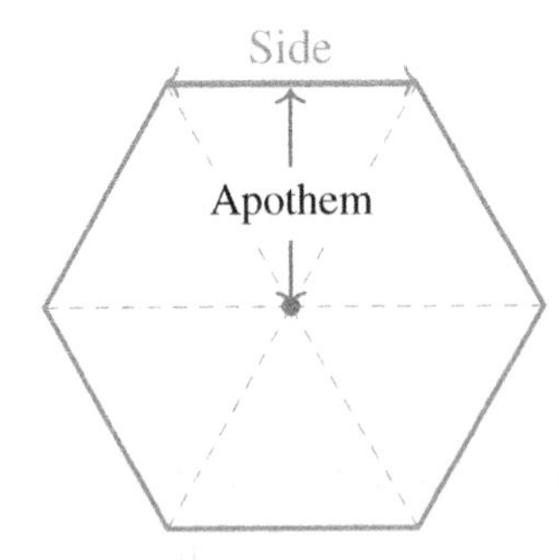


We can move on to an example of how to use this knowledge in a practical situation.

**Example** A park in the shape of a regular hexagon has a side length of 15 meters. Find the area of the park.

**Solution:** Knowing the side length, we can calculate the perimeter of the hexagon:

$$\text{perimeter} = 6 \times 15 = 90 \text{ meters}.$$

To find the apothem, we consider the regular hexagon split into equilateral triangles. The apothem is the height of one of these triangles. For an equilateral triangle with a side length ($s$) of 15 meters, the apothem ($a$) is: $a = \frac{s\sqrt{3}}{2} = \frac{15\sqrt{3}}{2}$.
Now, we use the area formula of a regular polygon:

$$\text{Area} = \frac{1}{2} \times \text{perimeter} \times \text{apothem} = \frac{1}{2} \times 90 \times \left(\frac{15\sqrt{3}}{2}\right) \approx 584.57\ m^2$$

Thus, the area of the park is approximately 584.57 square meters.

## 10.4 Nets of 3-D Figures

A net is a pattern that you can cut and fold to form a 3-D shape. They provide invaluable assistance in visualizing and comprehending the spatial relationships in geometric solids. It follows the logical progression from understanding two-dimensional shapes to appreciating the complexity of three-dimensional ones.

A net is composed of all the faces of a 3-$D$ figure laid out in a connected pattern.

**Key Point**

Nets help in calculating the surface area of a 3-$D$ figure by displaying all its faces in 2-$D$.

Imagine slicing through the edges of a 3-$D$ figure and laying it flat. The resulting shape is the net, a two-dimensional pattern of the figure's faces. Visualizing a net can be challenging, so let us discuss some strategies to construct nets for 3-$D$ shapes.

What is the net of a rectangular prism with length 4 units, width 3 units, and height 2 units?

**Solution:** A rectangular prism has 6 faces: 2 pairs of identical rectangles and another pair of squares. To draw its net, we would lay out 3 rectangles in a row for the front, bottom, and back, with 2 squares and an additional rectangle attached at the edges, representing the sides and top of the prism.

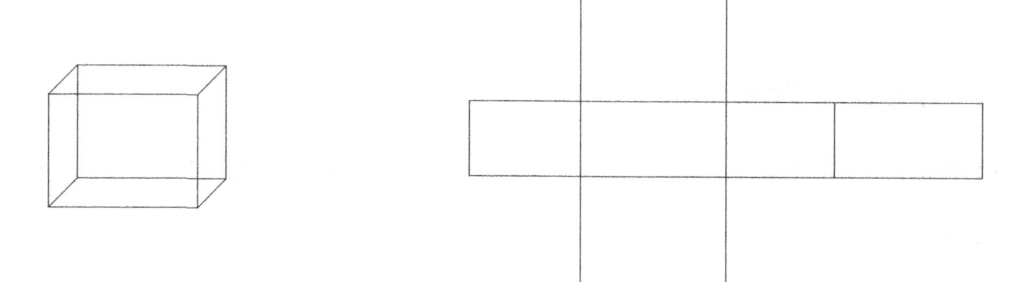

The rectangles represent the front, bottom, back, and top while the squares represent the sides. Note that in the net, lengths that were adjacent in the 3-$D$ figure are connected.

## 10.5 Cubes

A cube is a special kind of rectangular prism where all faces are squares. Since the lengths of all edges of a cube are equal, we typically refer to this common length as the "side" of the cube.

**Key Point**

The volume of a cube measures how much space is enclosed within it. The formula for the volume of a cube is: $V = a^3$, where $a$ is the length of a side of the cube.

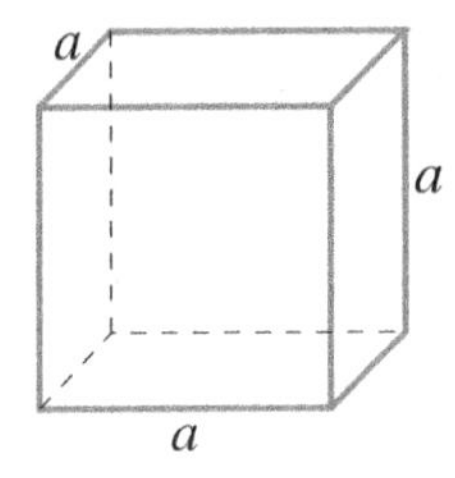


**Key Point**

The surface area of a cube measures the total area covered by all six faces. The formula for the surface area of a cube is: $SA = 6a^2$, where $a$ is the length of a side of the cube.

**Example** Find the volume and surface area of a cube with a side length of 4 cm.

**Solution:** Using the volume formula: $V = a^3 = (4\text{ cm})^3 = 64\text{ cm}^3$. And the surface area formula: $SA = 6a^2 = 6(4\text{ cm})^2 = 6(16\text{ cm}^2) = 96\text{ cm}^2$.

**Example** A cube has a side length of 7 inches. Calculate its volume and surface area.

**Solution:** Using the formula for the volume: $V = a^3 = (7\text{ in})^3 = 343\text{ in}^3$. Using the formula for the surface area: $SA = 6a^2 = 6(7\text{ in})^2 = 6(49\text{ in}^2) = 294\text{ in}^2$.

## 10.6 Rectangular Prisms

A rectangular prism, sometimes called a cuboid, is a solid three-dimensional object with six faces, each a rectangle.

**Key Point**

The volume of a rectangular prism is calculated by multiplying its length ($l$), width ($w$), and height ($h$):

$$\text{Volume} = l \times w \times h.$$

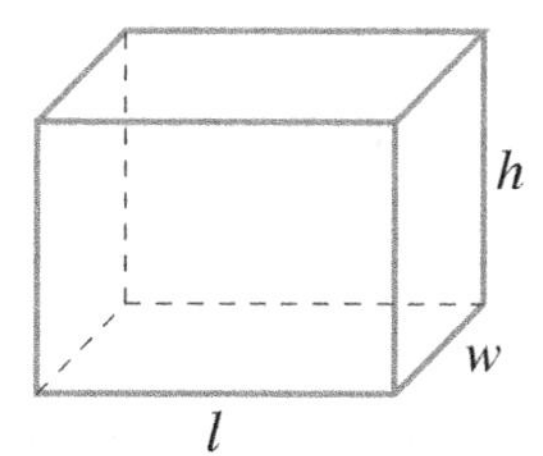


**Key Point**

The surface area of a rectangular prism is the sum of the areas of all six faces. Since opposite faces are equal, it can be simplified to: Surface area $= 2 \times (wh + lw + lh)$.

Let us examine an example.

**Example** Find the volume and surface area of a rectangular prism with a length of 7 m, a width of 5 m, and a height of 9 m.

**Solution:** The volume of this rectangular prism is:

$$\text{Volume} = 7 \times 5 \times 9 = 315 \text{ m}^3.$$

The surface area calculation gives us:

$$\text{Surface area} = 2 \times ((5 \times 9) + (7 \times 5) + (7 \times 9)) = 2 \times (45 + 35 + 63) = 2 \times 143 = 286 \text{ m}^2.$$

## 10.7 Cylinder

A cylinder is a three-dimensional shape with two parallel circular bases that are connected by a curved surface. Just like a prism, a cylinder has a base and a height. However, unlike a rectangular prism, the base of a cylinder is a circle, and its volume is calculated by taking the area of the base and multiplying it by the height.

**Key Point**

The volume of a cylinder can be calculated using the formula $V = \pi r^2 h$, where $r$ is the radius of the circular base and $h$ is the height of the cylinder.

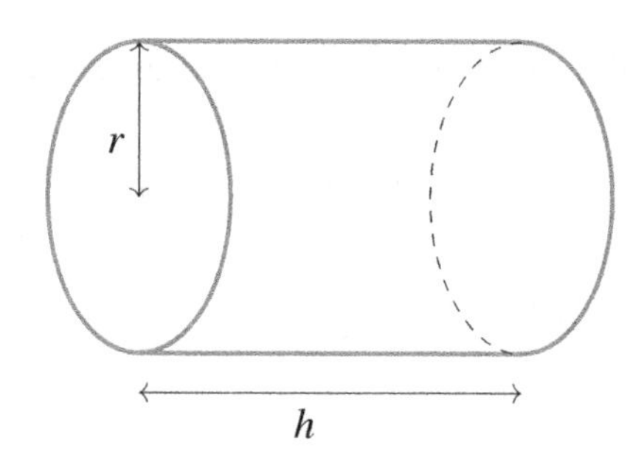

 Find the volume of a cylinder with a radius of 5 cm and a height of 12 cm.

**Solution:** The volume $V$ of the cylinder is given by the formula $V = \pi r^2 h$. Substituting the values of the radius and height, we get:

$$V = \pi(5\,\text{cm})^2 \times 12\,\text{cm} = 300\pi\,\text{cm}^3 \approx 942\,\text{cm}^3.$$

Therefore, the volume of the cylinder is approximately 942 cubic centimeters.

 A cylinder has a height of 15 cm and the area of its base is 50 cm$^2$. Calculate the volume.

**Solution:** Given the area of the base $A = 50\,\text{cm}^2$ and the height $h = 15\,\text{cm}$, the volume $V$ is:

$$V = A \times h = 50\,\text{cm}^2 \times 15\,\text{cm} = 750\,\text{cm}^3.$$

Thus, the volume of the cylinder is 750 cubic centimeters.

## 10.8 Surface Area of Prisms and Cylinders

**Prisms** are polyhedrons with two parallel and congruent bases connected by lateral faces that are parallelograms.

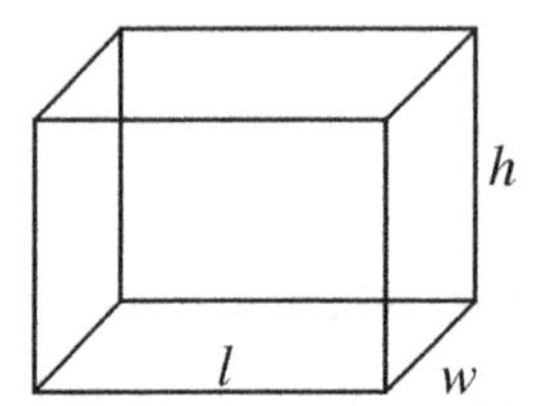

The surface area of a prism can be found using the formula:

$$\text{SA} = 2 \times \text{Area of Base} + \text{Perimeter of Base} \times \text{Height}.$$

For a *rectangular prism* specifically, the formula simplifies due to the rectangular nature of its faces:

$$\text{SA} = 2lw + 2lh + 2wh,$$

where $l$ is the length, $w$ is the width, and $h$ is the height.

A **cylinder** has two congruent and parallel bases (circles) and a curved surface connecting them.

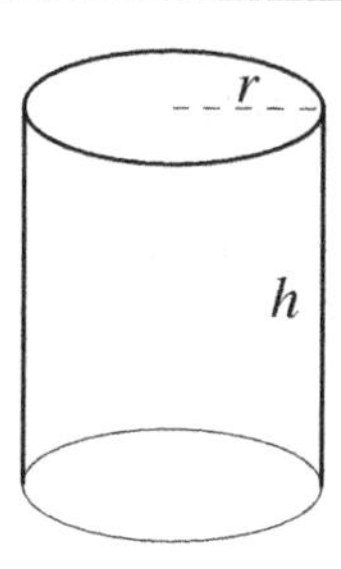


**Key Point**

The surface area of a cylinder includes the areas of the two bases and the curved surface, calculated as:

$$\text{SA} = 2\pi r^2 + 2\pi rh,$$

Where $r$ is the radius of the base and $h$ is the height of the cylinder.

**Example** Consider a rectangular prism with a length of 5 units, a width of 3 units, and a height of 7 units. Calculate its surface area.

**Solution:** Using the surface area formula for a rectangular prism:

$$\text{Surface Area} = 2lw + 2lh + 2wh,$$

substitute $l = 5$, $w = 3$, and $h = 7$:

$$\text{Surface Area} = (2 \times 5 \times 3) + (2 \times 5 \times 7) + (2 \times 3 \times 7) = 30 + 70 + 42 = 142.$$

Hence, the surface area is 142 square units.

**Example** A cylinder has a base radius of 4 units and a height of 10 units. What is the surface area of this cylinder?

**Solution:** Utilizing the cylinder surface area formula:

$$\text{Surface Area} = 2\pi r^2 + 2\pi rh,$$

inserting $r = 4$ and $h = 10$:

$$\text{Surface Area} = 2\pi(4)^2 + 2\pi(4) \times 10 = 32\pi + 80\pi = 112\pi \approx 351.68.$$

Therefore, the surface area of the cylinder is approximately 351.68 square units.

## 10.9 Volume of Cones and Pyramids

A *cone* is a solid that has a circular base connected by a curved surface to a single point called the vertex. The vertex is not in the same plane as the base.

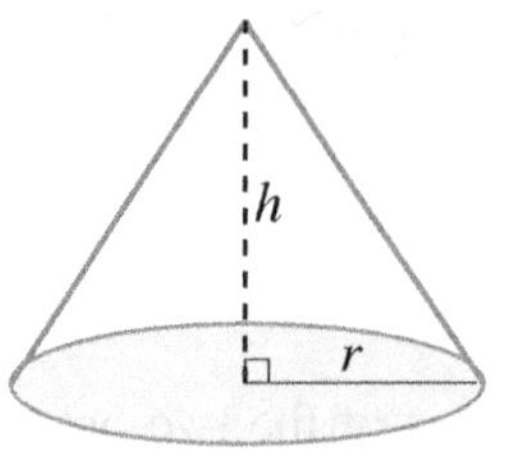


**Key Point**

The volume $V$ of a cone can be calculated using the formula: $V = \frac{1}{3}\pi r^2 h$, where $r$ is the radius of the base and $h$ is the height of the cone.

A *pyramid* is a solid that has a polygonal base and triangular faces that converge at a point called the apex or vertex. Unlike the cone, the base of a pyramid can have any polygonal shape.

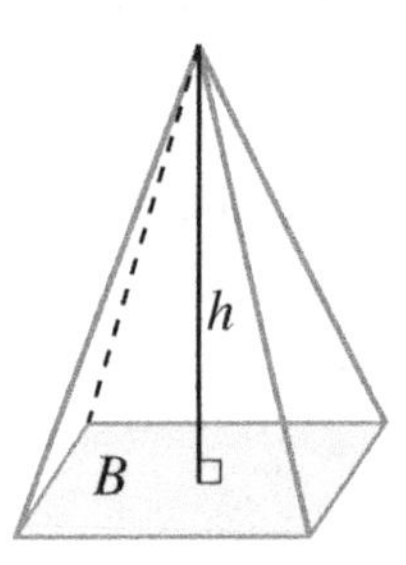


**Key Point**

The volume $V$ of a pyramid is given by the formula: $V = \frac{1}{3}Bh$, where $B$ is the area of the base and $h$ is the height from the base to the apex.

Let us delve into examples to solidify our understanding.

**Example** Find the volume of a cone with a base radius of 5 cm and a height of 12 cm. Use $\pi \approx 3.14$ for calculations.

**Solution:** Using the volume formula for cones:

$$V = \frac{1}{3}\pi r^2 h = \frac{1}{3}\pi (5\text{ cm})^2 (12\text{ cm}) \approx 314\text{ cm}^3.$$

**Example** A regular pyramid has a square base of side length 8 cm and a height of 15 cm. Calculate its volume.

**Solution:** First, calculate the area of the base $B$:

$$B = (\text{side length})^2 = (8 \text{ cm})^2 = 64 \text{ cm}^2.$$

Then, apply the pyramid volume formula:

$$V = \frac{1}{3}Bh = \frac{1}{3} \times 64 \text{ cm}^2 \times 15 \text{ cm} = 320 \text{ cm}^3.$$

## 10.10 Surface Area of Pyramids and Cones

A pyramid, regardless of its base shape, consists of a base and triangular lateral faces that converge at a point called the apex. The slant height ($l$) is the distance from the apex to the midpoint of a side of the base.

**Key Point**

The surface area ($SA$) of a pyramid can be calculated using the formula:

$$SA = B + \left(\frac{1}{2}\right) \times P \times l,$$

where $B$ is the area of the base, $P$ is the perimeter of the base, and $l$ is the slant height.

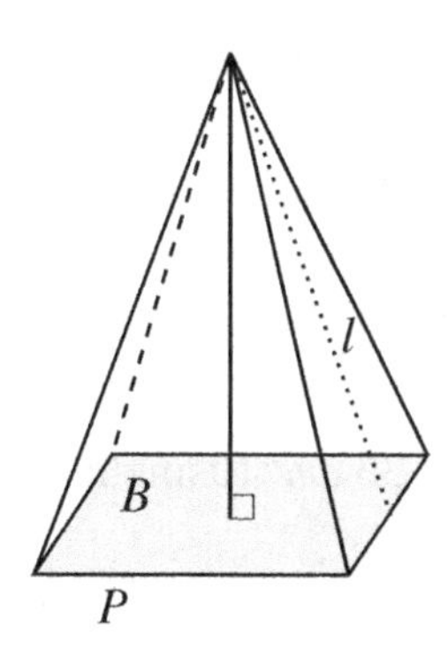


A cone has a circular base and a curved surface that tapers smoothly from the base to the apex. Like pyramids, the slant height ($l$) of a cone is the distance from the apex to any point on the edge of the base.

**Key Point**

The surface area ($SA$) of a cone can be found with the formula:

$$SA = \pi r(r + l),$$

where $r$ is the radius of the base, and $l$ is the slant height.

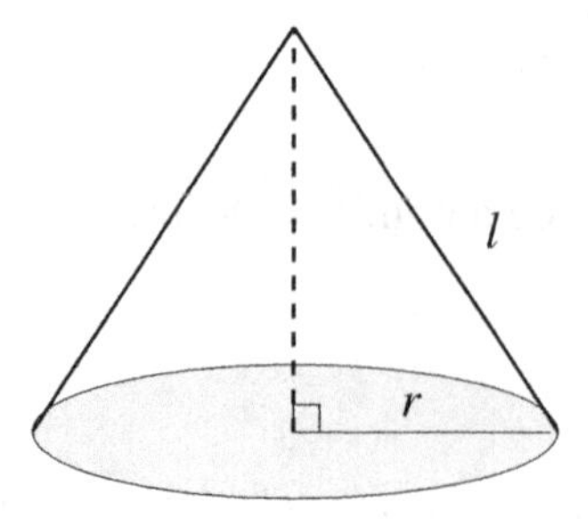


**Example** A square pyramid has a base side length $a$ of 10 $cm$ and a slant height $l$ of 12 $cm$. What is the surface area of the square pyramid?

**Solution:** Using the surface area formula for a pyramid:

$$B = a^2 = 100\ cm^2, \text{ and } P = 4a = 40\ cm.$$

So, the surface area will be:

$$SA = B + \left(\frac{1}{2}\right) \times P \times l = 100 + \left(\frac{1}{2}\right) \times 40 \times 12 = 340\ cm^2.$$

Therefore, the surface area of the pyramid is 340 $cm^2$.

**Example** Calculate the surface area of a cone with a radius $r$ of 3 $cm$ and a slant height $l$ of 5 $cm$.

**Solution:** Using the surface area formula for a cone:

$$SA = \pi r(r + l) = \pi(3)(3 + 5) = 24\pi\ cm^2.$$

The surface area of the cone is approximately 75.36 $cm^2$ (using $\pi \approx 3.14$).

## 10.11 Volume of Spheres

A sphere is a set of all points in space that are the same distance from a central point. This distance is known as the radius ($r$) of the sphere.

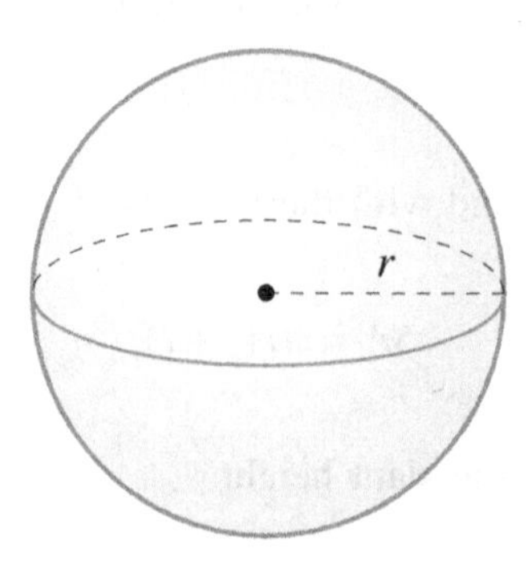

Key Point

The formula to calculate the volume of a sphere is expressed as $V = \frac{4}{3}\pi r^3$, where $r$ is the radius.

This formula tells us that the volume of a sphere grows rapidly with its radius since the radius is raised to the third power.

 Find the volume of a sphere with a radius of 5 cm. Use $\pi \approx 3.14$.

**Solution:** Given the radius ($r = 5\ cm$), we use the volume formula:

$$V = \frac{4}{3}\pi r^3 = \frac{4}{3}\pi(5\ cm)^3 = \frac{4}{3} \times 3.14 \times 125\ cm^3 \approx 523.33\ cm^3.$$

Hence, the volume of the sphere is approximately $523.33\ cm^3$.

 What is the volume of a spherical balloon with a diameter of 15 cm?

**Solution:** First, we find the radius of the sphere by halving the diameter:

$$\text{radius} = \frac{\text{diameter}}{2} = \frac{15\ cm}{2} = 7.5\ cm.$$

Now, we apply the volume formula:

$$V = \frac{4}{3}\pi r^3 = \frac{4}{3}\pi(7.5\ cm)^3 \approx 1766.25\ cm^3.$$

Therefore, the volume of the balloon is approximately $1766.25\ cm^3$.

## 10.12 Sphere Surface Area

The surface area of a sphere is a fundamental concept in geometry, representing the total area covered by the outer surface of the sphere.

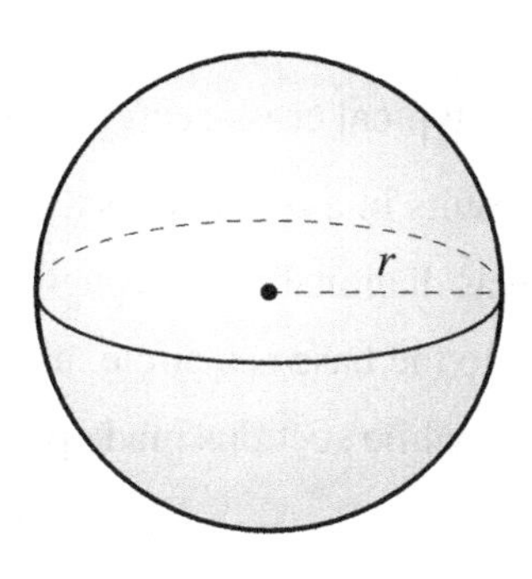

Key Point

The surface area $S$ of a sphere with radius $r$ is given by the formula $S = 4\pi r^2$. Alternatively, using the diameter $d$, the surface area is $S = \pi d^2$.

Calculate the surface area of a sphere with a diameter of 10 cm. Use $\pi \approx 3.14$.

**Solution:** First, we apply the surface area formula using the diameter:

$$S = \pi d^2 = 100\pi \approx 314\ cm^2.$$

Thus, the surface area of the sphere is approximately 314 $cm^2$.

Example A metallic sphere has a radius of 4 inches. Calculate both the surface area of the sphere and the amount of paint required to cover this area, knowing that the paint coverage per unit area is 0.1 ounce per square inch.

**Solution:** Using the radius to find the surface area, we have:

$$S = 4\pi r^2 = 4\pi(4 \text{ in})^2 = 64\pi \approx 200.96 \text{ in}^2.$$

To calculate the amount of paint needed, we multiply the surface area by the coverage rate:

$$\text{Paint required} = S \times \text{coverage rate} = 200.96 \times 0.1 \approx 20.1.$$

Thus, approximately 20.1 ounces of paint is required to cover the sphere.

## 10.13 Solids and Their Cross-Sections

When a solid is intersected by a plane, the shape that results on this plane is called a cross-section. The shape and size of this cross-section depend on the orientation and location of the plane that intersects the solid.

Every solid has a unique set of possible cross-sections that arise from various cuts at different angles.

Let us delve into the specific solids and their typical cross-sections resulting from particular cuts.

- **Cube:** Cutting perpendicular to a face results in a square cross-section. Diagonal slices that do not pass through vertices yield rectangular sections, whereas diagonal cuts through opposite vertices produce hexagonal sections.
- **Cylinder:** A cross-section perpendicular to the base is a circle, matching the base's shape. Angled cuts parallel to the base generate elliptical cross-sections, while sections made perpendicular to the cylinder's curved surface are rectangular.

- **Sphere:** Slicing through the sphere's center in any direction creates circular cross-sections, with sizes varying from smallest at the edges to largest at the center.
- **Pyramid:** Horizontal cuts produce scaled-down versions of the base polygon, decreasing in size towards the apex. A vertical cut through the center of a square or rectangular pyramid results in triangular sections.
- **Cone:** Cross-sections perpendicular to the base are circular and decrease in size as they approach the apex. Slanted cuts parallel to the base yield elliptical sections, while cuts perpendicular to the cone's slanted side form parabolic cross-sections.

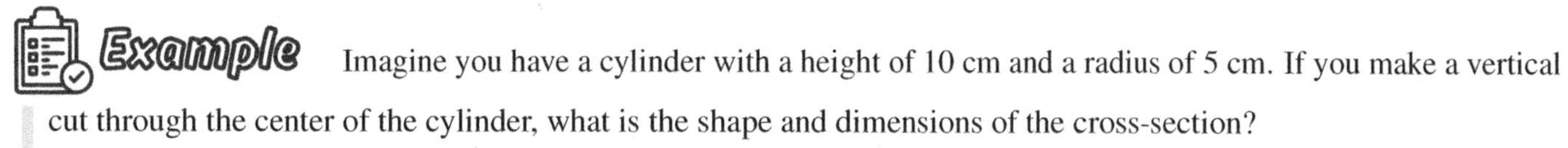

Example Imagine you have a cylinder with a height of 10 cm and a radius of 5 cm. If you make a vertical cut through the center of the cylinder, what is the shape and dimensions of the cross-section?

**Solution:** Such a cut would run along the length of the cylinder, intersecting its circular bases at the center.

Therefore, the cross-section will be a rectangle. Its length will be the same as the height of the cylinder, 10 cm, and its width will be twice the radius of the bases, which is $2 \times 5$ cm, so 10 cm.

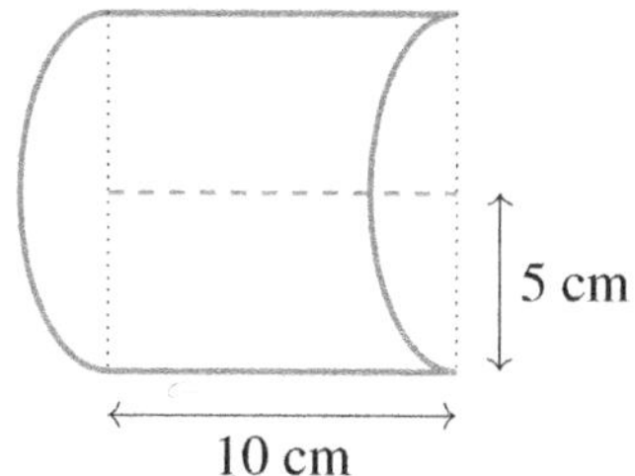


## 10.14 Volume of a Truncated Cone

A truncated cone is a shape resulting from slicing the top off a cone parallel to its base, creating two circular bases — a smaller top circle and a larger bottom circle — with a curved surface connecting them.

**Key Point**

The volume of a truncated cone, or frustum, is calculated using the formula:

$$V = \left(\frac{1}{3}\right) \pi h(R^2 + r^2 + Rr),$$

where $h$ is the height of the frustum, $R$ is the radius of the larger base, and $r$ is the radius of the smaller base.

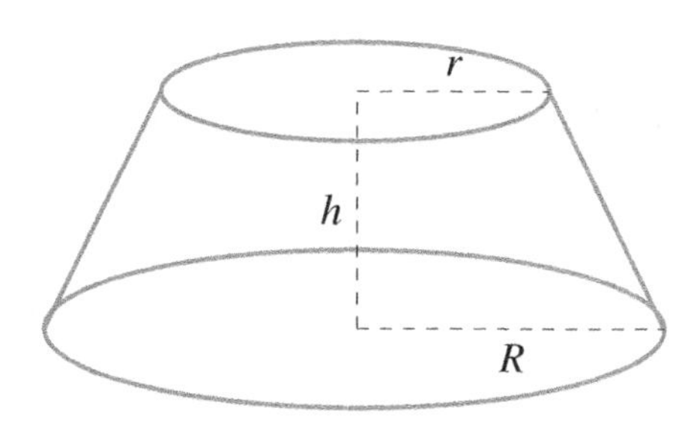


To better understand how to apply this formula, let us go through a couple of examples.

**Example** Given a truncated cone with a height $h = 10$ cm, a larger base radius $R = 8$ cm, and a smaller base radius $r = 5$ cm, calculate its volume.

**Solution:** Using the volume formula for a truncated cone:

$$V = \left(\frac{1}{3}\right)\pi(10)(8^2+5^2+(8)(5)) = \left(\frac{1}{3}\right)10\pi(129) = 430\pi \approx 1,350.2 \text{ cm}^3.$$

Thus, the volume of the truncated cone is approximately $1,350.2$ cm$^3$.

**Example** The volume of a truncated cone is $600\pi$ cm$^3$. The radii of the two circular bases are 5 cm and 3 cm. Find the height of the truncated cone.

**Solution:** The volume $V$ of a truncated cone is given by:

$$V = \left(\frac{1}{3}\right)\pi h(R^2+r^2+Rr)$$

where $R$ and $r$ are the radii of the larger and smaller bases, respectively, and $h$ is the height of the truncated cone. Given $R = 5$ cm, $r = 3$ cm, and $V = 600\pi$ cm$^3$, we can substitute these values into the formula to find $h$:

$$600\pi = \frac{1}{3}\pi h(5^2+3^2+5\times 3)$$

Simplifying the right side, we get:

$$600\pi = \frac{1}{3}\pi h(25+9+15) \Rightarrow 600\pi = \frac{1}{3}\pi h \times 49.$$

Solving for $h$, we have:

$$h = \frac{600\times 3}{49} \approx 36.73.$$

Therefore, the height of the truncated cone is approximately 36.73 cm.

## 10.15 Practices

**1)** Find the radius of each circle. Use your calculator's value of $\pi$. Round your answer to the nearest tenth.

1-1) Circumference $= 43.9$ *ft*

1-2) Circumference $= 75.4$ *yd*

1-3) Circumference $= 94.2$ *mi*

1-4) Circumference $= 12.6$ *yd*

**2)** Find the circumference of each circle.

2-1) Area $= 49\pi\ mi^2$

2-2) Area $= 81\pi\ in^2$

**3)** Find the area of each circle.

3-1) Circumference $= 12\pi\ yd$

3-2) Circumference $= 26\pi\ in$

**4)** Find the area of each trapezoid.

4-1)

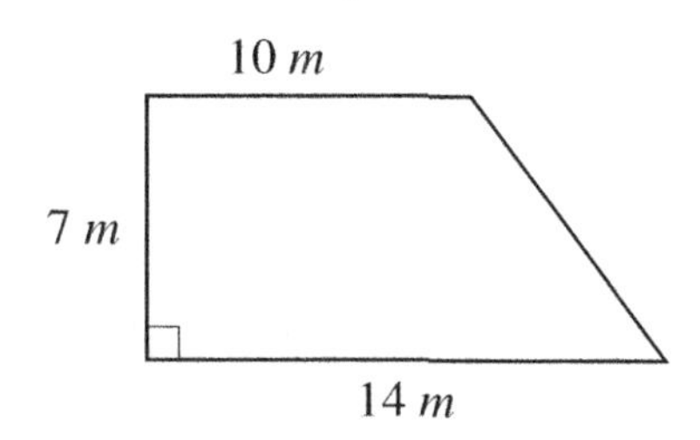


4-2)

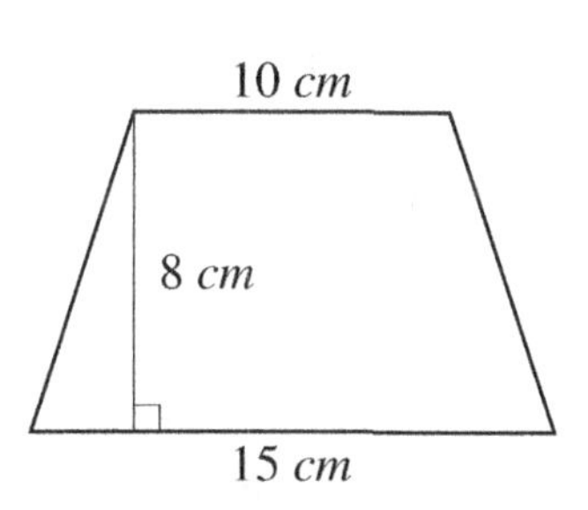


4-3)

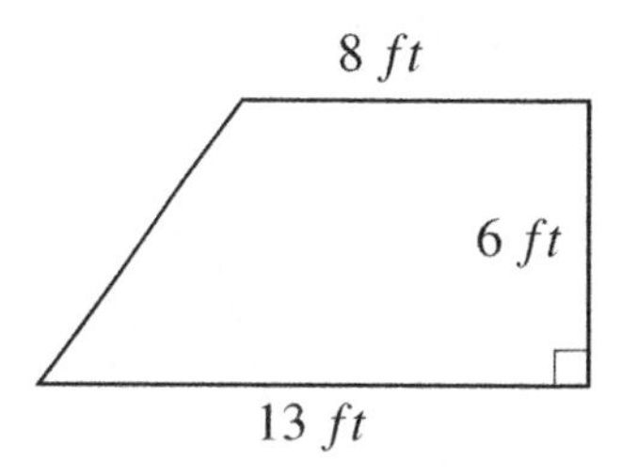


4-4)

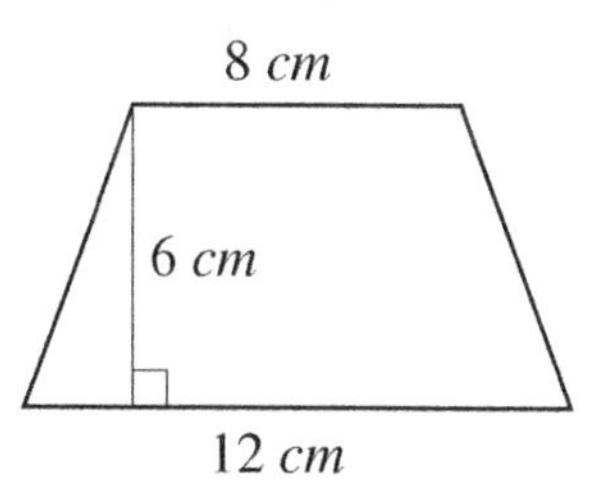


**5)** Find the area of each polygon. Round your answer to the nearest tenth.

5-1)

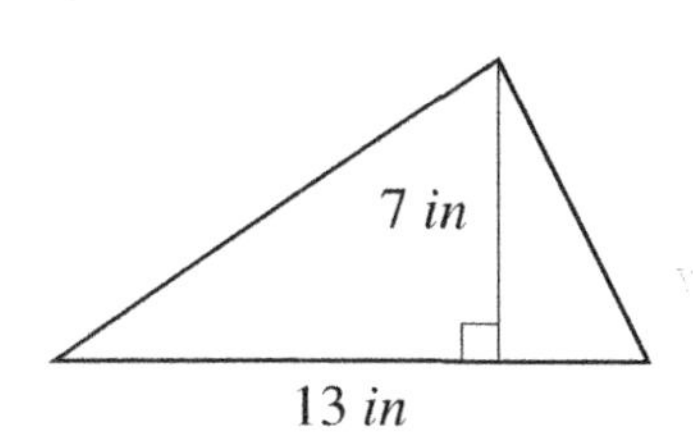


5-2)

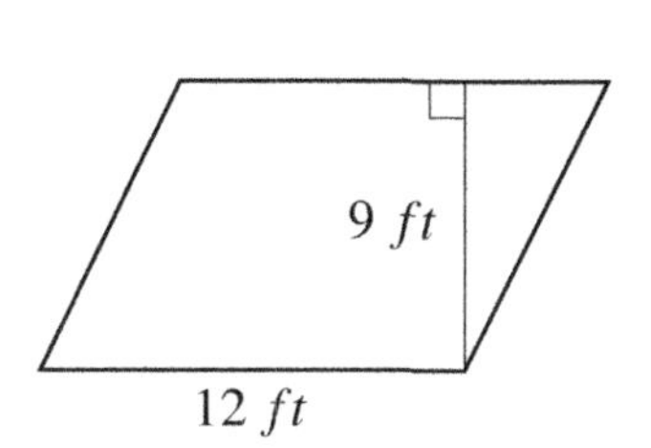


5-3)

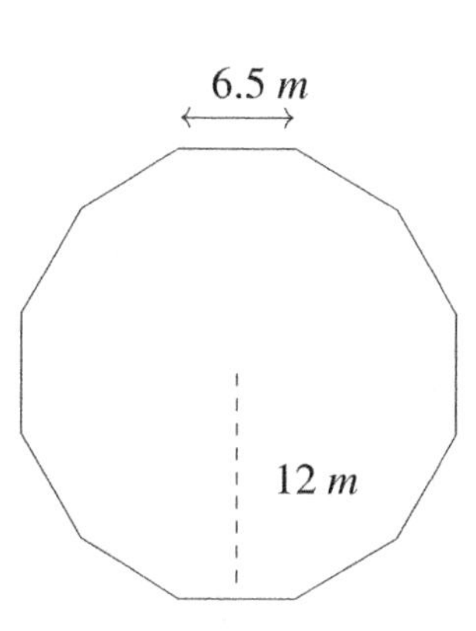

5-4)

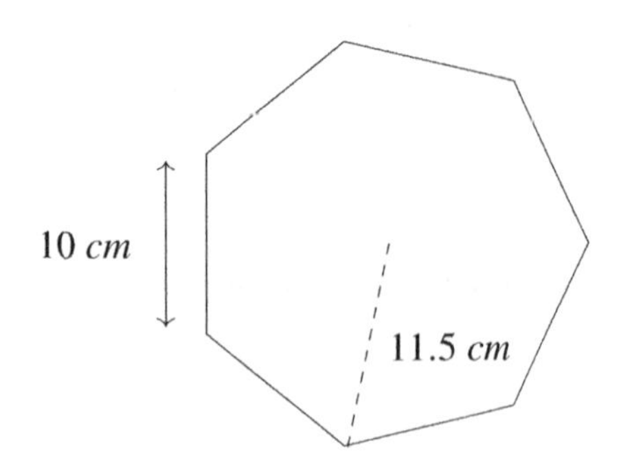


**6)** Create a net for the following figure.

6-1)

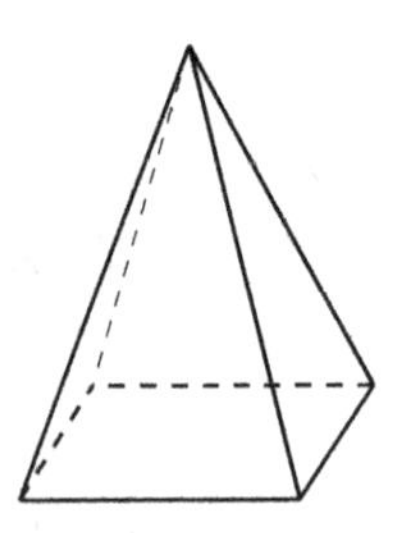

6-2)

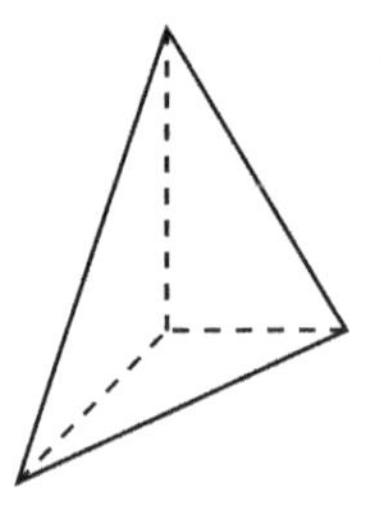

**7)** Create a figure for the following net.

7-1)

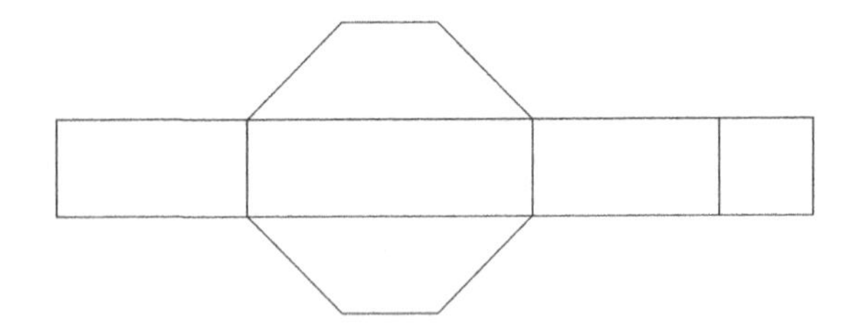

7-2)

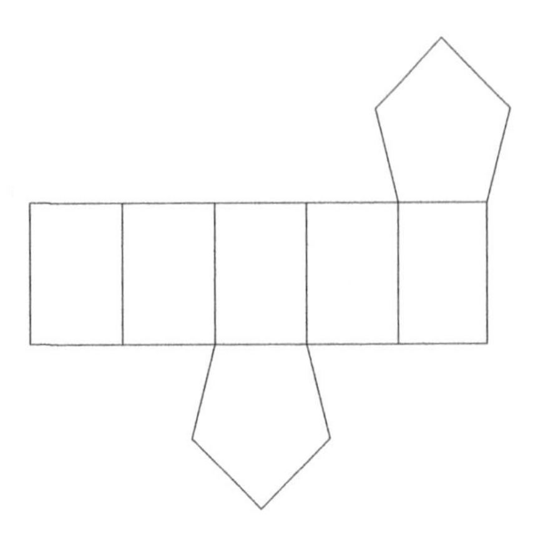

7-3)

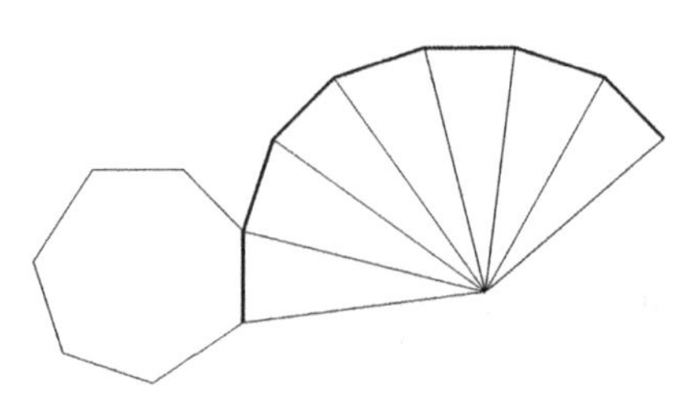

7-4)

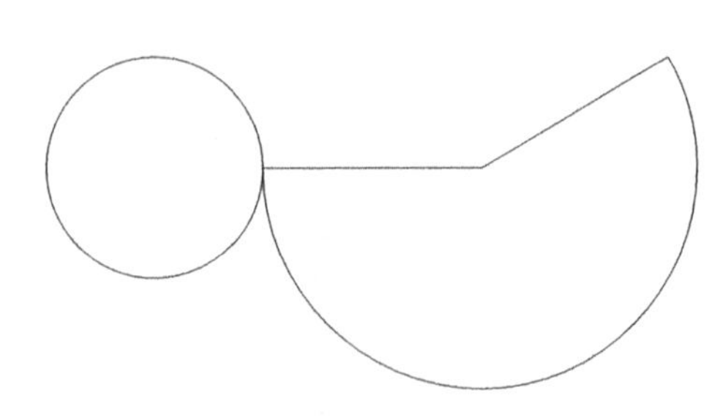

**8)** Name each figure.

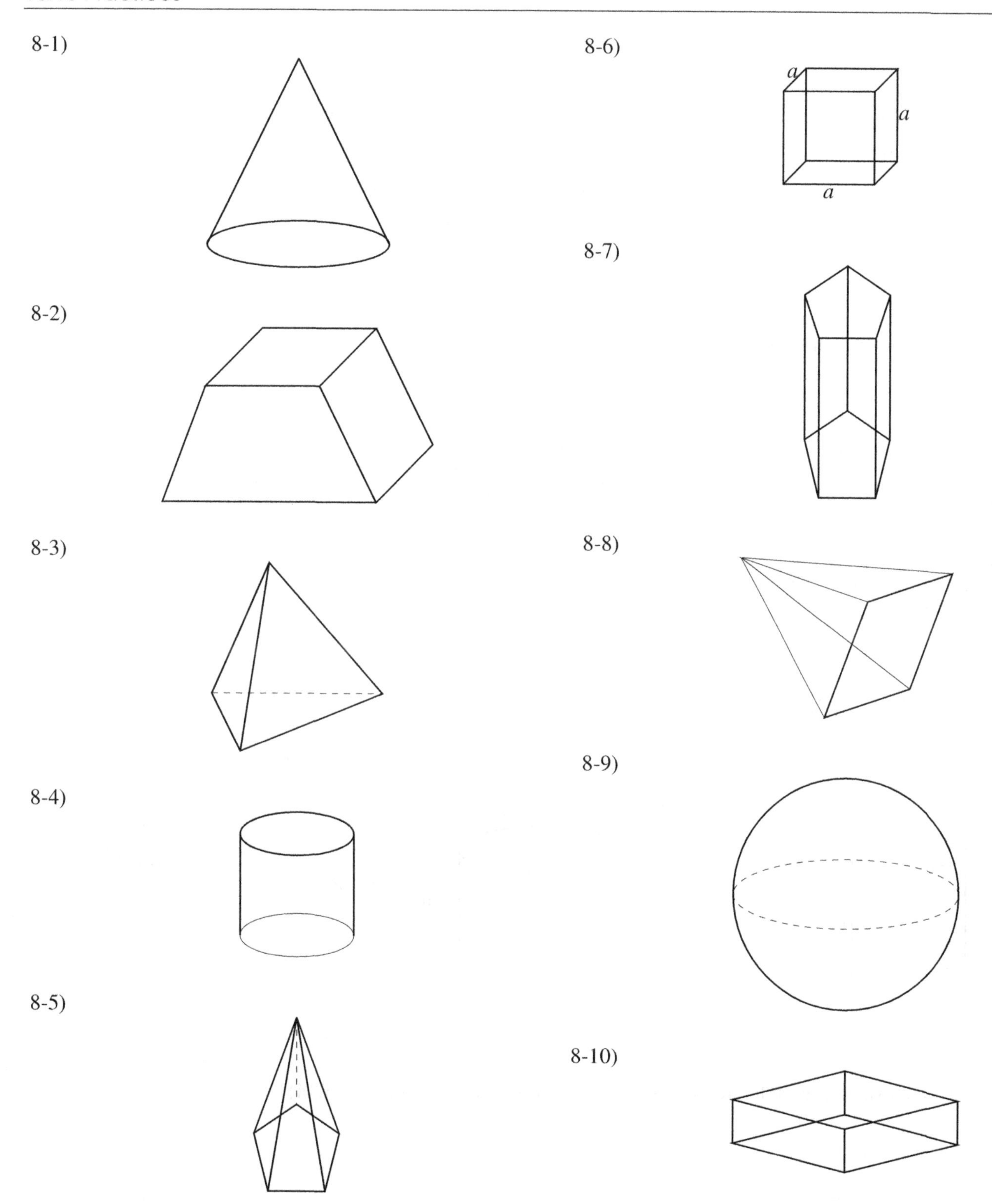


**9)** Find the volume of each cube.

9-1)

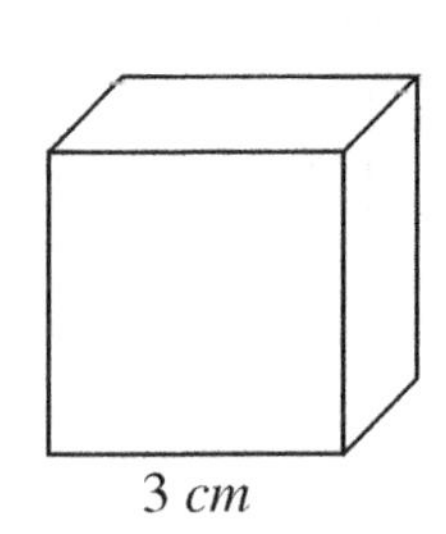


9-2)

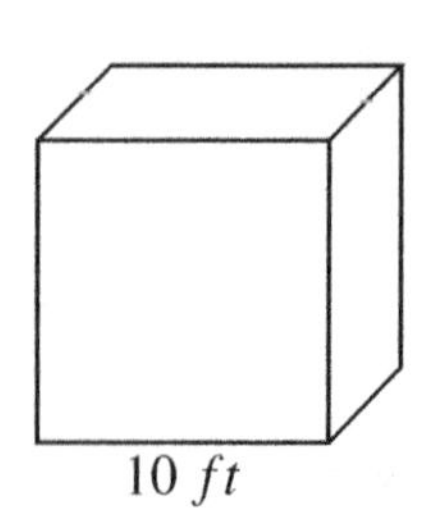


9-3)

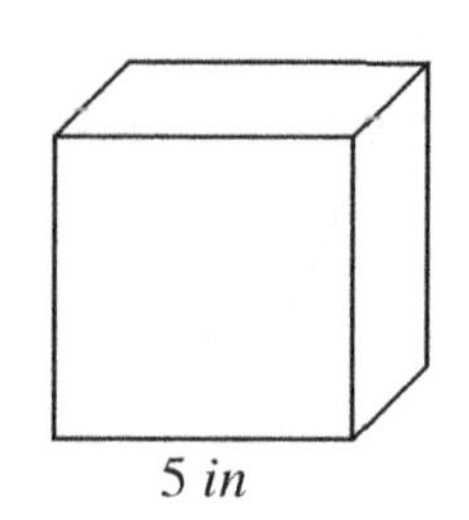


9-4)

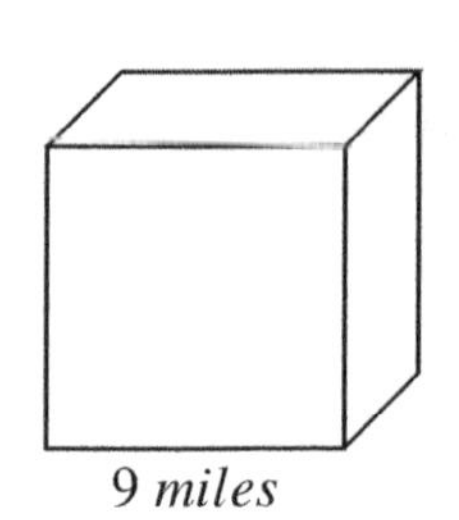


**10)** Find the volume of each Rectangular Prism.

10-1)

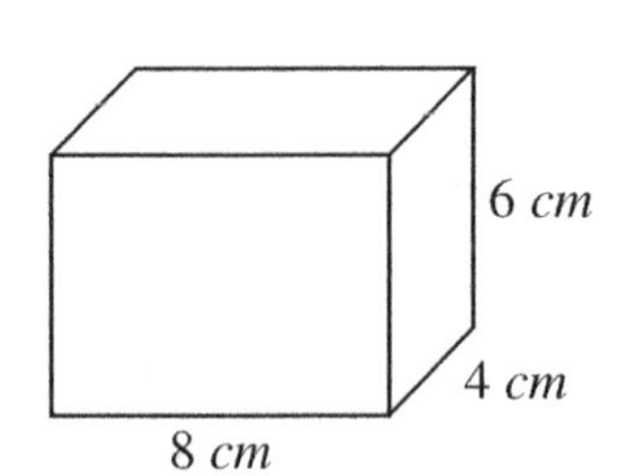


10-2)

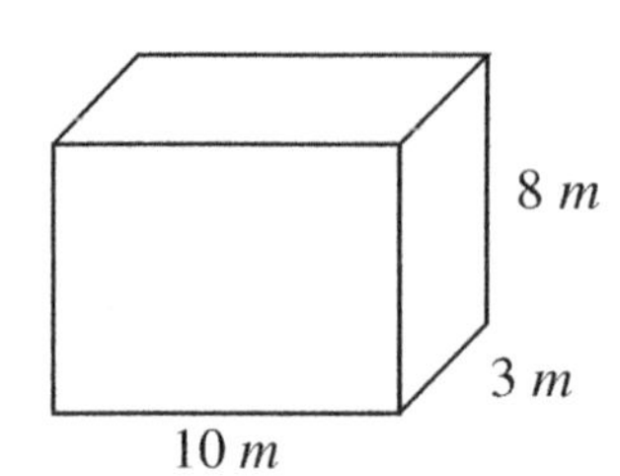


10-3)

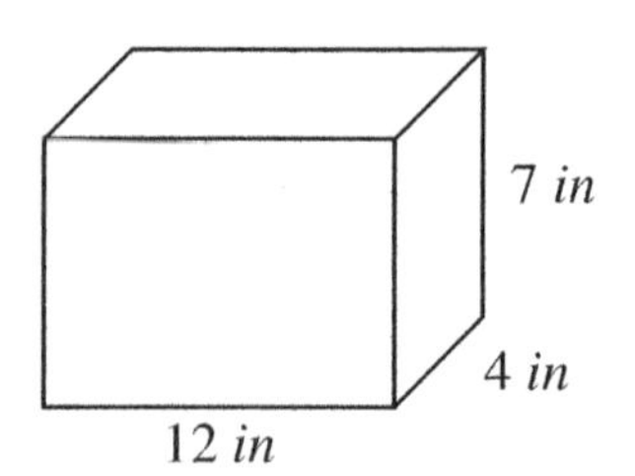


**11)** Find the volume of each Cylinder. Round your answer to the nearest tenth. ($\pi = 3.14$)

11-1)

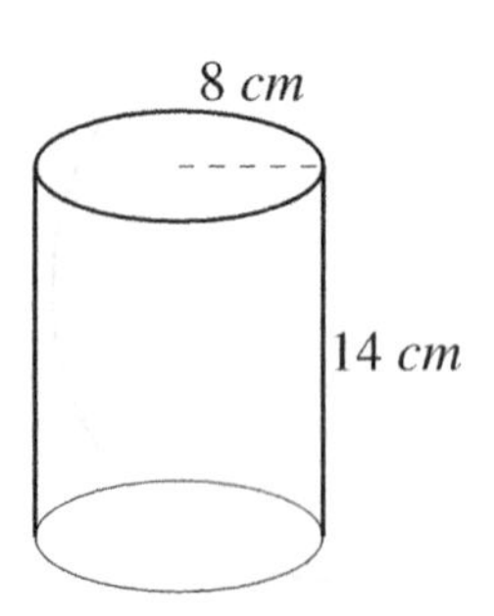


11-2)

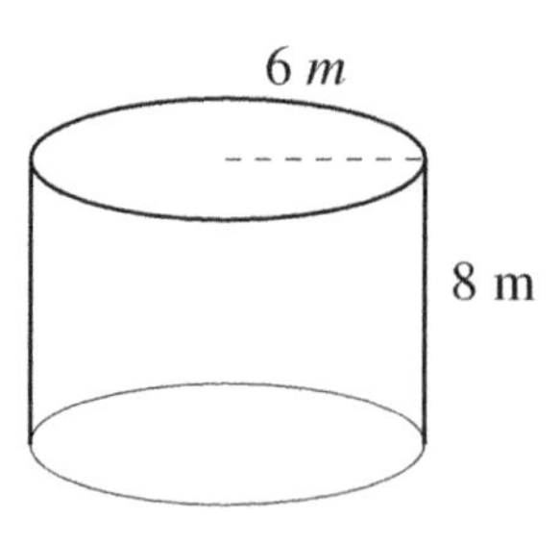


11-3)

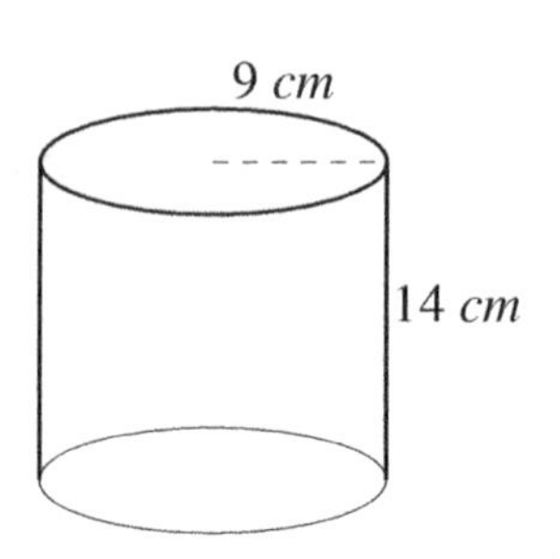


**12)** Find the area surface of each figure. Round your answer to nearest tenth. ($\pi = 3.14$)

12-1)

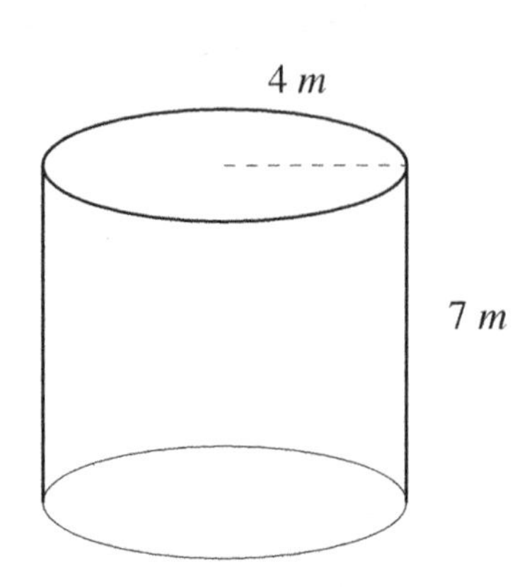

12-2)

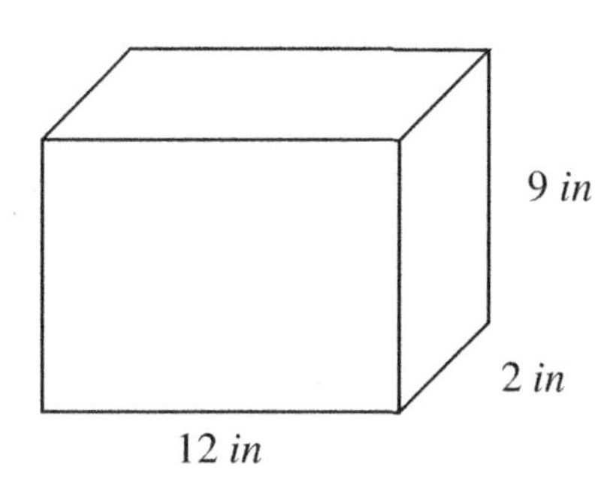


12-3)

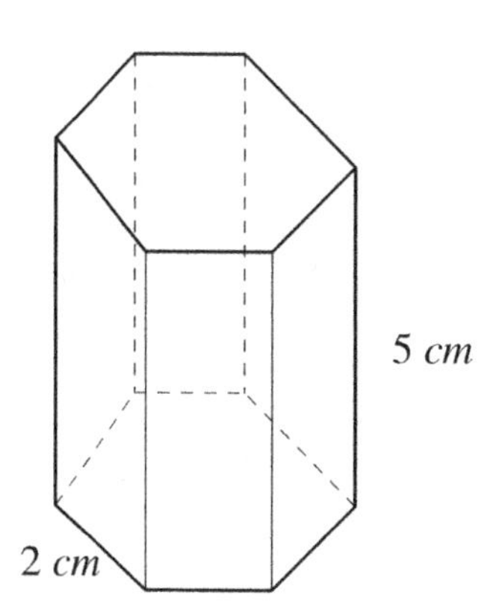


12-4)

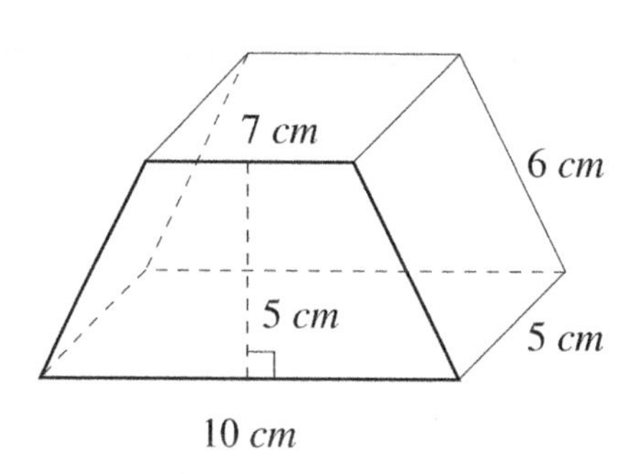


**13)** Find the volume of each figure. Round your answer to nearest hundredth. ($\pi = 3.14$)

13-1)

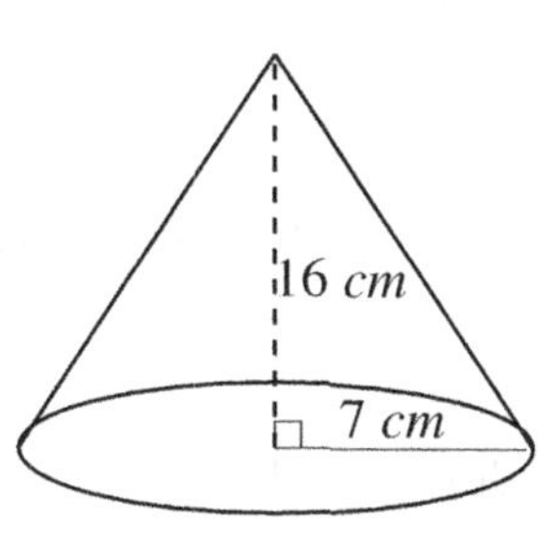


13-2)

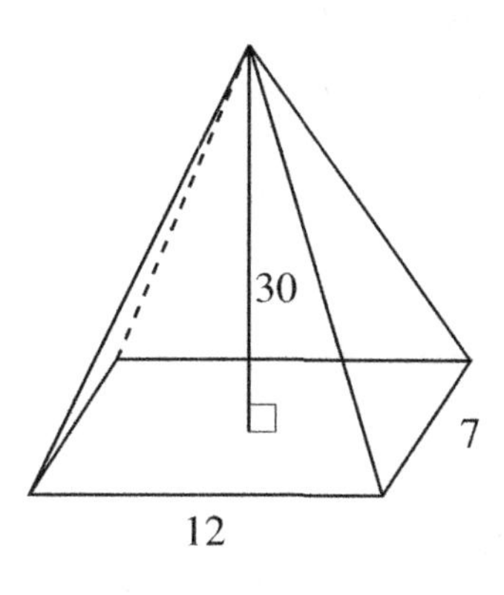


13-3)

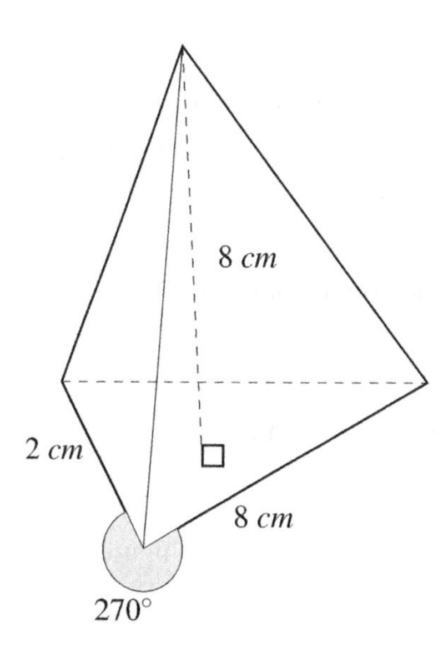

13-4)

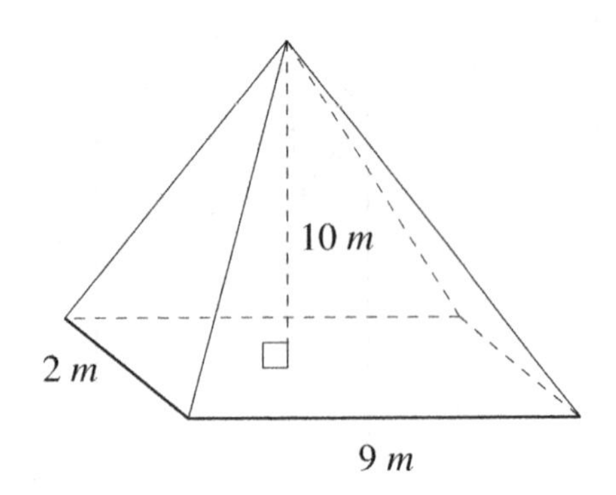


**14)** Find the area surface of each figure. Round your answer to nearest tenth. ($\pi = 3.14$)

14-1)

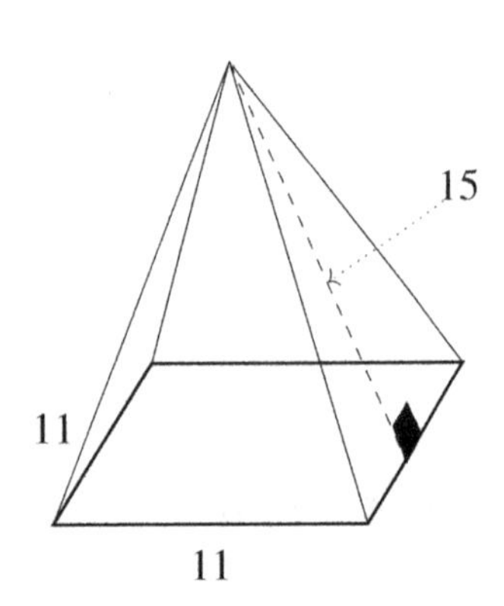


14-2)

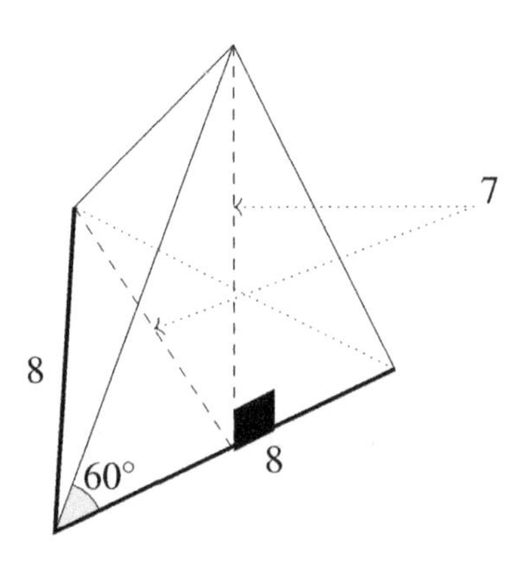


14-3) A cone with radius 6 $m$ and a slant height of 10 $m$.

14-4) A slant height of a hexagonal pyramid 14 $cm$ tall, with a regular base measuring 6 $cm$ on each side and an apothem of length 5.2 $cm$.

**15)** Find the volume of each Spheres. Round your answer to nearest hundredth. ($\pi = 3.14$)

15-1) Diameter $= 1.8\ ft$

15-2) Diameter $= 1.5\ cm$

15-3) Radius $= 10\ ft$

15-4) Radius $= 12\ ft$

**16)** Find the surface area each figure.

16-1)

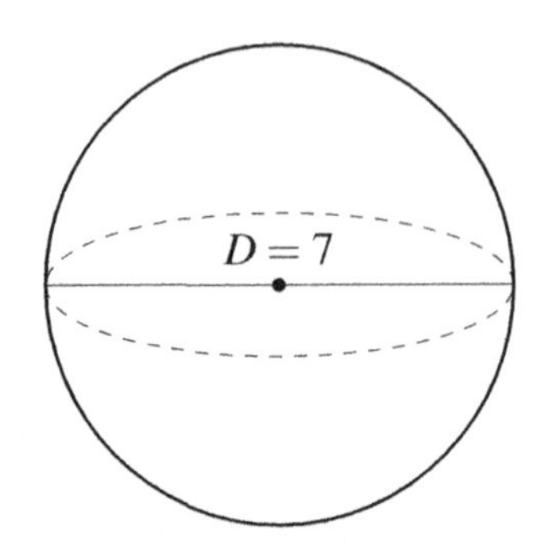


16-2)

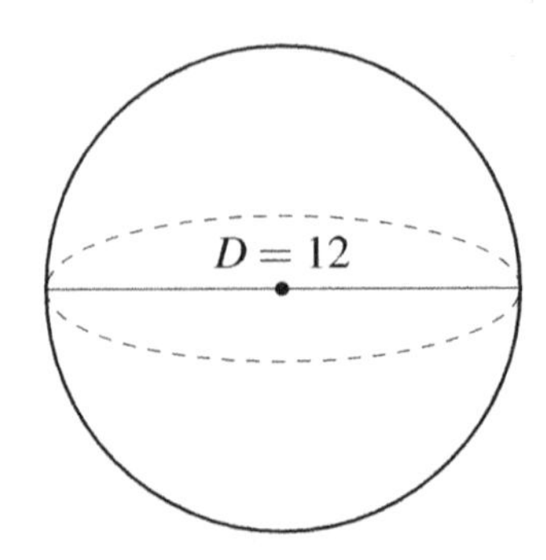

16-3)

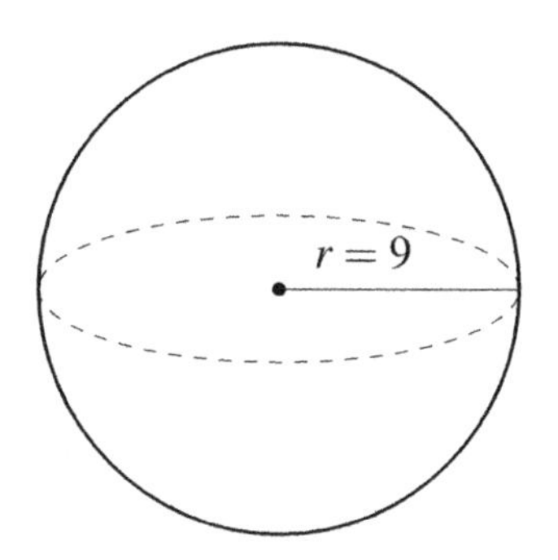


**17)** Solve.

17-1) If a rectangle is revolved around one of its sides, what is the solid produced?

17-2) If right triangle $ABC$, shown, was rotated around segment $BC$, what is the solid produced?

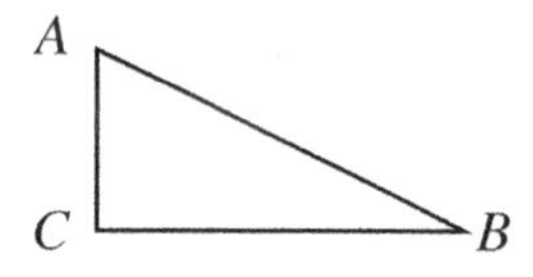


**18)** Find the volume of each figure. Round your answer to nearest hundredth. ($\pi = 3.14$)

18-1)

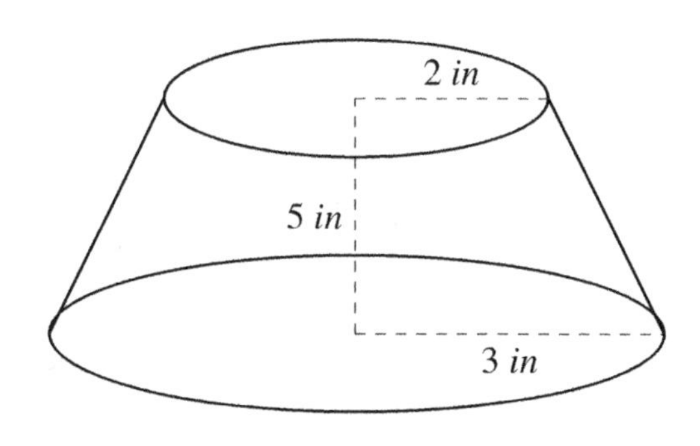


18-2)

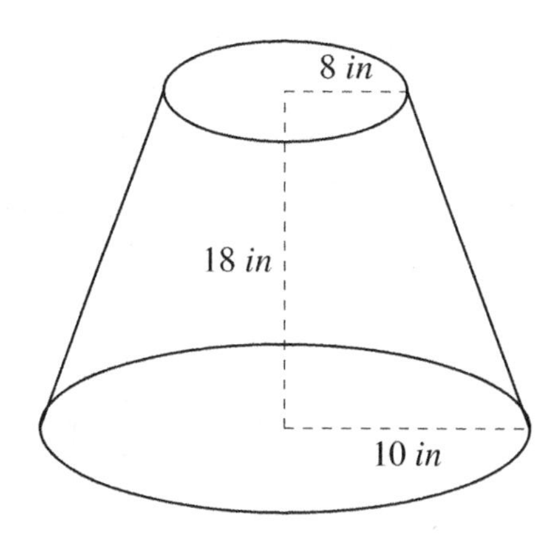


**19)** Find the volume of $x$. Round your answer to nearest tenth. ($\pi = 3.14$)

19-1)

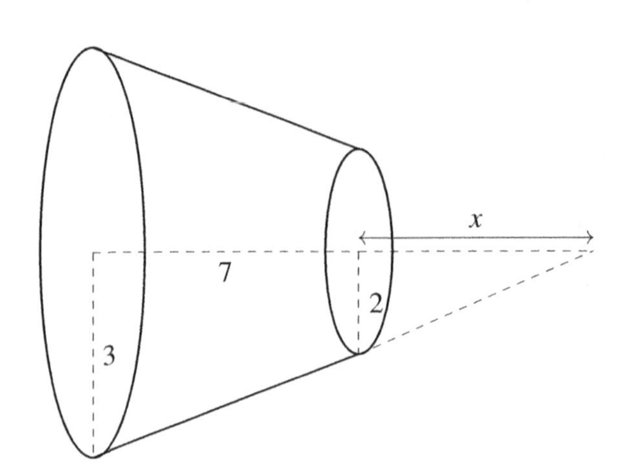


19-2)

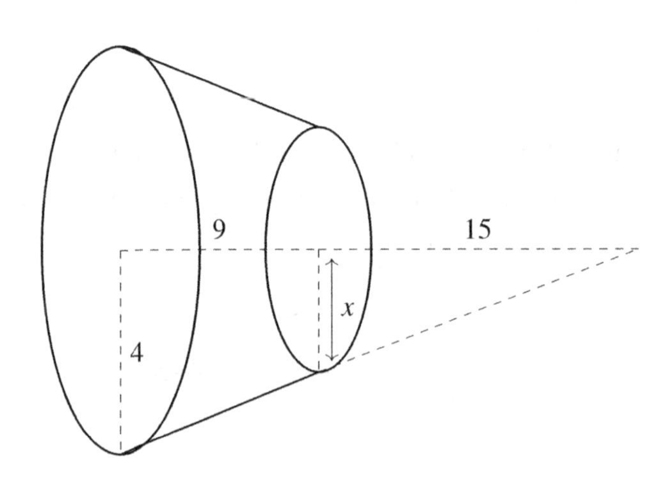


**20)** Draw the cross section of the given shapes.

20-1)

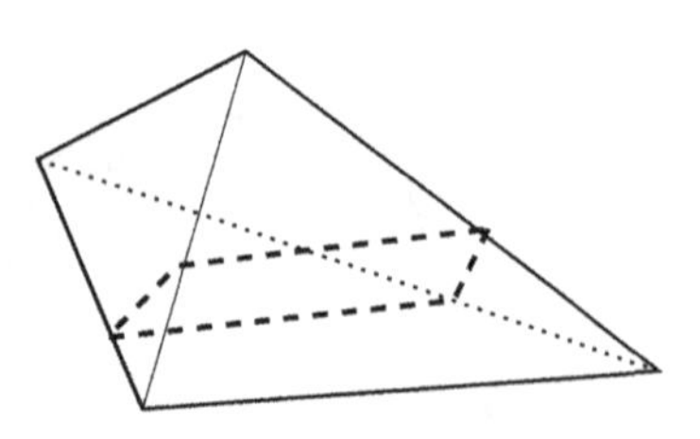

20-2)

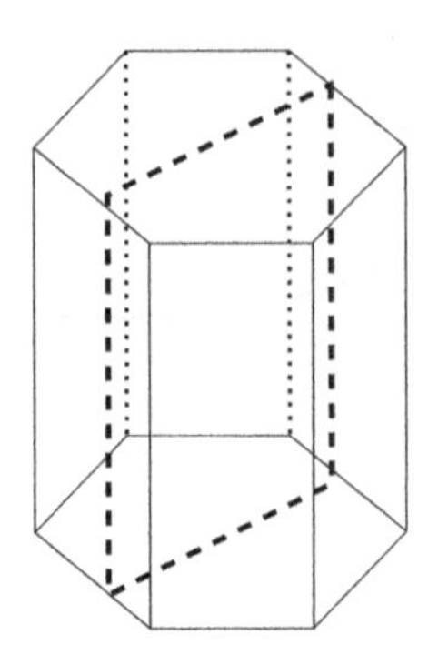

20-3)

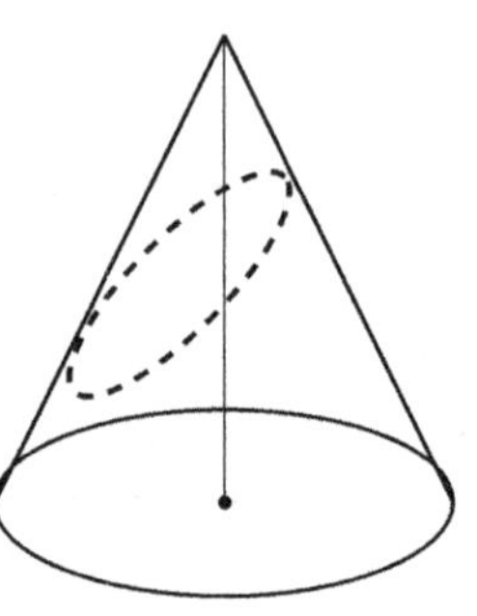

20-4)

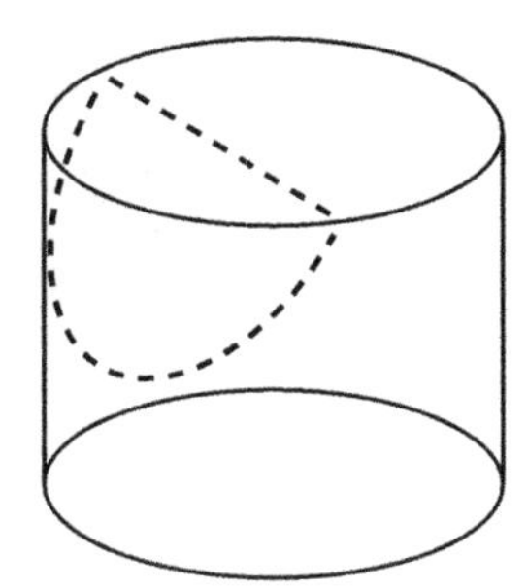

## 10.16 Answers

**1)**

1-1) $7\ ft$

1-2) $12\ yd$

1-3) $15\ mi$

1-4) $2\ yd$

**2)**

2-1) $14\pi\ mi$

2-2) $18\pi\ in$

**3)**

3-1) $36\pi\ yd^2$

3-2) $169\pi\ in^2$

**4)**

4-1) $84\ m^2$

4-2) $100\ cm^2$

4-3) $63\ ft^2$

4-4) $60\ cm^2$

**5)**

5-1) $45.5\ in^2$

5-2) $108\ ft^2$

5-3) $468\ m^2$

5-4) $362.5\ cm^2$

**6)**

6-1)

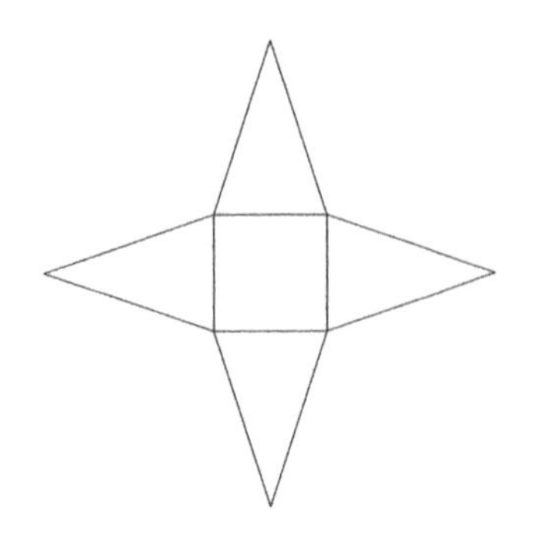

6-2)

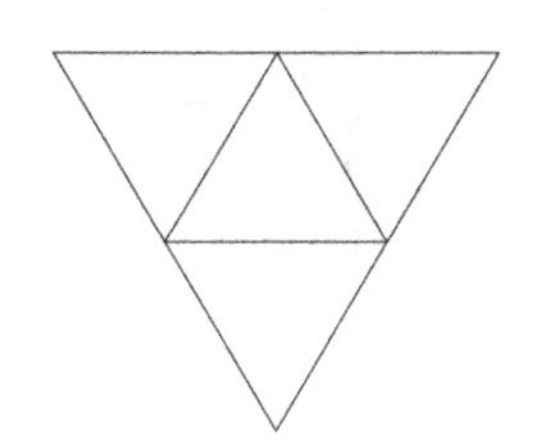

**7)**

7-1)

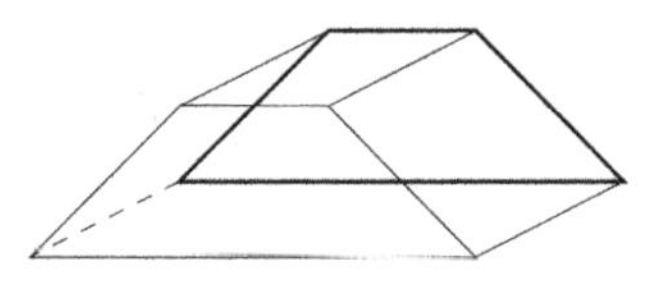

7-2)

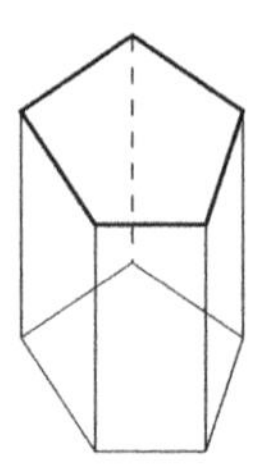

7-3)

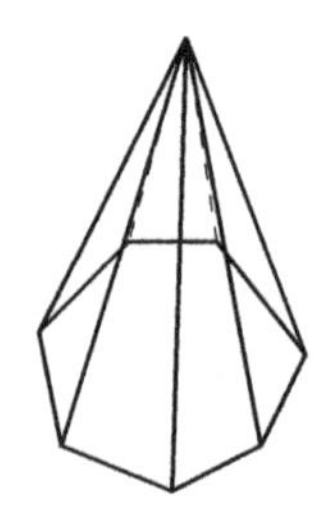

7-4)

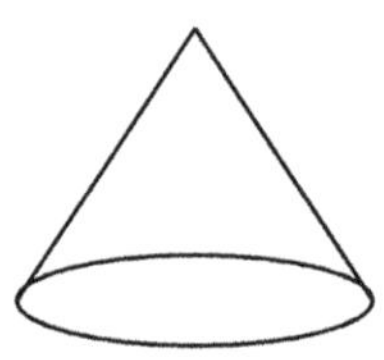

**8)**

8-1) Cone

8-2) Trapezoidal prism

8-3) Triangular pyramid

8-4) Cylinder

8-5) Pentagonal pyramid

8-6) Square prism

8-7) Pentagonal prism

8-8) Rectangular pyramid

8-9) Sphere

8-10) Rectangular prism

**9)**

9-1) 27 $cm^3$

9-2) 1000 $ft^3$

9-3) 125 $in^3$

9-4) 729 $mi^3$

**10)**

10-1) 192 $cm^3$

10-2) 240 $m^3$

10-3) 336 $in^3$

**11)**

11-1) 2813.44 $cm^3$

11-2) 904.32 $m^3$

11-3) 3560.76 $cm^3$

**12)**

12-1) 276.3 $m^2$

12-2) 300 $in^2$

12-3) 80.8 $cm^2$

12-4) 230 $cm^2$

**13)**

13-1) 821 $cm^3$

13-2) 840 *cubic units*

13-3) 21.3 $cm^3$

13-4) 60 $m^3$

**14)**

14-1) 451

14-2) 112

14-3) 301.4 $m^2$

14-4) 345.6 $cm^2$

**15)**

15-1) 3.05 $ft^3$

15-2) 1.77 $cm^3$

15-3) 4186.67 $ft^3$

15-4) 7234.56 $ft^3$

**16)**

16-1) 153.86

16-2) 452.16

16-3) 1017.36

**17)**

17-1) Cylinder

17-2) Cone

**18)**

18-1) 99.43 $in^3$

18-2) 4596.96 $in^3$

**19)**

19-1) 14

19-2) 2.5

**20)**

20-1)

20-2)

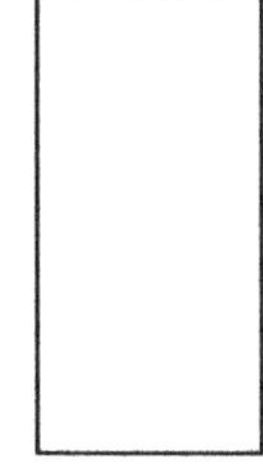

20-3)

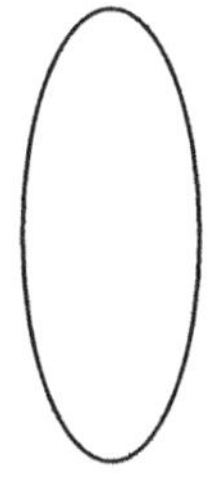

20-4)

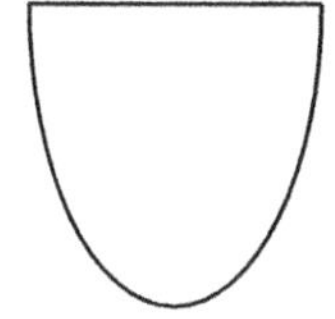

# 11. Practice Test 1

## 11.1 Practices

**1)** If angle $GQY = 130^\circ$, find the angle $CPQ$.

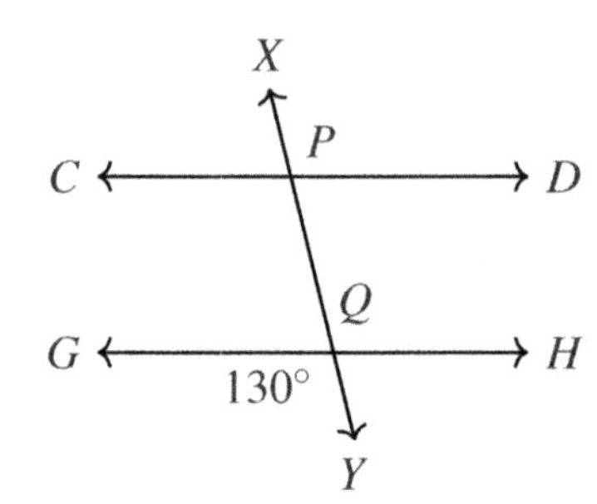

☐ A. 230°  ☐ C. 130°

☐ B. 30°  ☐ D. 50°

**2)** $QS$ bisects $\angle PQR$. If $\angle RQS = 55^\circ$, find measure of $\angle PQR$.

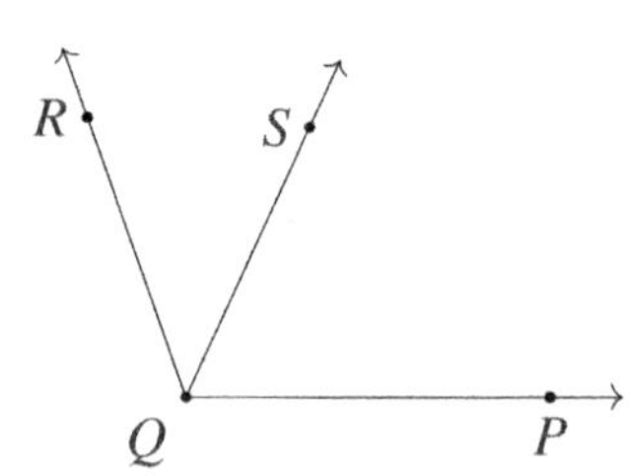

☐ A. 25°  ☐ C. 75°

☐ B. 50°  ☐ D. 110°

**3)** Which of the following best describes three points that lie on a curve?

☐ A. intersecting  ☐ C. non-coplanar

☐ B. collinear  ☐ D. non-collinear

**4)** The measure of $\angle XOY$ is 45°. $\angle XOY$ is:

☐ A. a right angle  ☐ C. a straight angle

☐ B. an obtuse angle  ☐ D. an acute angle

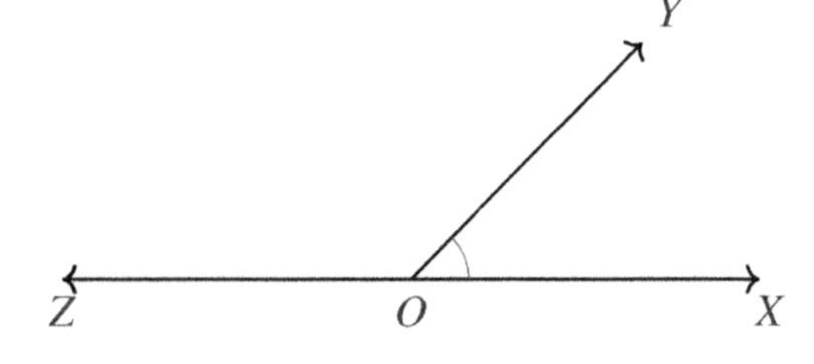

**5)** What is the equation of the line through $(2,-3)$ and perpendicular to $y=\frac{3}{4}x+\frac{1}{3}$?

☐ A. $y=-\frac{4}{3}x-\frac{1}{3}$

☐ B. $y=-\frac{4}{3}x+\frac{4}{3}$

☐ C. $y=\frac{3}{4}x-\frac{1}{4}$

☐ D. $y=-\frac{3}{4}x-\frac{3}{4}$

**6)** Find the slope of a line parallel to the line given by $y=\frac{4}{7}x-2$.

☐ A. $\frac{7}{4}$ ☐ B. $-\frac{7}{4}$ ☐ C. $-\frac{4}{7}$ ☐ D. $\frac{4}{7}$

**7)** Trapezoid $ABCD$ and $WXYZ$ are similar. Find $WZ$.

☐ A. 37

☐ B. 33

☐ C. 24

☐ D. 23

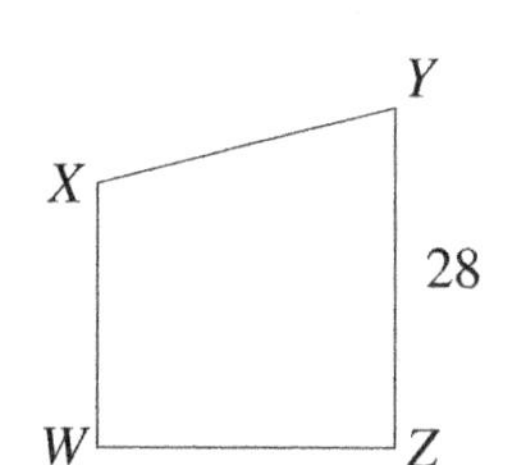


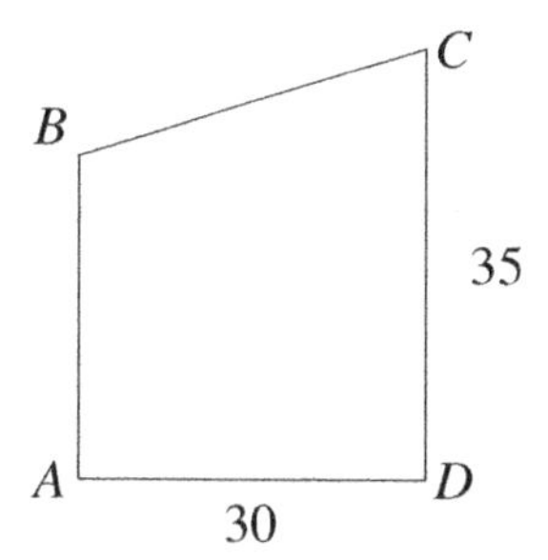


**8)** The volume of a cone is $200\,\text{cm}^3$. The height is $20\,\text{cm}$. Find the radius.

☐ A. $\sqrt{\frac{20}{\pi}}$ ☐ B. $\sqrt{\frac{10}{\pi}}$ ☐ C. $\sqrt{\frac{30}{\pi}}$ ☐ D. $\sqrt{\frac{40}{\pi}}$

**9)** Determine if the two triangles are congruent. If they are, state how you know.

☐ A. SSS

☐ B. ASS

☐ C. SAS

☐ D. ASA

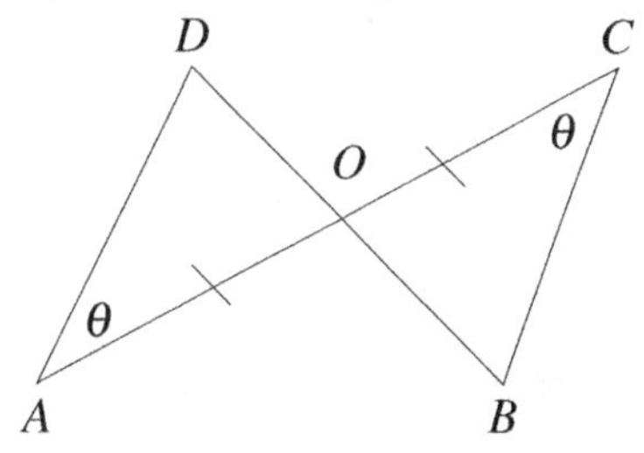


**10)** Find the volume of a sphere inscribed in a cube if each side of the cube is 6 m.

☐ A. $30\pi$

☐ B. $36\pi$

☐ C. $40\pi$

☐ D. $46\pi$

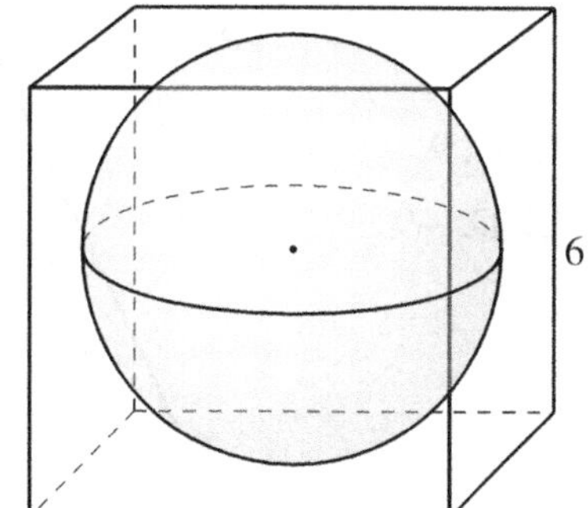


**11)** Find the length of $AB$ for the coordinates: $A(3,4)$, $B(9,16)$

☐ A. 14 ☐ B. $\sqrt{221}$ ☐ C. $\sqrt{180}$ ☐ D. $\sqrt{171}$

**12)** Find the slope of the line through each pair of points: $(5,7)$ and $(10,14)$.

☐ A. 1 ☐ B. $-1$ ☐ C. $-\frac{7}{5}$ ☐ D. $\frac{7}{5}$

**13)** Identify the type of triangle based on its angles and sides.

☐ A. Right Scalene
☐ B. Scalene isosceles
☐ C. Obtuse scalene
☐ D. Acute isosceles

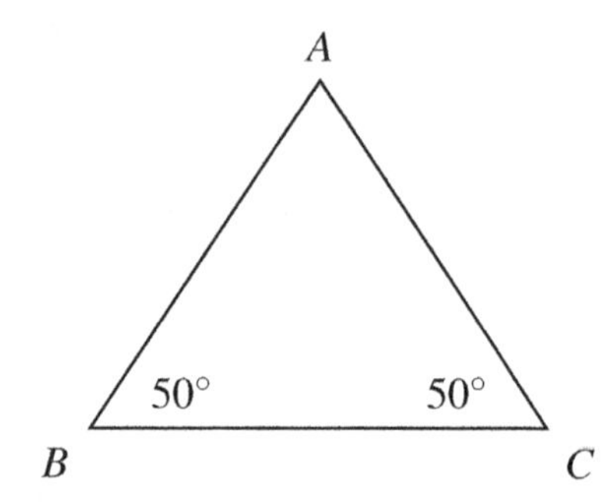


**14)** A cone has a radius of 6 in and a height of 14 in. A cylinder with a radius of 12 in has the same volume as the cone. What is the height of the cylinder?

☐ A. 7 ☐ B. $\frac{7}{6}$ ☐ C. 14 ☐ D. $\frac{6}{7}$

**15)** Order the sides of the triangle from shortest to longest.

☐ A. *GH*, *GI*, *HI*
☐ B. *HI*, *GI*, *GH*
☐ C. *GI*, *HI*, *GH*
☐ D. *HI*, *GH*, *GI*

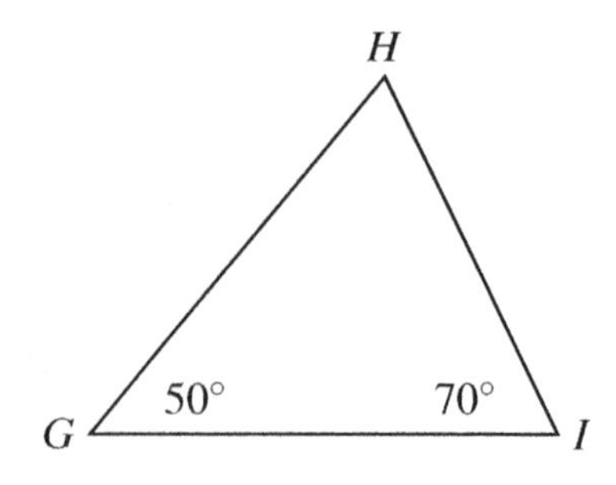


**16)** Find the value of $x$. If we extend $GH$ from point $G$, the measure of the exterior angle at $G$ is $200 + 10x$.

☐ A. 7
☐ B. $-7$
☐ C. 9
☐ D. $-9$

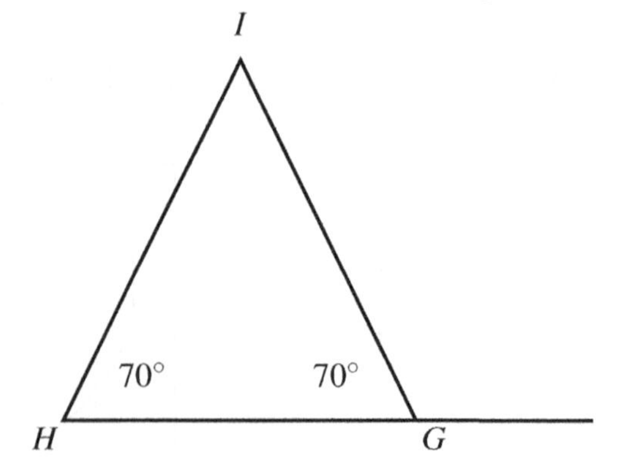


**17)** The water level of a swimming pool, 60 feet by 30 feet, is to be raised 4 inches. How many cubic feet of water must be added to achieve this?

☐ A. 600 ☐ B. 700 ☐ C. 820 ☐ D. 800

**18)** Which statement best describes the difference between a height and a bisector of a triangle?

☐ A. A height is a line segment drawn from a vertex perpendicular to the opposite side, while a bisector is a line segment that divides an angle into two equal parts.

☐ B. A height is a line segment that splits an angle of the triangle in half, while a bisector splits a side into two equal parts.

☐ C. A height divides an angle into two equal parts, while a bisector is always a segment connecting a vertex to the midpoint of the opposite side.

☐ D. A height divides the triangle into two triangles of equal area, while a bisector divides an angle into two equal angles.

**19)** Given the figure below, $SQ = 12$, $PS = 15$, and $RT = QT$. Find $QT$.

☐ A. $\sqrt{161}$ ☐ C. $\sqrt{162}$

☐ B. $\sqrt{164}$ ☐ D. $\sqrt{163}$

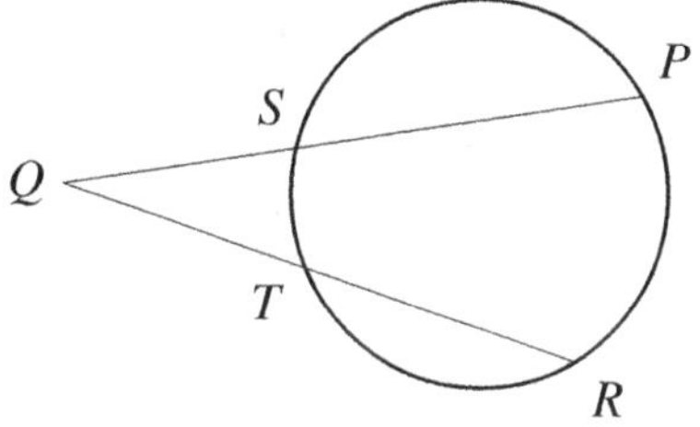


**20)** If the lengths of two sides of an isosceles triangle are 12 and 25, what is the perimeter of the triangle?

☐ A. 62 ☐ B. 74 ☐ C. 37 ☐ D. 49

**21)** Calculate the area between segment $QR$ and arc $QR$ in the figure below. $P$ is the center of the circle, $\angle QPR = 120°$, and $r = 8\,\text{cm}$.

☐ A. $\frac{32\pi-48\sqrt{3}}{3}\,\text{cm}^2$ ☐ C. $\frac{64\pi-48\sqrt{3}}{2}\,\text{cm}^2$

☐ B. $\frac{64\pi-48\sqrt{3}}{3}\,\text{cm}^2$ ☐ D. $\frac{64\pi-48\sqrt{2}}{3}\,\text{cm}^2$

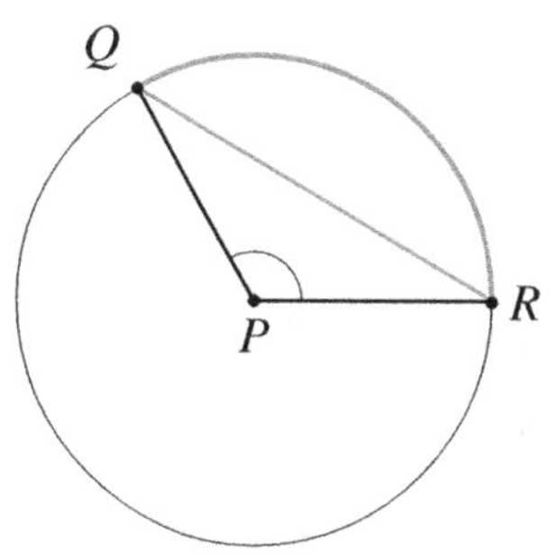


**22)** Determine the center and radius of the circle given by the equation:

$$(x-1)^2 + (y+2)^2 = 10$$

☐ A. Center: $(1,-2)$, Radius: 5 ☐ C. Center: $(-1,2)$, Radius: $\sqrt{10}$

☐ B. Center: $(1,2)$, Radius: $\sqrt{10}$ ☐ D. Center: $(1,-2)$, Radius: $\sqrt{10}$

**23)** A rectangular box has dimensions 5 feet wide, 6 feet high, and 8 feet in length. Calculate the surface area of the box.

☐ A. 60 ☐ B. 160 ☐ C. 236 ☐ D. 248

**24)** Find the length of the arc of a sector of 45° in a circle if the radius is 10. Find the area of the sector.

☐ A. $2.5\pi$, $12.5\pi$ ☐ C. $2.5\pi$, $5\pi$

☐ B. $5\pi$, $25\pi$ ☐ D. $5\pi$, $12.5\pi$

**25)** Find the area of an equilateral triangle if each side is 14.

☐ A. 49 ☐ B. $49\sqrt{3}$ ☐ C. 98 ☐ D. $98\sqrt{3}$

**26)** Each side of an equilateral triangle measures 14 units. Find the areas of the inscribed and circumscribed circles.

☐ A. $\frac{49\pi}{3}$, $\frac{196\pi}{3}$ ☐ C. $\frac{169\pi}{3}$, $\frac{94\pi}{3}$

☐ B. $\frac{193\pi}{3}$, $\frac{49\pi}{3}$ ☐ D. $\frac{193\pi}{3}$, $\frac{94\pi}{3}$

**27)** One acute angle of a right triangle has measure $m^\circ$. If $\cos(m^\circ) = \frac{3}{5}$, what is the value of $\cot(m^\circ)$?

☐ A. $\frac{3}{4}$ ☐ B. $\frac{4}{5}$ ☐ C. $\frac{5}{4}$ ☐ D. $\frac{4}{3}$

**28)** Solve for $PV$: if $UP = 5$, $WP = 4$, $PX = 9$.

☐ A. $\frac{63}{5}$ ☐ C. $\frac{54}{4}$

☐ B. $\frac{36}{5}$ ☐ D. $\frac{45}{4}$

X U P V W

**29)** What is the angle of elevation of the sun when a tree 25 meters tall casts a shadow 40 meters long? Round to the nearest degree.

☐ A. 30° ☐ B. 35° ☐ C. 38° ☐ D. 32°

**30)** In a right triangle $PQR$, side $PQ$ is 6 cm and angle $R = 35^\circ$. Find the length of side $QR$.

☐ A. 7.45 ☐ B. 8.57 ☐ C. 4.29 ☐ D. 10.12

**31)** Which conditional's inverse is true?

☐ A. If $\angle Y$ is less than 40°, then $\angle Y$ is acute.

☐ B. If two rays are adjacent rays, then they have a common endpoint.

☐ C. If point $M$ is the midpoint of $GH$, then points $G$, $M$, and $H$ are collinear.

☐ D. If two lines intersect to form an obtuse angle, then they are not perpendicular.

**32)** The sides of a rectangle are 24 and 32. What is the measure of the obtuse angle formed by the diagonals? Round to the nearest tenth.

☐ A. 25.2° ☐ B. 54.8° ☐ C. 125.2° ☐ D. 106.2

**33)** Which graph is correct for this equation $(x-1)^2+(y+1)^2=4$?

☐ A. ☐ B. ☐ C. ☐ D.

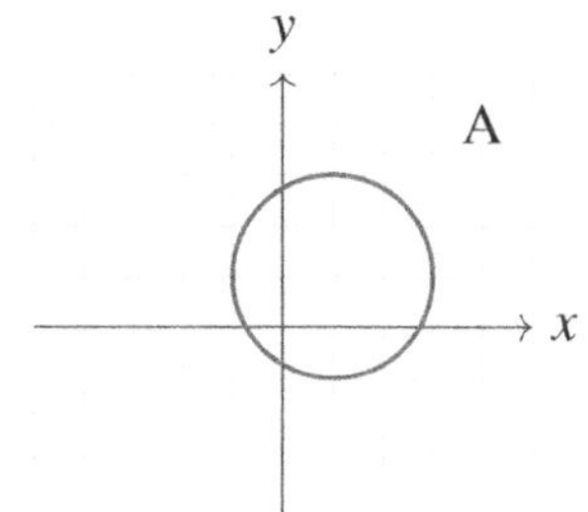


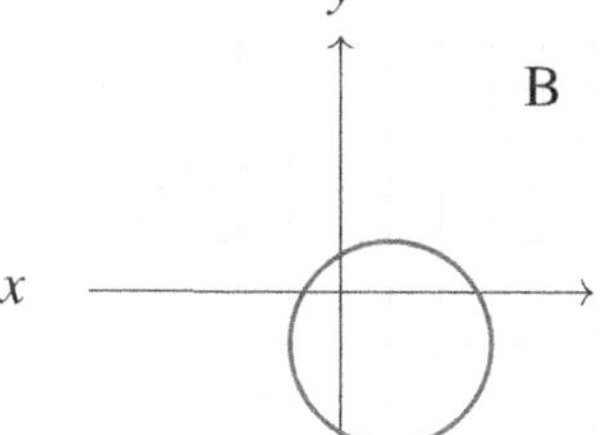


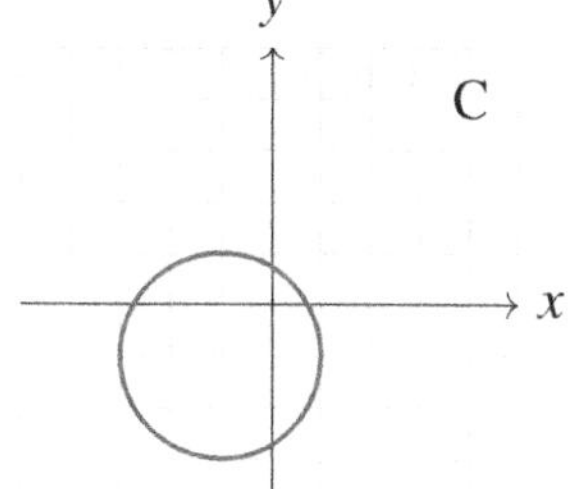


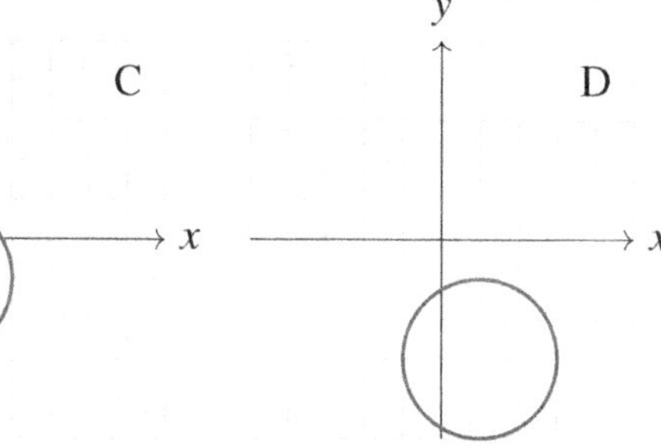


**34)** Given the figure $EFGH$ is a trapezoid with $EF \parallel GH$, and $EF = 14$, $FG = 14$, $\angle F = 60°$ and $\angle H = 30°$. Find the area and perimeter of $EFGH$.

☐ A. 169.68 and 84

☐ B. 254.52 and 80.25

☐ C. 254.52 and 84

☐ D. 486.45 and 80.25

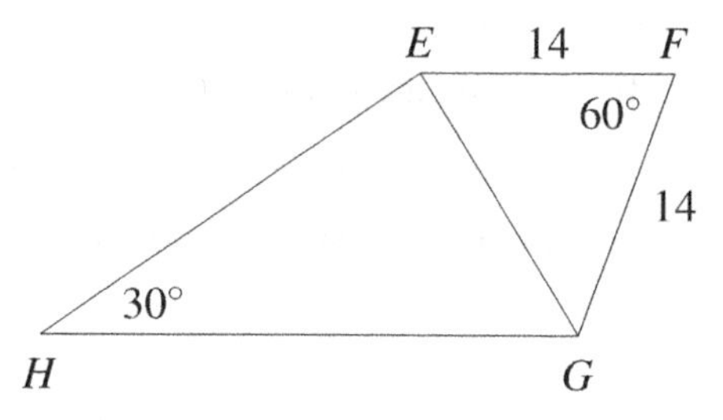


**35)** According to the given figure, find $OP$ and $MP$, when $MN = 20$.

☐ A. 8 and 6

☐ B. 6 and 8

☐ C. 6 and 12

☐ D. 30 and 20

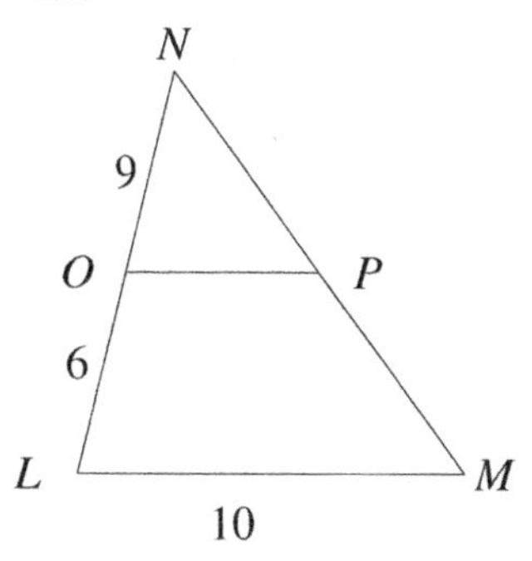


**36)** The apothem of a regular hexagon is $8\sqrt{3}$. Find the length of each side of the hexagon. Find the area of the hexagon.

☐ A. $288\sqrt{3}$ ☐ B. 384 ☐ C. $384\sqrt{3}$ ☐ D. 192

**37)** The lengths of the bases of a trapezoid are 12 and 28. What is the length of the median of the trapezoid?

☐ A. 15

☐ B. 20

☐ C. 18

☐ D. 19

**38)** Write a rule to describe the transformation.

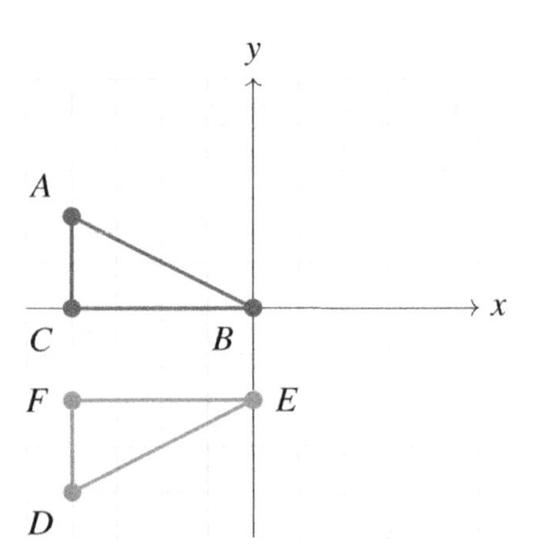


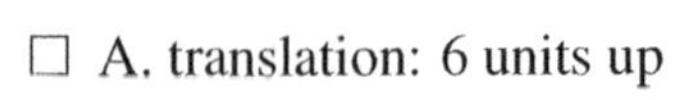

☐ A. translation: 6 units up

☐ B. reflection across $y = -x$

☐ C. translation: 2 units right and 2 units up

☐ D. reflection across $y = -1$

**39)** Is triangle $PQR$ congruent to triangle $STU$, given that the sides of $PQR$ measure 5 cm, 7 cm, and 10 cm, and the sides of $STU$ measure 10 cm, 14 cm, and 20 cm?

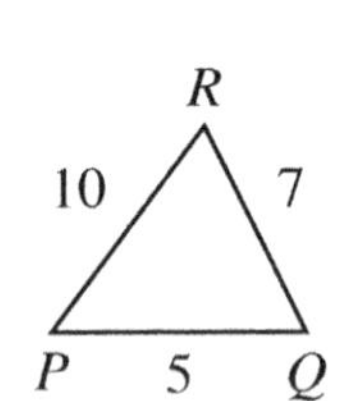


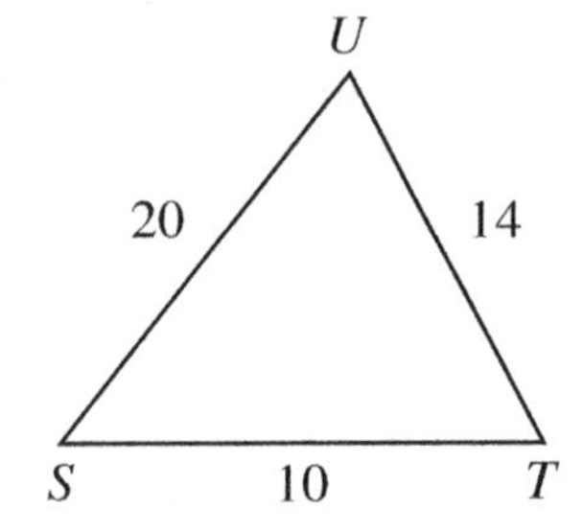


☐ A. Yes, they are congruent.

☐ B. No, they are not congruent but are similar.

☐ C. No, they are neither congruent nor similar.

☐ D. Yes, they are similar and congruent.

**40)** Among the following functions, which one is steeper than $h(x) = \frac{1}{5}x - 1$?

☐ A. $f(x) = \frac{2}{3}x + 2$ ☐ B. $g(x) = \frac{1}{6}x - 3$ ☐ C. $m(x) = \frac{2}{9}x + 1$ ☐ D. $k(x) = \frac{1}{8}x - 2$

## 11.2 Answer Keys

**1)** C. $130^\circ$

**2)** D. $110^\circ$

**3)** D. non-collinear

**4)** D. an acute angle

**5)** A. $y = -\frac{4}{3}x - \frac{1}{3}$

**6)** D. $\frac{4}{7}$

**7)** C. 24

**8)** C. $\sqrt{\frac{30}{\pi}}$

**9)** D. ASA

**10)** B. $36\pi$

**11)** C. $\sqrt{180}$

**12)** D. $\frac{7}{5}$

**13)** D. Acute isosceles

**14)** B. $\frac{7}{6}$

**15)** B. $HI$, $GI$, $GH$

**16)** D. $-9$

**17)** A. 600

**18)** A

**19)** C. $\sqrt{162}$

**20)** A. 62

**21)** B. $\frac{64\pi - 48\sqrt{3}}{3}$ cm$^2$

**22)** D. Center: $(1, -2)$, Radius: $\sqrt{10}$

**23)** C. 236

**24)** A. $2.5\pi$, $12.5\pi$

**25)** B. $49\sqrt{3}$

**26)** A. $\frac{49\pi}{3}$, $\frac{196\pi}{3}$

**27)** A. $\frac{3}{4}$

**28)** B. $\frac{36}{5}$

**29)** D. $32^\circ$

**30)** B. 8.57

**31)** B

**32)** D. 106.2

**33)** B

**34)** B. 254.52 and 80.25

**35)** B. 6 and 8

**36)** A. $288\sqrt{3}$

**37)** B. 20

**38)** D. reflection across $y = -1$

**39)** B. No, they are not congruent but are similar.

**40)** A. $f(x) = \frac{2}{3}x + 2$

## 11.3 Answers with Explanation

**1)** Since $CD$ is parallel to $GH$ and $XY$ is a transversal, angle $GQY$ and angle $CPQ$ are corresponding angles. Therefore, they are equal. Thus, $\angle CPQ = 130^\circ$.

**2)** Since $QS$ bisects $\angle PQR$, it follows that $\angle PQS = \angle RQS$. Given that $\angle RQS = 55^\circ$, we also have $\angle PQS = 55^\circ$. Therefore,

$$\angle PQR = \angle RQS + \angle PQS \Rightarrow \angle PQR = 55^\circ + 55^\circ \Rightarrow \angle PQR = 110^\circ.$$

**3)** Three points that lie on a curve are non-collinear, meaning they do not all lie on the same straight line. If they were collinear, they would be on a straight line.

**4)** The measure of $\angle XOY$ is given as $45^\circ$. An angle that measures $45^\circ$ is less than $90^\circ$, so it is an acute angle.

**5)** The slope of the given line is $\frac{3}{4}$. The slope of a line perpendicular to it will be the negative reciprocal, which is $m = -\frac{4}{3}$. Using the point-slope form of a line:

$$y - y_1 = m(x - x_1) \Rightarrow y + 3 = -\frac{4}{3}(x - 2) \Rightarrow y = -\frac{4}{3}x + \frac{8}{3} - 3 \Rightarrow y = -\frac{4}{3}x + \frac{8}{3} - \frac{9}{3} \Rightarrow y = -\frac{4}{3}x - \frac{1}{3}.$$

So, the equation of the line is $y = -\frac{4}{3}x - \frac{1}{3}$.

**6)** Parallel lines have the same slope. The slope of the given line is $\frac{4}{7}$. Therefore, the slope of a line parallel to it is also $\frac{4}{7}$.

**7)** Since the trapezoids are similar, the ratio of their corresponding sides is the same.

$$\frac{AD}{WZ} = \frac{CD}{YZ} \Rightarrow \frac{30}{WZ} = \frac{35}{28} \Rightarrow WZ = 30 \times \frac{28}{35} = 24.$$

So, $WZ = 24$ units.

**8)** The volume $V$ of a cone is given by: $V = \frac{1}{3}\pi r^2 h$, where $r$ is the radius, and $h$ is the height. Given that the volume is $200\,\text{cm}^3$, and the height is $20\,\text{cm}$, we can plug in the values: $200 = \frac{1}{3} \times \pi \times r^2 \times 20$. Simplifying:

$$200 = \frac{20}{3}\pi r^2 \Rightarrow r^2 = \frac{200 \times 3}{20\pi} \Rightarrow r^2 = \frac{30}{\pi} \Rightarrow r = \sqrt{\frac{30}{\pi}}.$$

So, the radius is $\sqrt{\frac{30}{\pi}}$.

**9)** Consider triangles $\triangle ADO$ and $\triangle OCB$: Both triangles share $\angle O$, which is a common angle. The angles $\angle A$ and $\angle C$ are congruent, denoted as $\theta$. The line segments $AO$ and $OC$ are equal in length. These conditions satisfy the criteria for the Angle-Side-Angle (ASA) postulate of congruence, which states that if two angles and the included side of one triangle are congruent to the corresponding two angles and included side of another triangle, then the two triangles are congruent. Therefore, $\triangle ADO$ is congruent to $\triangle OCB$, as proven by the ASA postulate.

**10)** For a sphere inscribed in a cube, the diameter of the sphere is equal to the side length of the cube. Given that the side of the cube is 6 m, the diameter of the sphere is also 6 m. Hence, the radius $r$ of the sphere is: $r = \frac{6}{2} = 3\,\text{m}$. The volume $V$ of a sphere is given by:

$$V = \frac{4}{3}\pi r^3 = \frac{4}{3}\pi(3)^3 = \frac{4}{3}\pi(27) = 36\pi.$$

**11)** Using the distance formula: $AB = \sqrt{(x_B - x_A)^2 + (y_B - y_A)^2}$. Plugging in the coordinates $A(3,4)$ and $B(9,16)$, we get:

$$AB = \sqrt{(9-3)^2 + (16-4)^2} = \sqrt{6^2 + 12^2} = \sqrt{36 + 144} = \sqrt{180}.$$

So, the length of $AB$ is $\sqrt{180}$.

**12)** The slope of a line through two points $(x_1, y_1)$ and $(x_2, y_2)$ is given by: $m = \frac{y_2 - y_1}{x_2 - x_1}$. For the points $(5,7)$ and $(10,14)$, the slope is: $m = \frac{14-7}{10-5} = \frac{7}{5}$.

**13)** The triangle has two equal angles, making it an isosceles triangle. Additionally, all the angles are less than 90°, so it is classified as an acute triangle. Therefore, triangle $ABC$ is an acute isosceles triangle.

**14)** Given: $r_{\text{cone}} = 6$ in, and $h_{\text{cone}} = 14$ in. The volume $V$ of a cone is given by:

$$V = \frac{1}{3}\pi r^2 h = \frac{1}{3}\pi \times 6^2 \times 14 = 168\pi\,\text{in}^3.$$

Given: $r_{\text{cylinder}} = 12$ in, and $V_{\text{cylinder}} = 168\pi\,\text{in}^3$ (since it is equal to the volume of the cone).
The volume $V$ of a cylinder is given by:

$$V = \pi r^2 h \Rightarrow 168\pi = \pi \times 12^2 \times h \Rightarrow h = \frac{168\pi}{144\pi} = \frac{7}{6}\,\text{in}.$$

Therefore, the height of the cylinder is $\frac{7}{6}$ inches.

**15)** In any triangle, the side opposite the largest angle is the longest, and the side opposite the smallest angle is the shortest. Given the angles:

Shortest side = Side opposite 50° = $HI$,
Middle side = Side opposite 60° = $GI$,
Longest side = Side opposite 70° = $GH$.

So, the order from shortest to longest is: $HI$, $GI$, $GH$.

**16)** In triangle $GHI$, the exterior angle at $G$ will be:

$$\text{Exterior } \angle G = 180° - 70° = 110°.$$

Setting this equal to the given expression:

$$200 + 10x = 110 \Rightarrow 10x = -90 \Rightarrow x = -9.$$

**17)** Given: Length $= 60$ feet, width $= 30$ feet, and the height increase $= 4$ inches $= \frac{4}{12}$ foot.

The volume of water needed is

$$V = 60 \times 30 \times \frac{4}{12} = 600 \text{ cubic feet.}$$

**18)** A height of a triangle is a segment drawn from a vertex that is perpendicular to the opposite side (or the line containing the opposite side). It represents the altitude of the triangle from that vertex. A bisector of a triangle is a segment drawn from a vertex that divides the angle at the vertex into two equal parts.

**19)** We have: $QP = PS + SQ = 15 + 12 = 27$. Since $RT = QT$, let us denote $QT$ as $x$. If two secant segments are drawn from a point outside a circle, then:

$$QP \times SQ = QR \times QT \Rightarrow 27 \times 12 = (2x) \times x \Rightarrow 324 = 2x^2 \Rightarrow x^2 = 162 \Rightarrow x = \sqrt{162}.$$

**20)** An isosceles triangle is defined by having two sides of equal length. Consider two different scenarios involving such triangles: In the first case, the lengths of the two equal sides are each 12 units, which sums up to 24 units. To satisfy the triangle inequality, this sum must be greater than the length of the third side, which is 25 units in this scenario; however, 24 is not greater than 25, indicating that this is not a valid triangle. In the second case, the two equal sides are each 25 units long, resulting in a sum of 50 units. With the third side being only 12 units, the sum of the lengths of the two equal sides is indeed greater than that of the third side, confirming this configuration as a valid triangle. Consequently, the perimeter of this valid triangle is $25 + 25 + 12 = 62$ units.

**21)** To find the area of the segment, we use the formula for the area of a segment of a circle:

$$A = \frac{1}{2} r^2 \left(\theta \left(\frac{\pi}{180^\circ}\right) - \sin\theta\right).$$

Here, $r = 8\,\text{cm}$ and $\theta = 120^\circ$. Plugging in the values:

$$A = \frac{1}{2}(8)^2 \left(120^\circ \left(\frac{\pi}{180^\circ}\right) - \sin 120^\circ\right).$$

Then, $\sin 120^\circ = \sin(180^\circ - 60^\circ) = \sin 60^\circ = \frac{\sqrt{3}}{2}$. Thus:

$$A = \frac{1}{2} \times 64 \times \left(\frac{2\pi}{3} - \frac{\sqrt{3}}{2}\right) = \frac{64\pi - 48\sqrt{3}}{3} \text{cm}^2.$$

**22)** The general form of a circle's equation is $(x-h)^2 + (y-k)^2 = r^2$, where $(h,k)$ is the center and $r$ is the radius.

Comparing $(x-1)^2+(y+2)^2=10$ with the general form: $h=1$, $k=-2$ and $r^2=10$ ($\Rightarrow r=\sqrt{10}$). Therefore, the center of the circle is $(1,-2)$ and the radius is $\sqrt{10}$.

**23)** The surface area ($SA$) of a rectangular box is given by the formula:

$$SA = 2lw + 2lh + 2wh,$$

where $l$ is the length, $w$ is the width, and $h$ is the height. Plugging in the given values:

$$SA = 2(8\times 5)+2(8\times 6)+2(5\times 6) = 80+96+60 = 236\,\text{ft}^2.$$

**24)** The formula for the arc length ($L$) of a sector is:

$$L = \frac{\theta}{360}\times 2\pi r,$$

and the formula for the area ($A$) of a sector is:

$$A = \frac{\theta}{360}\times \pi r^2.$$

Given $\theta = 45^\circ$ and $r = 10$: Calculate the arc length:

$$L = \frac{45}{360}\times 2\pi\times 10 = \frac{1}{8}\times 2\pi\times 10 = \frac{20\pi}{8} = 2.5\pi.$$

Calculate the area of the sector:

$$A = \frac{45}{360}\times \pi\times 10^2 = \frac{1}{8}\times \pi\times 100 = \frac{100\pi}{8} = 12.5\pi.$$

**25)** The formula for the area ($A$) of an equilateral triangle with side length $s$ is:

$$A = \frac{\sqrt{3}}{4}\times s^2.$$

Given $s = 14$, calculate the area:

$$A = \frac{\sqrt{3}}{4}\times 14^2 = \frac{\sqrt{3}}{4}\times 196 = 49\sqrt{3}.$$

**26)** For an equilateral triangle with each side measuring 14 units, we can find the areas of the inscribed and circumscribed circles as follows:

Inscribed circle: The radius $r$ of the inscribed circle (inradius) in an equilateral triangle is given by: $r = \frac{a}{2\sqrt{3}}$, where $a$ is the side length of the triangle. Plugging in the given side length:

$$r = \frac{14}{2\sqrt{3}} = \frac{7}{\sqrt{3}} = \frac{7\sqrt{3}}{3}\text{ units}.$$

Now, the area of the inscribed circle is:

$$A = \pi r^2 = \pi \left(\frac{7\sqrt{3}}{3}\right)^2 = \frac{49\pi}{3} \text{ square units.}$$

Circumscribed circle: The radius $R$ of the circumscribed circle (circumradius) for an equilateral triangle is given by: $R = \frac{a}{\sqrt{3}}$, using the given side length:

$$R = \frac{14}{\sqrt{3}} = \frac{14\sqrt{3}}{3} \text{ units.}$$

The area of the circumscribed circle is:

$$A = \pi R^2 = \pi \left(\frac{14\sqrt{3}}{3}\right)^2 = \frac{196\pi}{3} \text{ square units.}$$

**27)** Using the Pythagorean identity:

$$\sin(m^\circ) = \sqrt{1 - \cos^2(m^\circ)} = \sqrt{1 - \left(\frac{3}{5}\right)^2} = \sqrt{\frac{16}{25}} = \frac{4}{5}.$$

Note that $\cot(m^\circ) = \frac{\cos(m^\circ)}{\sin(m^\circ)}$, so

$$\cot(m^\circ) = \frac{\frac{3}{5}}{\frac{4}{5}} = \frac{3}{4}.$$

**28)** Using the property of intersecting chords:

$$UP \times PV = WP \times PX.$$

Substituting the given values:

$$5 \times PV = 4 \times 9 \Rightarrow 5 \times PV = 36 \Rightarrow PV = \frac{36}{5}.$$

**29)** Height of the tree: $h = 25$ meters

Length of the shadow: $s = 40$ meters

The angle of elevation, $\theta$, can be found using the tangent function:

$$\tan\theta = \frac{h}{s} \Rightarrow \tan\theta = \frac{25}{40} \Rightarrow \tan\theta = \frac{5}{8}.$$

Using the inverse tangent function:

$$\theta = \tan^{-1}\left(\frac{5}{8}\right).$$

Using a calculator:

$$\theta \approx 32.005^\circ,$$

rounding to the nearest degree, the angle of elevation $\theta$ is $32^\circ$.

**30)** Using the tangent function:

$$\tan(R) = \frac{PQ}{QR} \Rightarrow \tan(35^\circ) = \frac{6}{QR} \Rightarrow QR = \frac{6}{\tan(35^\circ)}.$$

Using a calculator to find $\tan(35^\circ) \approx 0.7002$:

$$QR = \frac{6}{0.7002} \approx 8.57.$$

Therefore, the length of side $QR$ is approximately 8.57 cm.

**31)** To determine the inverse of a conditional statement, both the hypothesis and the conclusion must be negated. Let us evaluate each option:

A) Inverse: If $\angle Y$ is not less than $40^\circ$, then $\angle Y$ is not acute.

This statement is not necessarily true. An angle less than $90^\circ$ is considered acute. If $\angle Y$ is not less than $40^\circ$, it could be either an acute angle (between $40^\circ$ and $90^\circ$) or not acute ($90^\circ$ and above).

B) Inverse: If two rays are not adjacent rays, then they do not have a common endpoint.

This is true. Two rays that are not adjacent do not share a common endpoint.

C) Inverse: If point $M$ is not the midpoint of $GH$, then points $G$, $M$, and $H$ are not collinear.

This is not necessarily true. Even if $M$ is not the midpoint, points $G$, $M$, and $H$ could still be collinear.

D) Inverse: If two lines do not intersect to form an obtuse angle, then they are perpendicular.

This statement is not universally true. Two lines that do not intersect to form an obtuse angle might form an acute angle instead.

Therefore, the true inverse statement is option B.

**32)** The diagonal, $d$, can be found using the Pythagorean theorem:

$$d = \sqrt{l^2 + w^2} = \sqrt{32^2 + 24^2} = \sqrt{1024 + 576} = \sqrt{1600} = 40.$$

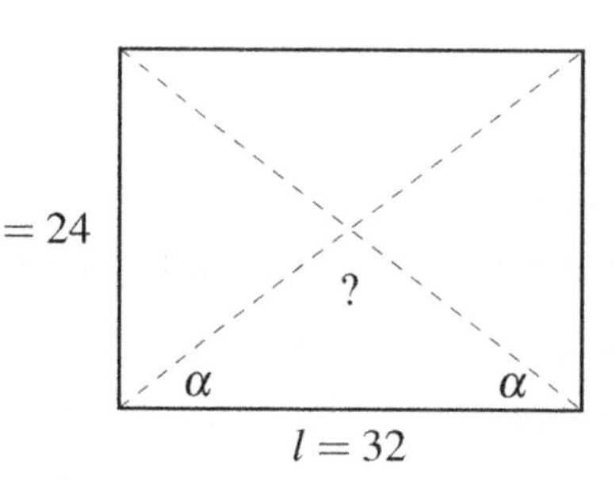


The cosine of the angle between the diagonal and the length is:

$$\cos\alpha = \frac{l}{d} = \frac{32}{40} = 0.8.$$

Using the inverse cosine function:

$$\alpha = \cos^{-1}(0.8) \Rightarrow \alpha \approx 36.9^\circ.$$

The obtuse angle formed by the diagonals is:

$$180^\circ - 36.9^\circ - 36.9^\circ = 106.2^\circ.$$

**33)** The given equation $(x-1)^2 + (y+1)^2 = 4$ represents a circle with center $(1,-1)$ and radius 2. Let us analyze each graph:

Graph A: Center $(1,1)$, incorrect center.

Graph B: Center $(1,-1)$, correct center.

Graph C: Center $(-1,-1)$, incorrect center.

Graph D: Center $(1,-3)$, incorrect center.

Therefore, the correct graph is B.

**34)** To solve this problem, we divide the trapezoid into two triangles: $\triangle EFG$ and $\triangle EGH$. First, let us solve for $\triangle EFG$: since $\triangle EFG$ has two equal sides, it is an isosceles triangle. To find $\angle E'$ and $\angle G'$, we know that in an isosceles triangle, the angles opposite the equal sides are equal. Therefore, $\angle E'$ and $\angle G'$ are congruent and both are equal to $60^\circ$.

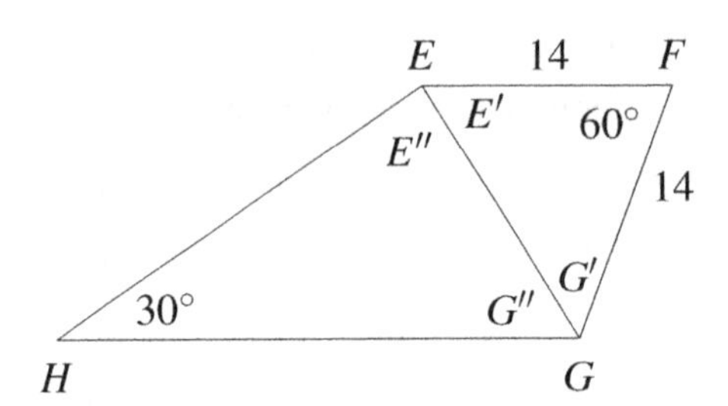


To find the side $EG$: since $\triangle EFG$ is equilateral, all sides are equal, so we have $EG = EF = FG = 14$ cm.
Now, let us solve for $\triangle EGH$: since $EF$ is parallel to $GH$, $\angle G = 180^\circ - 60^\circ = 120^\circ$ and $\angle E = 180^\circ - 30^\circ = 150^\circ$. Therefore:

$$\angle G = \angle G' + \angle G'' \Rightarrow 120^\circ = 60^\circ + \angle G'' \Rightarrow \angle G'' = 60^\circ,$$

$$\angle E = \angle E' + \angle E'' \Rightarrow 150^\circ = 60^\circ + \angle E'' \Rightarrow \angle E'' = 90^\circ.$$

Thus, $\triangle EGH$ is a 30-60-90 right triangle. For a 30-60-90 triangle, the side ratios are:
1. Side opposite $30^\circ$ is half the hypotenuse.
2. Side opposite $60^\circ$ is $\sqrt{3}$ times the side opposite the $30^\circ$ angle.
3. Side opposite $90^\circ$ (hypotenuse) is twice the side opposite the $30^\circ$ angle.

Given $EG$ (side opposite the $30^\circ$ angle) is 14 cm:
1. $HG$ (hypotenuse): $HG = 2 \times 14 \Rightarrow HG = 28$ cm.
2. $EH$ (side opposite the $60^\circ$ angle): $EH = 14 \times \sqrt{3} \Rightarrow EH \approx 24.25$ cm.

So, $EH$ is approximately 24.25 cm, and $HG = 28$ cm. Now, let us find the perimeter:

$$P = EF + FG + HG + EH = 14 + 14 + 28 + 24.25 \Rightarrow P = 80.25 \text{ cm}.$$

To find the area of the trapezoid, we need to calculate the height. The height of the trapezoid is equal to the height of $\triangle EFG$:

Drop a perpendicular from $G$ to $EF$, splitting $EF$ into two equal parts, each 7 cm. Call the point where the perpendicular meets $EF$ as $P$.

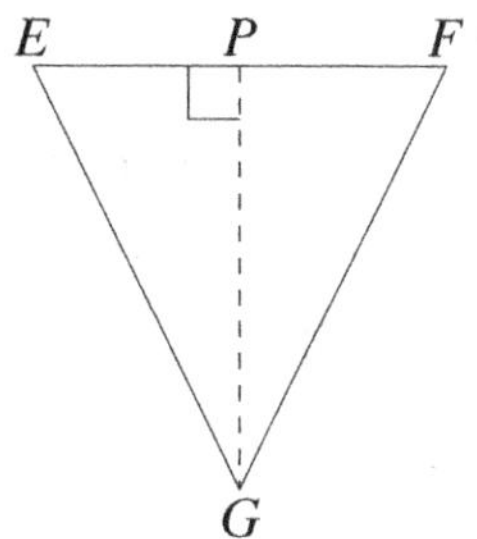


$\triangle GPE$ is a 30-60-90 right triangle, so:

$$GP = 7 \times \sqrt{3} \Rightarrow GP \approx 12.12 \text{ cm}.$$

The height (altitude) $GP$ of the triangle is approximately 12.12 cm. Now, we can calculate the area of the trapezoid:

$$A = \frac{1}{2}(EF + HG) \times h = \frac{1}{2}(14 + 28) \times 12.12 = \frac{42 \times 12.12}{2} = \frac{509.04}{2} \Rightarrow A = 254.52 \text{ cm}^2.$$

Therefore, the area is 254.52 cm$^2$ and the perimeter is 80.25 cm.

**35)** Given $OP \parallel LM$, we can use similar triangles to find the lengths of $OP$ and $MP$. Using the properties of similar triangles:

$$\frac{OP}{LM} = \frac{ON}{LN} \Rightarrow \frac{OP}{10} = \frac{9}{9+6} \Rightarrow OP = \frac{90}{15} \Rightarrow OP = 6.$$

Similarly:

$$\frac{MP}{MN} = \frac{LO}{LN} \Rightarrow \frac{MP}{20} = \frac{6}{9+6} \Rightarrow MP = \frac{120}{15} \Rightarrow MP = 8.$$

Therefore, the length of $OP$ is 6 units, and $MP$ is 8 units.

**36)** For a regular hexagon with apothem $a$:

$$\text{Side length: } s = a\sqrt{3}, \qquad \text{Area: } A = \frac{3}{2} \times s \times a.$$

Then, using $a = 8\sqrt{3}$:

$$s = 8\sqrt{3} \times \sqrt{3} = 24 \text{ units}, \qquad A = \frac{3}{2} \times 24 \times 8\sqrt{3} = 288\sqrt{3} \text{ square units}.$$

**37)** The median (or mid-segment) of a trapezoid is the segment that connects the midpoints of the non-parallel sides. Its length is the average of the lengths of the bases. Given $b_1 = 12$ and $b_2 = 28$:

$$\text{Median } (M) = \frac{b_1 + b_2}{2} = \frac{12 + 28}{2} = 20.$$

So, the length of the median of the trapezoid is 20 units.

**38)** The given transformation moves triangle $ABC$ to triangle $DEF$. By observing the coordinates of the vertices of the original and transformed triangles, we notice that the transformation involves reflecting across the line $y = -1$.

**39)** The triangles are not congruent because their corresponding sides are not equal in length. However, the sides of $STU$ are exactly twice the length of the corresponding sides of $PQR$. This means that the triangles have the same shape but different sizes. As the ratios of the corresponding sides are equal, the triangles are similar.

**40)** To determine which function is steeper, we compare the slopes of the given functions. The slope of $h(x)$ is $\frac{1}{5}$. The slopes of the other functions are as follows:

$f(x)$: $\frac{2}{3}$ $g(x)$: $\frac{1}{6}$ $m(x)$: $\frac{2}{9}$ $k(x)$: $\frac{1}{8}$

Since $\frac{2}{3}$ is greater than $\frac{1}{5}$, function $f(x)$ is steeper than $h(x)$.

# 12. Practice Test 2

## 12.1 Practices

**1)** The areas of two similar triangles are 198 and 324. If a side of the smaller triangle is 11, how long is the corresponding side of the larger triangle?

☐ A. 13.05 ☐ B. 15.06 ☐ C. 17.07 ☐ D. 14.07

**2)** If $D$ is the midpoint of $EF$, $ED = 6x + 10$, and $DF = 9x - 20$, then what is $EF$?

☐ A. 120 ☐ B. 60 ☐ C. 140 ☐ D. 15

**3)** Points $(2, 1)$ and $(8, w)$ lie on a line with slope $\frac{3}{2}$. What is the value of $w$?

☐ A. 9 ☐ B. 10 ☐ C. 5 ☐ D. 4

**4)** Which of the following best describes two lines in different planes that never intersect?

☐ A. collinear
☐ B. parallel
☐ C. perpendicular
☐ D. skew

**5)** The area of an equilateral triangle is $45\sqrt{3}$. Find the length of its sides and altitudes.

☐ A. $3\sqrt{10}, 2\sqrt{30}$
☐ B. $3\sqrt{5}, 6\sqrt{15}$
☐ C. $6\sqrt{5}, 3\sqrt{15}$
☐ D. $3\sqrt{30}, 2\sqrt{10}$

**6)** What is the pre-image, $P$, of $P'(16, -10)$ using the transformation $(x, y) \rightarrow (2x, 3y)$?

☐ A. $(-8, \frac{10}{3})$ ☐ B. $(8, -\frac{10}{3})$ ☐ C. $(\frac{16}{3}, -\frac{10}{3})$ ☐ D. $(4, -10)$

**7)** Solve for the variables to make the lines parallel.

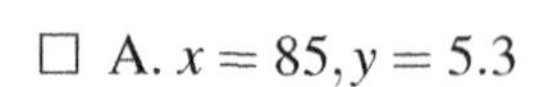

☐ A. $x = 85, y = 5.3$

☐ B. $x = 18, y = 9.5$

☐ C. $x = 5, y = 18$

☐ D. $x = 18, y = 5$

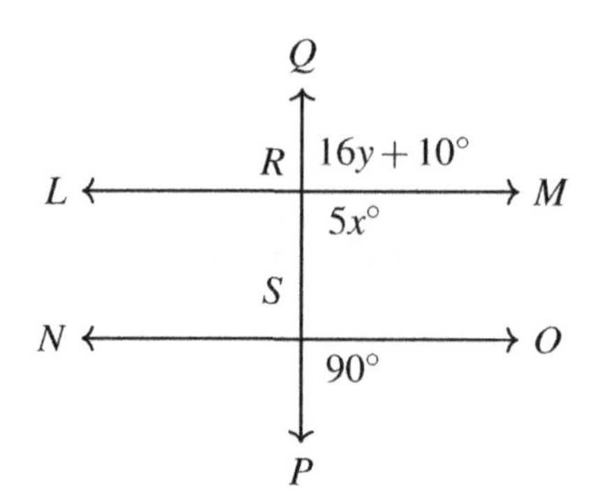


**8)** Find the slope of a line perpendicular to the given line: $y = \frac{1}{3}x + 2$

☐ A. $\frac{1}{3}$ ☐ B. $-3$ ☐ C. 3 ☐ D. $-\frac{1}{3}$

**9)** The angle of depression from the top of a tower to a point $P$ is 30°. The distance from $P$ to the base, $Q$, of the tower is 80 m. How tall is the tower to the nearest meter?

☐ A. 56 m ☐ C. 46 m

☐ B. 80 m ☐ D. 138 m

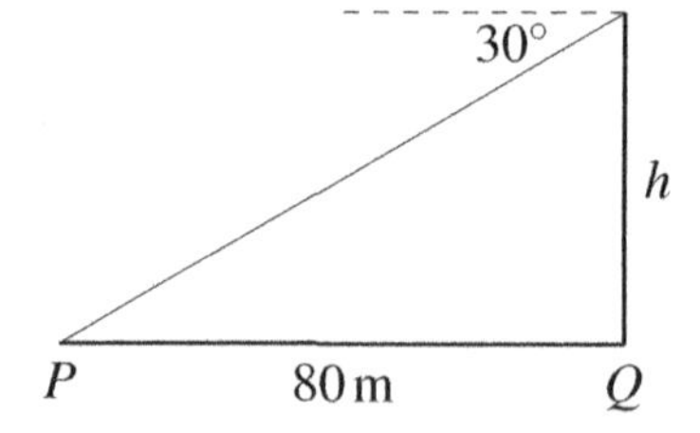


**10)** Determine if the two triangles are congruent. If they are, state how you know.

☐ A. SSS ☐ C. HL

☐ B. ASA ☐ D. AAS

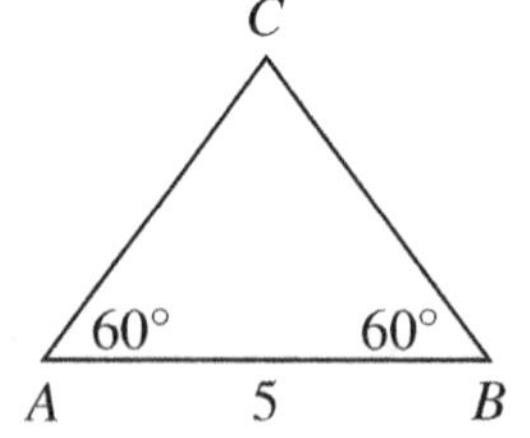


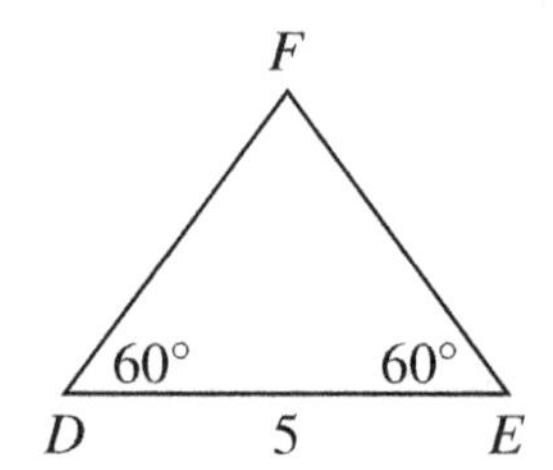


**11)** Write the circle equation given the center $(-3, 5)$ and radius 7.

☐ A. $(x+3)^2 + (y+5)^2 = 49$ ☐ C. $(x-3)^2 + (y+5)^2 = 7$

☐ B. $(x+3)^2 + (y-5)^2 = 49$ ☐ D. $(x-3)^2 + (y-5)^2 = 7^2$

**12)** How do you determine one interior angle of a regular hexagon? And one exterior angle?

☐ A. 120°, 60° ☐ C. 60°, 120°

☐ B. 90°, 60° ☐ D. 120°, 120°

**13)** In triangle $ABC$, it is known that $AB > AC$ and $AB < BC$. Order the angles from least to greatest measure.

☐ A. $\angle B < \angle C < \angle A$

☐ B. $\angle C < \angle B < \angle A$

☐ C. $\angle A < \angle C < \angle B$

☐ D. $\angle B < \angle A < \angle C$

**14)** If $GHI$ is a right triangle: $\angle I = 90°$, $\angle H = 45°$, $IH = 10\sqrt{2}$cm. Find hypotenuse $GH$ and side $GI$.

☐ A. 15.05, 14.14

☐ B. 15.05, 5.15

☐ C. $10\sqrt{2}$, 12.25

☐ D. 20, $10\sqrt{2}$

H

G I

**15)** In a circle whose radius is 8, the area of a sector is $18\pi$. Find the measure of the central angle of the sector and the length of the arc of the sector.

☐ A. 101.25°, 14.13

☐ B. 48°, 27.8

☐ C. 48°, 28

☐ D. 35.3°, $2\sqrt{193}$

**16)** Given the statement "All Professors are at least 40 years old," and the information that "John is 32 years old," which of the following conclusions is true?

☐ A. John is a Professor.

☐ B. John is not a Professor.

☐ C. John might be a Professor.

☐ D. John's age is unrelated to being a Professor.

**17)** Which of the following statements is logically equivalent to "If it snows, the roads become slippery"?

☐ A. If the roads do not become slippery, then it does not snow.

☐ B. If the roads become slippery, then it snows.

☐ C. If it does not snow, the roads do not become slippery.

☐ D. If it snows, the roads do not become slippery.

**18)** Given the following argument pattern:

$$Q \to P, \quad R \to Q.$$

Which conclusion logically follows from the given statements?

☐ A. $P \to R$

☐ B. $P \to Q$

☐ C. $R \to P$

☐ D. $Q \to R$

**19)** Two sides of a triangle are 10 and 24. Determine the range of possible values for the third side, $z$.

☐ A. $10 < z < 24$

☐ B. $10 \leq z \leq 24$

☐ C. $14 \leq z \leq 34$

☐ D. $14 < z < 34$

**20)** Given the truth values $p =$ true, $q =$ false, and $r =$ true, determine the truth value of the statement: $(\neg q \wedge p) \vee \neg r$.

☐ A. False

☐ B. True

☐ C. Neither true nor false

☐ D. Cannot be determined

**21)** A square pyramid with a base edge of 5 is inscribed in a cone with a height of 7. Determine the volume of both the pyramid and the cone. Round your answer to the nearest tenth.

☐ A. $\frac{175}{3}$, $29.17\pi$

☐ B. $\frac{175}{2}$, $28.17\pi$

☐ C. $\frac{175}{2}$, $30.15\pi$

☐ D. $\frac{175}{6}$, $21.15\pi$

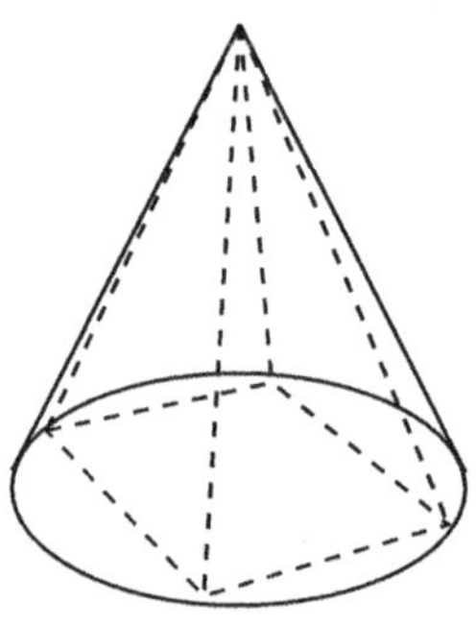

**22)** The volume of a cone is $450\,\text{cm}^3$ and the radius of the base is $6\,\text{cm}$. Find its altitude. Round your answer to the nearest centimeter.

☐ A. 10 cm

☐ B. 12 cm

☐ C. 14 cm

☐ D. 16 cm

**23)** A hemisphere is placed on top of a cylinder with the same radius of $7\,\text{cm}$, and the height of the cylinder is $5\,\text{cm}$. Find the surface area and volume. Round your answers to the nearest whole number.

☐ A. $217\pi$, 474

☐ B. $315\pi$, $702\pi$

☐ C. $266\pi$, 474

☐ D. $217\pi$, 702

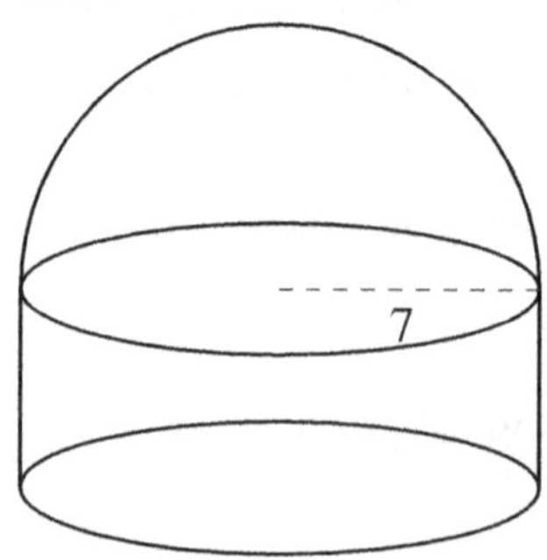


**24)** Given the figure, $IJK$ is a right triangle, and $KL \perp IJ$, $JK \perp KI$, with sides as marked. Find the length of $KL$, $JL$, and the area of triangle $IJK$.

☐ A. 26.67, 35, 96
☐ B. 9.6, 12.8, 96
☐ C. 26.67, 7.2, 96
☐ D. 96, 12.8, 9.6

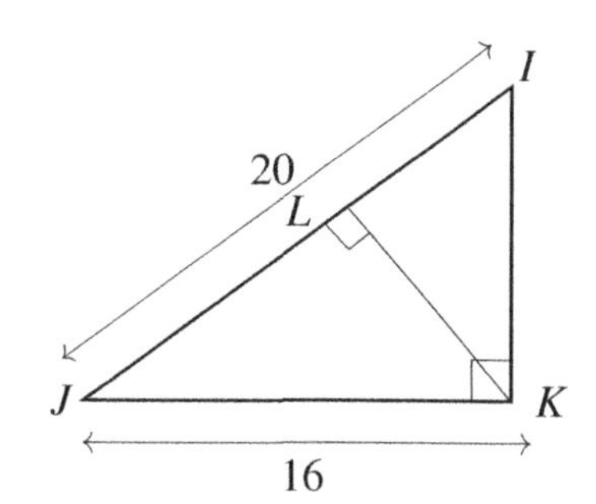


**25)** In $\triangle XYZ$, the bisector of $\angle X$ (line $XW$) creates $\angle 1$ and $\angle 2$. The sides are marked as follows. Determine the lengths of $YW$ and $ZW$.

☐ A. 12.75, 8.25
☐ B. 10.5, 10.5
☐ C. 8.25, 12.75
☐ D. 7.4, 13.6

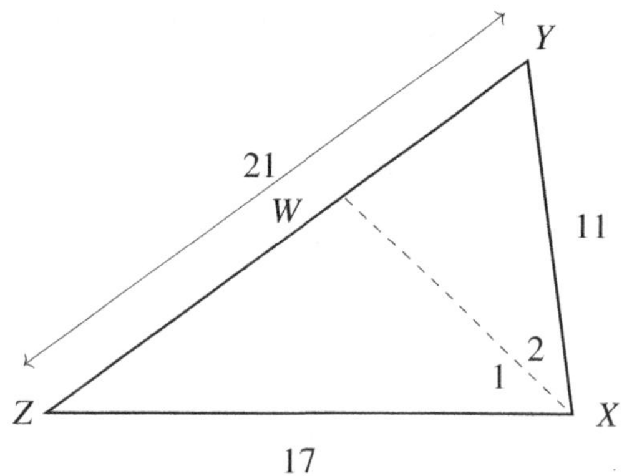


**26)** $CD$ from $\angle ECD$ is tangent to the circle. $CE$ intersects the circle creating $EF = 10$ and $CF = 7$. Determine the length of $CD$:

☐ A. $\sqrt{14}$
☐ B. $\sqrt{70}$
☐ C. 35
☐ D. 70

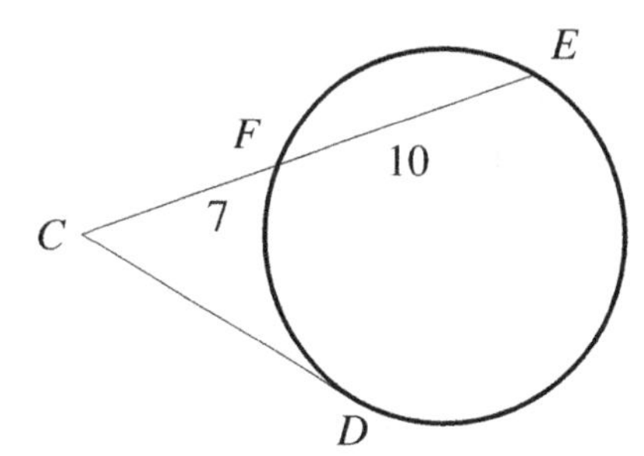


**27)** Classify the special quadrilateral. Then find the values of $x$ and $y$.

☐ A. 110°, 3
☐ B. 60°, 3
☐ C. 70°, 6
☐ D. 110°, 9

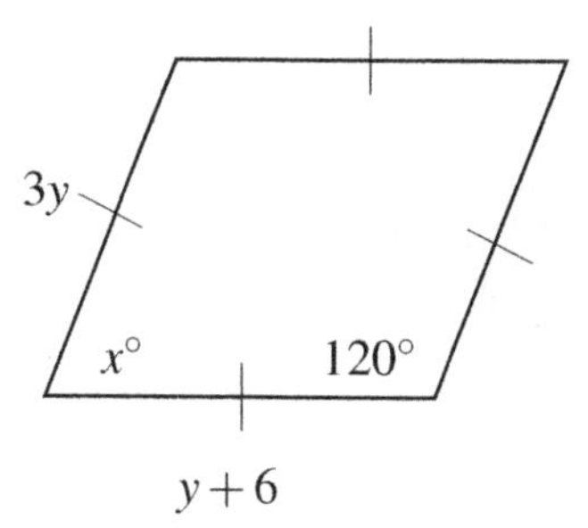


**28)** $U$, $V$, and $W$ are all points of tangency. What is the perimeter of $\triangle ABC$?

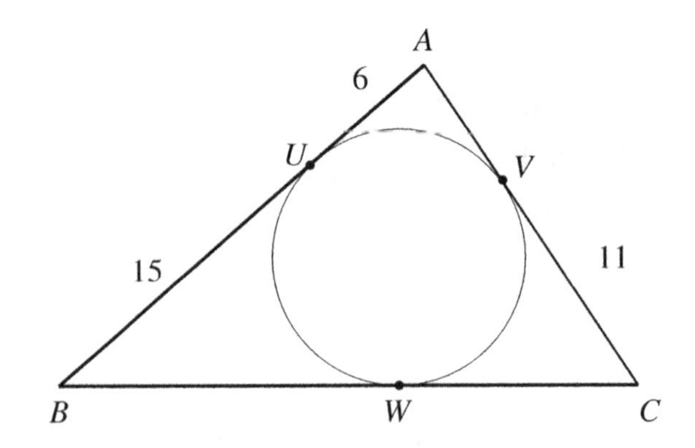


☐ A. 56
☐ B. 32
☐ C. 64
☐ D. 45

**29)** *LMNO* is a parallelogram. Given $LN = 7$, $LP = 5$, and $PM = 3$, determine the lengths of $LQ$ and $QN$.

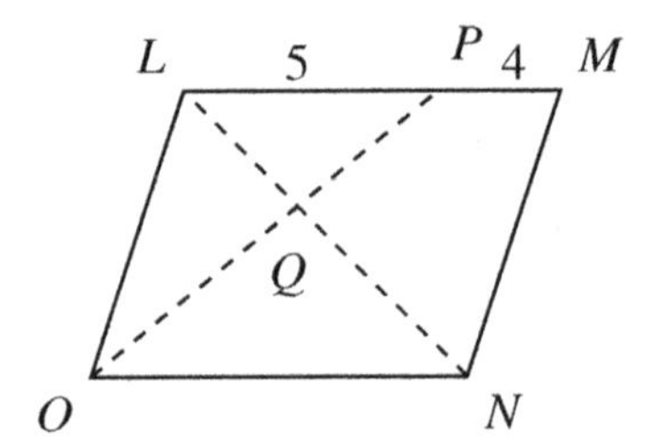


☐ A. 3, 4
☐ B. 1.5, 5.5
☐ C. 2.5, 4.5
☐ D. 5, 2

**30)** $EFGH$ is a kite, with $EF = EH = 18$, $PG = 24$, and $PH = 14$. Find the lengths of $EG$, $FH$, and $FG$.

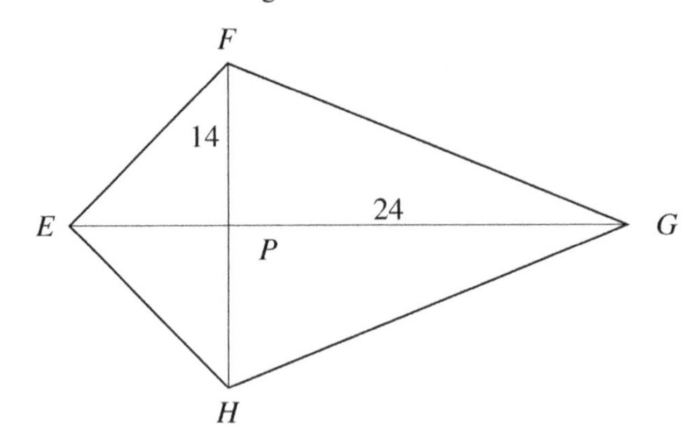


☐ A. 35.3, 28, $2\sqrt{193}$
☐ B. 48, 14, 27.8
☐ C. 48, 28, 27.8
☐ D. 46.8, 28, $2\sqrt{95}$

**31)** In $\triangle FGH$, given $FI = 6$, $IG = 4$, and $GH = 24$, determine the lengths of $FH$, $HJ$, and $IJ$.

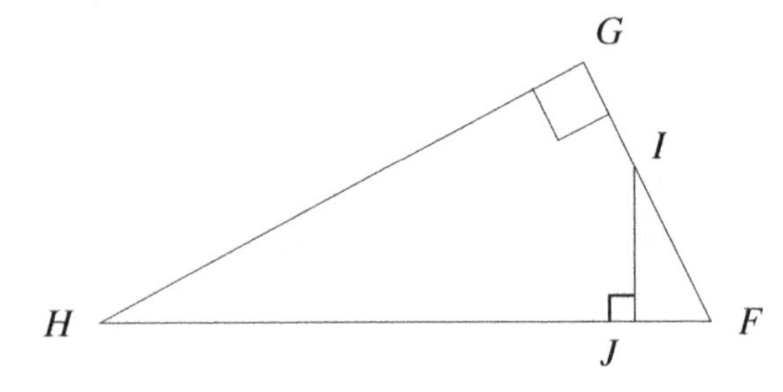


☐ A. 26, $2\sqrt{119}$, $\frac{72}{13}$
☐ B. $2\sqrt{119}$, $\frac{30}{13}$, $\frac{72}{13}$
☐ C. 30, $\frac{30}{13}$, $\frac{308}{13}$
☐ D. 26, $\frac{308}{13}$, $\frac{72}{13}$

**32)** Find the measure of the given arc or variable.

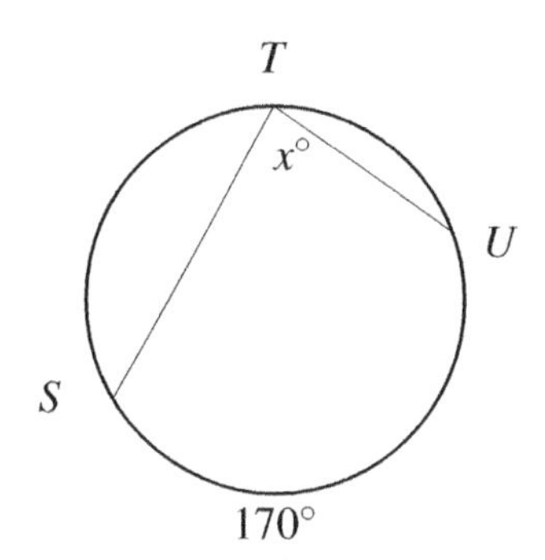


☐ A. 170° ☐ C. 85°

☐ B. 100° ☐ D. 60°

**33)** What are the coordinates of $\triangle ABC$ after a reflection in the line $y = -x$?

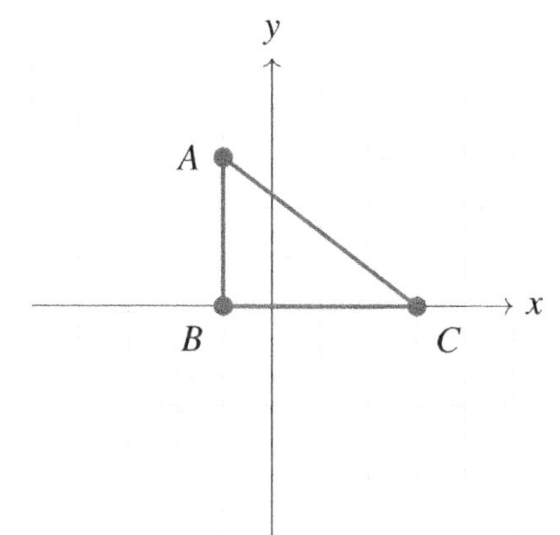


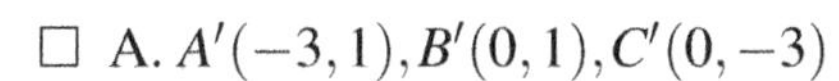

☐ A. $A'(-3,1), B'(0,1), C'(0,-3)$

☐ B. $A'(-3,-1), B'(0,0), C'(-3,0)$

☐ C. $A'(-3,1), B'(0,1), C'(0,3)$

☐ D. $A'(0,-3), B'(0,-1), C'(0,-3)$

**34)** What are the coordinates of $AB$ after a $45^\circ$ rotation (counterclockwise) about the origin?

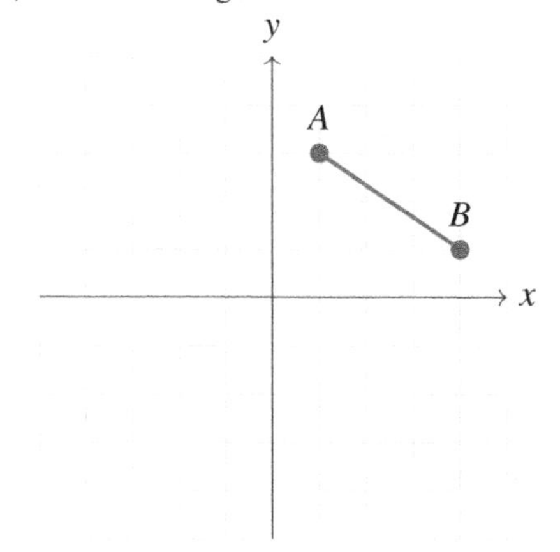


☐ A. $A'(-1,3), B'(-4,1)$

☐ B. $A'(2,1), B'(1,4)$

☐ C. $A'(-\sqrt{2}, 2\sqrt{2}),\ B'(\frac{3\sqrt{2}}{2}, \frac{5\sqrt{2}}{2})$

☐ D. $A'(2\sqrt{2}, -\sqrt{2}), B'(5\sqrt{2}, \frac{5\sqrt{2}}{2})$

**35)** According to the figure below, $E \perp AB$, and $D$ and $F$ are midpoints. Find $AC$ and $GC$.

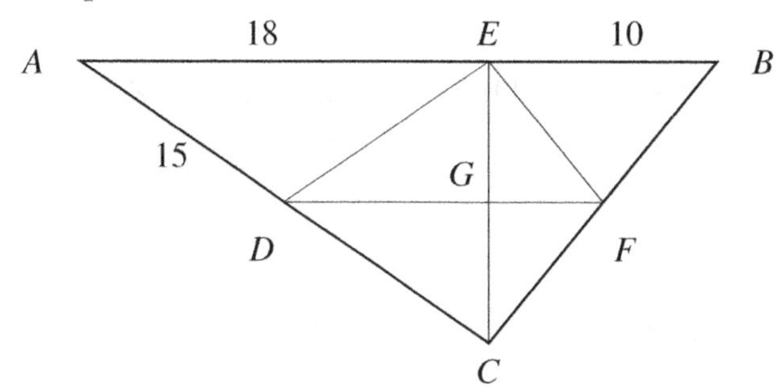


☐ A. 30, 12 ☐ C. 15, 13

☐ B. 30, 24 ☐ D. 15, 24

**36)** The radius of a circle is 22. The length of chord $AB$ is 26. How far is $AB$ from the center of the circle?

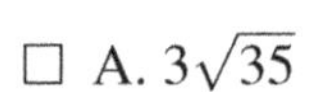

☐ A. $3\sqrt{35}$ ☐ C. $3\sqrt{7}$

☐ B. 22 ☐ D. $15\sqrt{7}$

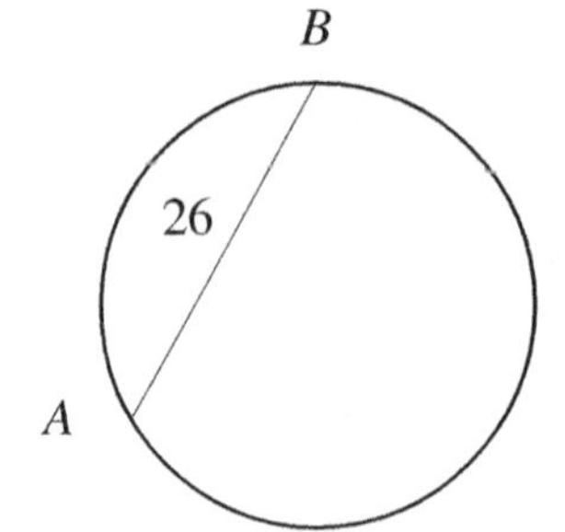


**37)** Order the sides of each triangle from shortest to longest.

☐ A. $HI, GH, GI$ ☐ C. $GH, HI, GI$

☐ B. $HI, GI, GH$ ☐ D. $GI, GH, HI$

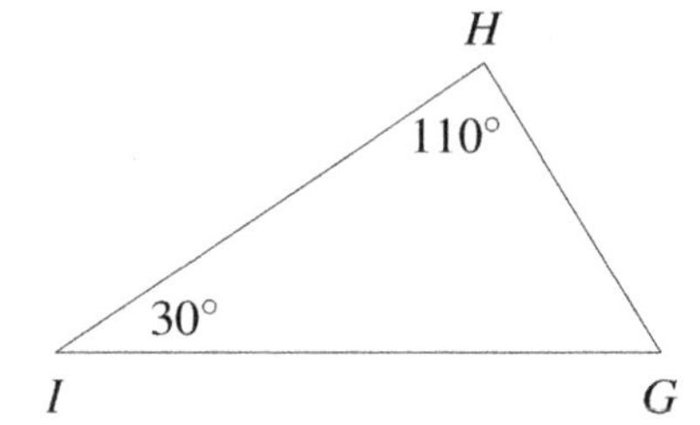


**38)** $IJKL$ is a parallelogram in the figure below and $JK = 7$. Given the information in the figure, find $KM$ and the ratio of the areas of $\triangle IJN$ and $\triangle MNK$.

☐ A. 2, 5 ☐ C. 3.6, 2.5

☐ B. 3.6, 6.25 ☐ D. 2, 6.25

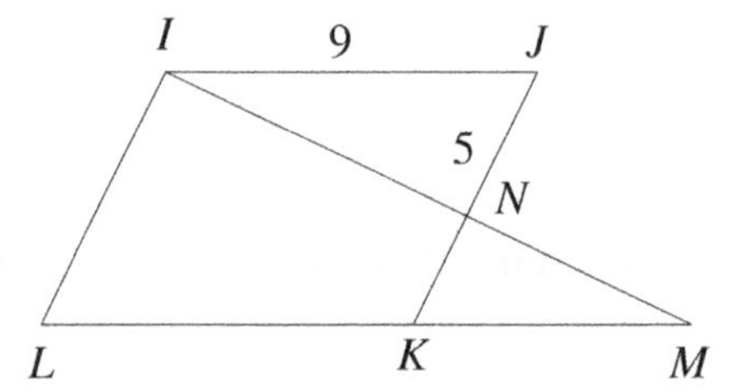


**39)** $GH, IJ$, and $KL$ intersect at $Q$. If $\angle G = 95°$, $\angle H = 55°$, $\angle K = 65°$, $\angle J = 50°$, and $\angle I = 55°$, what is the value of $\angle L$?

☐ A. 30° ☐ C. 65°

☐ B. 40° ☐ D. 15°

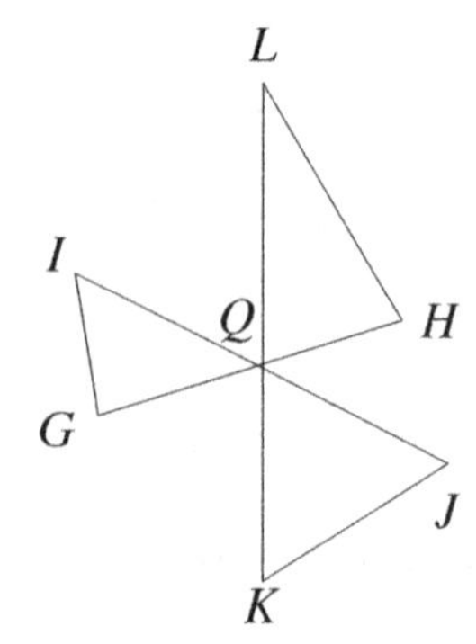


**40)** Triangles $ADC$ and $WZY$ are similar. Find $WZ$.

☐ A. 37 ☐ C. 24

☐ B. 33 ☐ D. 23

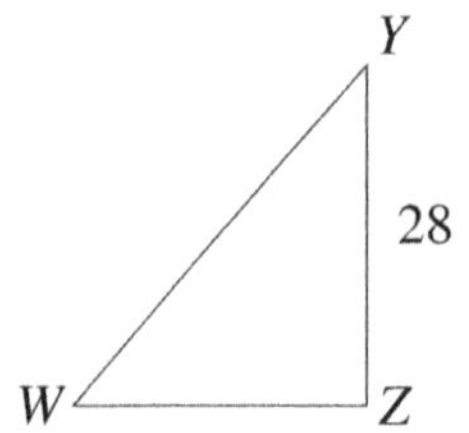


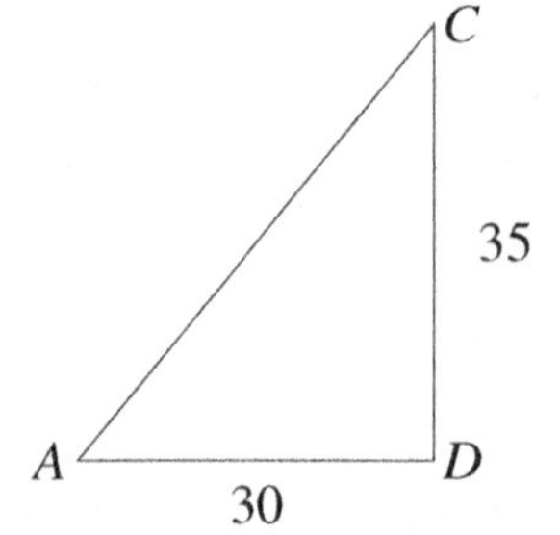

## 12.2 Answer Keys

**1)** D. 14.07

**2)** C. 140

**3)** B. 10

**4)** D. skew

**5)** C. $6\sqrt{5}, 3\sqrt{15}$

**6)** B. $(8, -\frac{10}{3})$

**7)** D. $x = 18, y = 5$

**8)** B. $-3$

**9)** C. 46 m

**10)** B. ASA

**11)** B. $(x+3)^2 + (y-5)^2 = 49$

**12)** A. $120°, 60°$

**13)** A. $\angle B < \angle C < \angle A$

**14)** D. 20, $10\sqrt{2}$

**15)** A. $101.25°$, 14.13

**16)** B. John is not a Professor.

**17)** A. If the roads do not become slippery, then it does not snow.

**18)** C. $R \rightarrow P$

**19)** D. $14 < z < 34$

**20)** B. True

**21)** A. $\frac{175}{3}, 29.17\pi$

**22)** B. 12 cm

**23)** A. $217\pi, 474$

**24)** B. 9.6, 12.8, 96

**25)** C. 8.25, 12.75

**26)** B. $\sqrt{70}$

**27)** B. $60°$, 3

**28)** C. 64

**29)** C. 2.5, 4.5

**30)** A. 35.3, 28, $2\sqrt{193}$

**31)** D. 26, $\frac{308}{13}$, $\frac{72}{13}$

**32)** C. $85°$

**33)** A. $A'(-3,1), B'(0,1), C'(0,-3)$

**34)** C. $A'(-\sqrt{2}, 2\sqrt{2})$, $B'(\frac{3\sqrt{2}}{2}, \frac{5\sqrt{2}}{2})$

**35)** A. 30, 12

**36)** A. $3\sqrt{35}$

**37)** C. $GH, HI, GI$

**38)** B. 3.6, 6.25

**39)** B. $40°$

**40)** C. 24

## 12.3 Answers with Explanation

**1)** The areas of similar triangles are proportional to the square of the ratio of their corresponding sides. If the areas of the two similar triangles are $A_1 = 198$ and $A_2 = 324$, and a side of the smaller triangle $s_1 = 11$, then the corresponding side of the larger triangle $s_2$ can be found using the following formula:

$$\frac{A_1}{A_2} = \left(\frac{s_1}{s_2}\right)^2.$$

Substituting the given values, we have:

$$\frac{198}{324} = \left(\frac{11}{s_2}\right)^2 \implies \frac{198}{324} = \frac{121}{s_2^2}.$$

Solving for $s_2$:

$$s_2^2 = \frac{324 \times 121}{198} = \frac{39204}{198} = 198.$$

Finally, take the square root to find $s_2$:

$$s_2 = \sqrt{198} \approx 14.07.$$

**2)** Since $D$ is the midpoint of $EF$, $ED = DF$. Thus:

$$6x + 10 = 9x - 20.$$

Solving for $x$, we get:

$$6x + 10 = 9x - 20 \implies 3x = 30 \implies x = 10.$$

Now, $EF = ED + DF$:

$$EF = (6(10) + 10) + (9(10) - 20) \implies EF = (60 + 10) + (90 - 20) \implies EF = 70 + 70 \implies EF = 140.$$

Therefore, the length of $EF$ is 140.

**3)** Using the formula for slope: $m = \frac{y_2 - y_1}{x_2 - x_1}$. Given: $m = \frac{3}{2}$, $x_1 = 2$, $y_1 = 1$, $x_2 = 8$. Now, substitute in the slope formula:

$$\frac{3}{2} = \frac{w - 1}{8 - 2} = \frac{w - 1}{6}.$$

Solving for $w$:

$$3 \times 6 = 2(w - 1) \implies 18 = 2w - 2 \implies 18 + 2 = 2w \implies 20 = 2w \implies w = 10.$$

Therefore, the value of $w$ is 10.

**4)** Lines that are in different planes and never intersect are called skew lines. They are neither parallel nor perpendicular because those concepts apply to lines within the same plane.

**5)** For an equilateral triangle with area $A$, the side length $s$ can be found using the formula $A = \frac{\sqrt{3}}{4}s^2$. Given $A = 45\sqrt{3}$:

$$45\sqrt{3} = \frac{\sqrt{3}}{4}s^2 \implies 180\sqrt{3} = \sqrt{3}s^2.$$

Divide both sides by $\sqrt{3}$:

$$s^2 = 180 \implies s = \sqrt{180} = 6\sqrt{5}.$$

The altitude $h$ of an equilateral triangle is related to the side length $s$ through the formula $h = \frac{\sqrt{3}}{2}s$:

$$h = \frac{\sqrt{3}}{2} \times 6\sqrt{5} = 3\sqrt{15}.$$

Therefore, the side length of the equilateral triangle is $6\sqrt{5}$ units and its altitude is $3\sqrt{15}$ units.

**6)** Given the transformation $(x,y) \to (2x,3y)$, we need to find the pre-image $P$ of $P'(16,-10)$. From the transformation, we know:

$$x' = 2x \quad \text{and} \quad y' = 3y.$$

Given $x' = 16$ and $y' = -10$:

$$2x = 16 \implies x = \frac{16}{2} = 8,$$

$$3y = -10 \implies y = -\frac{10}{3}.$$

Therefore, the pre-image $P$ is $(8, -\frac{10}{3})$.

**7)** Given that lines $LM$ and $NO$ are parallel and line $PQ$ intersects them, angles $QRM$ and $SRM$ will be supplementary because they are consecutive interior angles. The sum of supplementary angles is $180°$. Since angles $QRM$ and $SRM$ are supplementary, we have:

$$QRM + SRM = 180° \implies 16y + 10 + 5x = 180 \implies 16y + 5x = 170. \quad (1)$$

Since $LM$ and $NO$ are parallel and $PQ$ is a transversal, angles $SRM$ and $QSO$ are alternate interior angles. Given that $QSO = 90°$, for the lines to remain parallel, $SRM$ must also be $90°$:

$$5x = 90 \implies x = 18. \quad (2)$$

Using the value of $x$ from (2) in (1):

$$16y + 5(18) = 170 \implies 16y + 90 = 170 \implies 16y = 80 \implies y = 5.$$

Therefore, for the lines $LM$ and $NO$ to be parallel, $x = 18$ and $y = 5$.

**8)** To find the slope of a line perpendicular to a given line, you need to determine the negative reciprocal of the slope of the given line. Given the equation $y = \frac{1}{3}x + 2$, the slope $(m)$ is $\frac{1}{3}$. The negative reciprocal of $\frac{1}{3}$ is $-3$. Therefore, the slope of a line perpendicular to $y = \frac{1}{3}x + 2$ is $-3$.

**9)** The angle of elevation from point $P$ to the top of the tower is also 30°. Using the tangent function:

$$\tan(30°) = \frac{\text{height of the tower, } h}{PQ},$$

Therefore:

$$h = PQ \times \tan(30°) \implies h = 80 \times 0.577 \implies h \approx 46\,\text{m}.$$

Thus, the height of the tower is approximately 46 meters.

**10)** We can use the Angle-Side-Angle (ASA) congruence postulate to determine if the triangles are congruent. In this case:

- Side (Given): $AB = DE$
- Angle (Given): $\angle BAC = \angle EDF = 60°$
- Angle (Given): $\angle BCA = \angle EFD = 60°$

This fits the Angle-Side-Angle (ASA) postulate. Therefore, the two triangles are congruent by ASA.

**11)** Using the general equation of a circle: $(x - x_1)^2 + (y - y_1)^2 = r^2$, we substitute the given center $(-3, 5)$ and radius 7:

$$(x - (-3))^2 + (y - 5)^2 = 7^2.$$

Simplifying, we get:

$$(x + 3)^2 + (y - 5)^2 = 49.$$

So, the equation of the circle is $(x + 3)^2 + (y - 5)^2 = 49$.

**12)** The formula for the measure of each interior angle of a regular polygon with $n$ sides is:

$$\text{Interior angle} = \frac{(n-2) \times 180°}{n}.$$

The formula for the measure of each exterior angle of a regular polygon with $n$ sides is:

$$\text{Exterior angle} = \frac{360°}{n}.$$

For a hexagon $(n = 6)$:

$$\text{Interior angle} = \frac{(6-2) \times 180°}{6} = \frac{4 \times 180°}{6} = \frac{720°}{6} = 120°,$$

$$\text{Exterior angle} = \frac{360^\circ}{6} = 60^\circ.$$

Therefore, one interior angle of a regular hexagon is $120^\circ$, and one exterior angle is $60^\circ$.

**13)** From the properties of triangles, the size of an angle is directly proportional to the length of the side opposite it. Therefore: Since $AB > AC$, $\angle C > \angle B$. Since $AB < BC$, $\angle C < \angle A$. Combining these results, the order from least to greatest measure is: $\angle B < \angle C < \angle A$.

**14)** Using the cosine function:

$$\cos H = \frac{IH}{GH} \Longrightarrow \cos 45^\circ = \frac{10\sqrt{2}}{GH}.$$

Solving for $GH$:

$$GH = \frac{10\sqrt{2}}{\cos 45^\circ} = \frac{10\sqrt{2}}{\frac{\sqrt{2}}{2}} = 20\,\text{cm}.$$

Using the sine function:

$$\sin H = \frac{GI}{GH} \Longrightarrow \sin 45^\circ = \frac{GI}{20}.$$

Solving for $GI$:

$$GI = \sin 45^\circ \times 20 = \frac{\sqrt{2}}{2} \times 20 = 10\sqrt{2}\,\text{cm}.$$

Therefore, $GH$ is $20\,\text{cm}$ and $GI$ is $10\sqrt{2}\,\text{cm}$.

**15)** The area of a sector of a circle is given by:

$$A = \frac{\theta}{360^\circ} \times \pi r^2,$$

Given the area $A = 18\pi$ and radius $r = 8$, we can find the central angle $\theta$:

$$18\pi = \frac{\theta}{360^\circ} \times \pi \times 8^2 \Longrightarrow 18 = \frac{\theta}{360^\circ} \times 64.$$

Solving for $\theta$:

$$\theta = \frac{18 \times 360}{64} = 101.25^\circ.$$

The length of the arc $L$ is given by:

$$L = \frac{\theta}{360^\circ} \times 2\pi r.$$

Substituting the known values:

$$L = \frac{101.25^\circ}{360^\circ} \times 2\pi \times 8 = \frac{101.25}{360} \times 16\pi \approx 14.13\,\text{units}.$$

Therefore, the measure of the central angle is $101.25^\circ$, and the length of the arc is approximately 14.13 units.

**16)** The statement "All Professors are at least 40 years old" can be written in if-then form as: "If someone is a Professor, then she/he is at least 40 years old.."

Given this conditional statement, if John is 32 years old, he cannot be a Professor since he is not at least 40 years old.

**17)** The given statement "If it snows, the roads become slippery" is a conditional. The logically equivalent statement in its contrapositive form is "If the roads do not become slippery, then it does not snow."

The contrapositive inverts and negates both the hypothesis and the conclusion of the original statement, making option A the correct answer.

**18)** This is an example of a valid argument pattern known as "Hypothetical Syllogism" or "Chain Argument." If $Q \to P$, and $R \to Q$, then it logically follows that $R \to P$. Thus, the correct conclusion is $R \to P$.

**19)** For any triangle, the sum of the lengths of any two sides must be greater than the length of the third side. Using the given sides:

$$10 + 24 > z \implies 34 > z,$$

$$24 - 10 < z \implies 14 < z.$$

Thus, the range for $z$ is $14 < z < 34$. Therefore, the correct option is:

**20)** To determine the truth value of the statement, follow these steps:

1. $\neg q$ (not $q$): Since $q$ is false, $\neg q$ is true.
2. $(\neg q \wedge p)$: Combining the true value of $\neg q$ with the true value of $p$ using "and" ($\wedge$) results in true.
3. $\neg r$ (not $r$): Since $r$ is true, $\neg r$ is false.
4. $(\neg q \wedge p) \vee \neg r$: Combining the true value from step 2 with the false value from step 3 using "or" ($\vee$) results in true.

Therefore, the truth value of the statement $(\neg q \wedge p) \vee \neg r$ is true.

**21)** The formula for the volume $(V)$ of a square pyramid is:

$$V = \frac{1}{3} \times \text{base area} \times \text{height}.$$

For a square pyramid, the base area $(A)$ is $s^2$, where $s$ is the side length. Given the side length $s$ of the base of the pyramid is 5, the base area $A$ is:

$$A = 5^2 = 25.$$

Since the pyramid is inside the cone, both have the same height of 7. Then

$$V = \frac{1}{3} \times \text{base area} \times \text{height} = \frac{1}{3} \times 25 \times 7 = \frac{175}{3}.$$

The volume $(V)$ of a cone is given by:

$$V = \frac{1}{3} \times \text{base area} \times \text{height}.$$

Since the square is inscribed in a circle, the diagonal of the square is the diameter of the circle. To find the area of the circle, we first need to find the diameter using the Pythagorean theorem. The diagonal ($d$) of the square (which is the diameter of the circle) is:

$$d^2 = s^2 + s^2 \implies d^2 = 25 + 25 \implies d = \sqrt{50} = 5\sqrt{2}.$$

The radius ($r$) is half the diameter:

$$r = \frac{1}{2}d = \frac{1}{2} \times 5\sqrt{2} = \frac{5\sqrt{2}}{2}.$$

Now, the formula for the area ($A$) of a circle is:

$$A = \pi r^2 = \pi\left(\frac{5\sqrt{2}}{2}\right)^2 = \pi\left(\frac{25 \times 2}{4}\right) = 12.5\pi.$$

We can now find the volume of the cone (Base area is $12.5\pi$, and height is 7):

$$V = \frac{1}{3} \times 12.5\pi \times 7 \approx 29.17\pi.$$

Therefore, the volume of the pyramid is $\frac{175}{3}$ cubic units, and the volume of the cone is approximately $29.17\pi$ cubic units.

**22)** We know: $V_{\text{cone}} = 450\,\text{cm}^3$, $r_{\text{cone}} = 6\,\text{cm}$, and the formula for the volume of a cone is:

$$V_{\text{cone}} = \frac{1}{3}\pi r^2 h.$$

Now, calculate the height of the cone:

$$450 = \frac{1}{3}\pi(6^2)h \implies 450 = \frac{1}{3}\pi(36)h \implies 450 = 12\pi h.$$

Solving for $h$:

$$h = \frac{450}{12\pi} \implies h = \frac{75}{2\pi} \approx 12.$$

So, the altitude of the cone is approximately $12\,\text{cm}$.

**23)** The surface area of the combination includes the curved surface area of the cylinder and the curved surface area of the hemisphere (excluding the base):

$$\text{Surface Area} = \pi r^2 + 2\pi rh + 2\pi r^2.$$

For the given values, $r = 7\,\text{cm}$ and $h = 5\,\text{cm}$:

$$\text{SA} = \pi(7^2) + 2\pi(7)(5) + 2\pi(7^2) = 49\pi + 70\pi + 98\pi = 217\pi\,\text{cm}^2.$$

The volume of the combination includes the volume of the cylinder and the volume of the hemisphere:

$$\text{Volume} = \pi r^2 h + \frac{2}{3}\pi r^3.$$

For the given values, $r = 7\,\text{cm}$ and $h = 5\,\text{cm}$:

$$\text{Volume} = \pi(7^2)(5) + \frac{2}{3}\pi(7^3) = 245\pi + \frac{2}{3}(343\pi) \approx 245\pi + 228.67\pi \approx 473.67\pi\,\text{cm}^3.$$

Therefore, the surface area is $217\pi\,\text{cm}^2$, and the volume is approximately $474\,\text{cm}^3$.

**24)** Using the Pythagorean theorem for the right triangle $IJK$:

$$KI^2 = IJ^2 - JK^2 \implies KI^2 = 20^2 - 16^2 \implies KI^2 = 400 - 256 \implies KI = \sqrt{144} \implies KI = 12.$$

Now, triangle $ILK$ is similar to triangle $IJK$. Using the properties of similar triangles:

$$\frac{KI}{IJ} = \frac{LK}{JK} \implies \frac{12}{20} = \frac{LK}{16} \implies LK = \frac{12 \times 16}{20} \implies LK = 9.6.$$

Now, using the property of similar triangles $\triangle JLK \sim \triangle IJK$, we can find $JL$:

$$\frac{JL}{JK} = \frac{LK}{KI} \implies \frac{JL}{16} = \frac{9.6}{12} \implies JL = \frac{9.6 \times 16}{12} \implies JL = 12.8.$$

The area of triangle $IJK$ is:

$$\text{Area} = \frac{1}{2} \times JK \times KI \implies \text{Area} = \frac{1}{2} \times 16 \times 12 \implies \text{Area} = 96\,\text{square units}.$$

Therefore, the answer is $9.6, 12.8, 96$.

**25)** Using the angle bisector theorem, we have:

$$\frac{XY}{XZ} = \frac{YW}{ZW}.$$

Given that $ZY = ZW + YW$, we can write:

$$ZW = 21 - YW.$$

From this, we can find $YW$ and $ZW$:

$$\frac{11}{17} = \frac{YW}{21 - YW} \implies 17 \times YW = 231 - 11 \times YW \implies 28YW = 231 \implies YW = 8.25\,\text{units}.$$

Substituting the value of $YW$ into the equation:

$$ZW = 21 - YW \implies ZW = 21 - 8.25 \implies ZW = 12.75\,\text{units}.$$

Therefore, the correct answer is option C: 8.25 and 12.75.

**26)** Using the property that the tangent from an external point is perpendicular to the radius at the point of tangency, we have:

$$CD^2 = CF \times FE.$$

Given $CF = 7$ and $EF = 10$. Now, calculating $CD$:

$$CD^2 = 7 \times 10 = 70 \implies CD = \sqrt{70}.$$

Therefore, the length of $CD$ is $\sqrt{70}$ units.

**27)** Since all four sides are equal, the quadrilateral is a rhombus. For the angles, the opposite angles of a rhombus are supplementary. So:

$$120^\circ + x^\circ = 180^\circ \implies x = 60^\circ.$$

For the equation involving $y$:

$$3y = y + 6 \implies 2y = 6 \implies y = 3.$$

Therefore, the values of $x$ and $y$ are $60^\circ$ and 3, respectively.

**28)** In a triangle with an inscribed circle, the tangent segments from each vertex to the points of tangency are equal. Given:

$$AU = AV = 6, \quad CV = CW = 11, \quad BU = BW = 15.$$

We find the sides of the triangle as follows:

$$AB = AU + BU = 6 + 15 = 21, \quad AC = AV + VC = 6 + 11 = 17, \quad BC = BW + WC = 15 + 11 = 26.$$

Therefore, the perimeter of the triangle is:

$$AB + AC + BC = 21 + 17 + 26 = 64.$$

Thus, the perimeter of $\triangle ABC$ is 64.

**29)** From the properties of a parallelogram, we know that $LM \parallel ON$. Using the given conditions and properties of a parallelogram, let's analyze the situation: When $OP$ intersects $LN$ at point $Q$, we use the Vertical Angles Theorem which states that the angles opposite each other are equal. In our case, $\angle LQP = \angle NQO$. Additionally, using the Alternate Interior Angles theorem, we have congruent angles when two parallel lines are intersected by a transversal. Thus, $\angle LPQ = \angle QON$ and $\angle QLP = \angle QNO$. With these angle relationships, we can conclude that triangles $LQP$ and $NQO$ are similar by the AAA (Angle-Angle-Angle) criterion. Since triangles $LQP$ and $NQO$ are similar, the ratio of their corresponding sides will be equal:

$$\frac{LP}{NO} = \frac{LQ}{QN}.$$

Given:

$$NO = LP + PM = 5 + 4 = 9,$$

and:

$$LQ = LN - QN \implies LQ = 7 - QN.$$

Using the side ratios from the similar triangles:

$$\frac{5}{9} = \frac{7 - QN}{QN}.$$

Solving for $QN$:

$$5QN = 63 - 9QN \implies 14QN = 63 \implies QN = 4.5.$$

Then, substituting $QN$ back into the equation for $LQ$:

$$LQ = 7 - 4.5 = 2.5.$$

Therefore, $LQ$ is 2.5 units, and $QN$ is 4.5 units.

**30)** In a kite, the diagonals are perpendicular, and one of the diagonals bisects the other. Thus, $FP = \frac{1}{2}FH$. Therefore:

$$FH = 2 \times 14 = 28.$$

To find $EG$, note that:

$$EG = EP + PG.$$

First, find $EP$. In the right triangle $EFP$, using the Pythagorean theorem:

$$EF^2 = FP^2 + EP^2 \implies 18^2 = 14^2 + EP^2 \implies 324 = 196 + EP^2 \implies EP^2 = 128 \implies EP = \sqrt{128} \approx 11.3.$$

Then:

$$EG = EP + 24 \implies EG = 11.3 + 24 \implies EG = 35.3.$$

To find $FG$, use the Pythagorean theorem in triangle $FPG$:

$$FG^2 = PG^2 + FP^2 \implies FG^2 = 24^2 + 14^2 \implies FG^2 = 576 + 196 \implies FG = \sqrt{772} = 2\sqrt{193}$$

So, $EG = 35.3$, $FH = 28$, and $FG = 2\sqrt{193}$.

**31)** First, since $GH = 24$ and $FG = 10$, and given that $\angle G = 90°$, we can use the Pythagorean theorem to find the length of $FH$:

$$FH^2 = GH^2 + FG^2 \implies FH^2 = 24^2 + 10^2 \implies FH = \sqrt{676} \implies FH = 26\text{units}.$$

Next, we note that $\triangle FIJ$ is similar to $\triangle FGH$ based on the AA similarity criterion since both are right triangles and share

$\angle F$. Let us use the ratios from these similar triangles to find $IJ$:

$$\frac{IJ}{GH} = \frac{FI}{FH} \Longrightarrow \frac{IJ}{24} = \frac{6}{26} \Longrightarrow 26 \times IJ = 24 \times 6 \Longrightarrow IJ = \frac{144}{26} \Longrightarrow IJ = \frac{72}{13} \text{ units.}$$

Finally, to find $HJ$, use the property of the similar triangles again:

$$\frac{FJ}{FG} = \frac{FI}{FH} \Longrightarrow \frac{FJ}{10} = \frac{6}{26} \Longrightarrow 26 \times FJ = 10 \times 6 \Longrightarrow FJ = \frac{60}{26} \Longrightarrow FJ = \frac{30}{13} \text{ units.}$$

Then, substitute $FJ$ into the equation:

$$HJ = FH - FJ \Longrightarrow HJ = 26 - \frac{30}{13} \Longrightarrow HJ = \frac{308}{13} \text{ units.}$$

In summary:

$$FH = 26,\ HJ = \frac{308}{13},\ IJ = \frac{72}{13}.$$

**32)** The measure of an angle formed by two chords inside a circle is half the sum of the measures of the arcs intercepted by the angle and its vertical angle.

$$\angle STU = \frac{1}{2} \times SU = \frac{1}{2} \times 170^\circ = 85^\circ.$$

**33)** To reflect a point over the line $y = -x$, you switch the $x$ and $y$ coordinates and change the signs. So, the reflection of a point $(a,b)$ over the line $y = -x$ is $(-b,-a)$.

Given the points:

$$A(-1,3) \to A'(-3,1),$$

$$B(-1,0) \to B'(0,1),$$

$$C(3,0) \to C'(0,-3).$$

After reflection in the line $y = -x$, the coordinates of $\triangle A'B'C'$ are: $A'(-3,1), B'(0,1), C'(0,-3)$.

**34)** To rotate a point $(x,y)$ around the origin by $45^\circ$ counterclockwise, apply the following transformations:

$$x' = x\cos 45^\circ - y\sin 45^\circ, \quad \text{and} \quad y' = x\sin 45^\circ + y\cos 45^\circ.$$

Given $\cos 45^\circ = \frac{\sqrt{2}}{2}$ and $\sin 45^\circ = \frac{\sqrt{2}}{2}$:

For $A(1,3)$:

$$x' = 1 \times \frac{\sqrt{2}}{2} - 3 \times \frac{\sqrt{2}}{2} \Longrightarrow x' = \frac{\sqrt{2}}{2} - \frac{3\sqrt{2}}{2} \Longrightarrow x' = -\sqrt{2},$$

$$y' = 1 \times \frac{\sqrt{2}}{2} + 3 \times \frac{\sqrt{2}}{2} \Longrightarrow y' = \frac{\sqrt{2}}{2} + \frac{3\sqrt{2}}{2} \Longrightarrow y' = 2\sqrt{2}.$$

For $B(4,1)$:

$$x' = 4 \times \frac{\sqrt{2}}{2} - 1 \times \frac{\sqrt{2}}{2} \Longrightarrow x' = \frac{4\sqrt{2}}{2} - \frac{\sqrt{2}}{2} \Longrightarrow x' = \frac{3\sqrt{2}}{2},$$

$$y' = 4 \times \frac{\sqrt{2}}{2} + 1 \times \frac{\sqrt{2}}{2} \Longrightarrow y' = \frac{4\sqrt{2}}{2} + \frac{\sqrt{2}}{2} \Longrightarrow y' = \frac{5\sqrt{2}}{2}.$$

Therefore, after a 45° rotation, the new coordinates are: $A'(-\sqrt{2}, 2\sqrt{2})$, $B'(\frac{3\sqrt{2}}{2}, \frac{5\sqrt{2}}{2})$.

**35)** Since $D$ is the midpoint of $AC$ and $AD = 15$, then $CD = 15$ as well because midpoints divide segments into two equal parts. Therefore:

$$AC = AD + CD = 15 + 15 = 30\,\text{units}.$$

Next, using the Pythagorean theorem in triangle $AEC$:

$$EC^2 = AE^2 + EC^2 \Longrightarrow 30^2 = 18^2 + EC^2 \Longrightarrow 900 = 324 + EC^2 \Longrightarrow EC^2 = 576 \Longrightarrow EC = 24\,\text{units}.$$

To find $GC$: Since $CF = \frac{1}{2}BC$, $DC = \frac{1}{2}AC$, and $DF = \frac{1}{2}AB$, then $GC = \frac{1}{2}EC$. Therefore:

$$GC = \frac{1}{2} \times 24 \Longrightarrow GC = 12.$$

Hence, $AC = 30$ and $GC = 12$.

**36)** To determine the distance of chord $AB$ from the center of the circle, we can use the properties of right triangles and the Pythagorean theorem. Let's name the circle's center as $O$ and the midpoint of the chord $AB$ as $M$. When we draw a radius from $O$ to the midpoint $M$ of the chord, we form a right triangle with the radius, half of the chord, and the segment from the chord's midpoint to the center. Therefore, half of the chord $AM = \frac{26}{2} = 13$. In the right triangle $OAM$:

$$OA^2 = OM^2 + AM^2 \Longrightarrow 22^2 = OM^2 + 13^2 \Longrightarrow 484 = OM^2 + 169 \Longrightarrow OM^2 = 315 \Longrightarrow OM = \sqrt{315} = 3\sqrt{35}.$$

Thus, chord $AB$ is $3\sqrt{35}$ units away from the center of the circle.

**37)** In triangle $GHI$, the side opposite the largest angle is the longest, and the side opposite the smallest angle is the shortest. Given the angles, the order of sides from shortest to longest is:

$$\text{Shortest} = GH\,(\text{opposite } 30^\circ),$$

$$\text{Middle} = HI\,(\text{opposite } 180^\circ - 110^\circ - 30^\circ = 40^\circ),$$

$$\text{Longest} = GI\,(\text{opposite } 110^\circ).$$

Therefore, the correct answer is option C: $GH, HI, GI$.

**38)** To find $NK$ and $KM$, we first note that if $IN$ extends through $M$ to the other side of $JK$, the segment beyond $M$, which we'll call $NK$, is what we need to determine for the length $KM$. Given that $JN = 5$ and $JK = 7$:

$$NK = JK - JN = 7 - 5 = 2.$$

Given that triangles $IJN$ and $KNM$ are similar by the AA postulate, the sides are proportional. Using the proportionality:

$$\frac{IJ}{KM} = \frac{JN}{NK} \implies \frac{9}{KM} = \frac{5}{2} \implies KM = \frac{18}{5} \implies KM = 3.6.$$

Now, let us find the ratio of the areas of triangles $IJN$ and $KNM$. Using the property of similar triangles:

$$\frac{\text{Area}_{IJN}}{\text{Area}_{KNM}} = \left(\frac{IJ}{KM}\right)^2 = \left(\frac{9}{3.6}\right)^2 = (2.5)^2 = 6.25.$$

This means the area of triangle $IJN$ is 6.25 times larger than the area of triangle $KNM$.

**39**) We are tasked with finding the measure of $\angle L$. By the vertical angle theorem, $\angle JQK = \angle IQL$. Since $GH = 180°$, we have:

$$\angle IQG + \angle IQL + \angle HQL = 180°. \quad (1)$$

Let us calculate the measures of $\angle IQG$, $\angle IQL$, and $\angle HQL$ to determine $\angle L$. First, consider triangle $IQG$, the sum of angles in a triangle is $180°$. Thus:

$$\angle IQG + \angle I + \angle G = 180° \implies \angle IQG + 55° + 95° = 180° \implies \angle IQG = 30°.$$

Next, consider triangle $JQK$, the sum of angles in a triangle is $180°$. So:

$$\angle JQK + \angle J + \angle K = 180° \implies \angle JQK + 50° + 65° = 180° \implies \angle JQK = 65°$$

Therefore, $\angle IQL = \angle JQK = 65°$. Now, consider triangle $HQL$, the sum of angles in a triangle is $180°$. Hence:

$$\angle HQL + \angle H + \angle L = 180° \implies \angle HQL + 55° + \angle L = 180° \implies \angle HQL = 125° - \angle L.$$

Plugging this into our equation (1):

$$30° + 65° + 125° - \angle L = 180° \implies -\angle L = 180° - 30° - 65° - 125° \implies \angle L = 40°.$$

So, $\angle L$ is $40°$. The answer is option B.

**40**) Since the triangles are similar, the ratio of their corresponding sides is the same.

$$\frac{AD}{WZ} = \frac{CD}{YZ} \Rightarrow \frac{30}{WZ} = \frac{35}{28} \Rightarrow WZ = 30 \times \frac{28}{35} = 24.$$

Therefore, $WZ = 24$ units.

## Author's Final Note

I hope you enjoyed this book as much as I enjoyed writing it. I have tried to make it as easy to understand as possible. I have also tried to make it fun. I hope I have succeeded. If you have any suggestions for improvement, please let me know. I would love to hear from you.

The accuracy of examples and practice is very important to me. We have done our best. But I also expect that I have made some minor errors. Constant improvement is the name of the game. If you find any errors, please let me know. I will fix them in the next edition.

Your learning journey does not end here. I have written a series of books to help you learn math. Make sure you browse through them. I especially recommend workbooks and practice tests to help you prepare for your exams.

I also enjoy reading your reviews. If you have a moment, please leave a review on Amazon. It will help other students find this book.

If you have any questions or comments, please feel free to contact me at
drNazari@effortlessmath.com.

And one last thing: Remember to use online resources for additional help. I recommend using the resources on `https://effortlessmath.com`. There are many great videos on YouTube.

Good luck with your studies!

Dr. Abolfazl Nazari

Made in the USA
Middletown, DE
30 August 2024

60078386R00166